"From Red Buttons to Smart Technology Support"

Part 1

Kevin Doughty, Ph.D.

Chairman - T-Cubed Ltd.

Director – i-Centre for Usable Home Technology

Visiting Professor in Digital Transformation of Care Services, University of Cumbria

Contents

Acknowledgements

There are many people that I need to thank for providing me with the inspiration to write this book. It begins with the residents of the former Bryn Idan Residential Care Home in Caernarfon, who shared with me, as the proprietor, their frustrations that they had to give up their homes to move into somewhere to be watched over 24 hours a day, 7 days a week. They may not have seen the potential for technology that I was predicting, but they did know that support in the home is what they wanted if they were to be given the opportunity to remain independent.

Secondly, there are the many researchers both at the University of Wales in Bangor, where I formed the Telecare Research Group in 1993, and the scientists and engineers at BT Labs in Martlesham outside Ipswich. In particular, BT managers, such as Keith Cameron, not only provided the funding and industrial input to the challenge of monitoring people in their own homes, but also offered the commercial credibility to a business opportunity that will grow steadily and be worth hundreds of millions of pounds as the population ages, and as life expectancy increases.

Students including Gareth Williams, Peter Smith and Peter King became important players in my decision to spin out the Technology in Healthcare company in 1998, and to develop sensors that would integrate with the community alarm systems that had been the mainstay of community support for decades. These enabled passive monitoring to detect the emergency situations that people might not be aware of. Gareth Williams, after the award of his Ph.D., went on to become a telecare specialist in his own right, and has collaborated with me for many years on a range of projects including system design, product evaluation and service audit and improvements for national governments. The work that we have done on behalf of T-Cubed, the company that we founded, has provided diagrams, models and other content for this book, for which I am extremely grateful. He is also a source of reasoned arguments, publications, reviews, and of innovative ideas, which should enable him to continue to prosper in this field for many years after I have been forced into retirement for whatever reason.

I acknowledge Professor Andrew Monk for his leading role in the establishment of the Centre for Usable Home Technology (CUHTec) at the University of York in 2003 (and its subsequent relocation to Newcastle and then Coventry). This became the platform for collaboration with over 100 local authorities, housing associations and third sector organisations, and which provided me with the opportunity to develop new models of connected support. CUHTec became the vehicle for questioning support arrangements, assessment models, integrated delivery, and a reason for developing training courses for all levels of interest. Many current members of iCUHTec, the virtual successor to CUHTec, have helped me to plan this book and to approve its content

Finally, I must thank my family, friends and collaborators who have both worked with me in gaining a better understanding of what technology works, what doesn't work, and how services need to change in order to realise the potential of telecare and AT services. I should not forget to mention Jan Costa, both as a co-author of many papers dating back to the 1990s and for providing a practical view on many ideas, based on her professional skills as an Occupational Therapists. Others have shared the authorship of papers with me or have collaborated closely in developing new ideas. They are named below; apologies for any missing names – my memory is not perfect!

Kathryn Andrew
Alistair Appleby
Mark Bannister
Steve Barnard
Karen Bradshaw
Kay Board
Keith Cameron
Alan Capper
Julia Champion
Joanna Clark
Paul Clitheroe
Kath Cooper
Jan Costa
Ellen ap Dafydd
Helen Deakin
Chris Doughty
Barbara Dunk
Diane Edge
Malcolm Fisk
David Godfrey (R.I.P.)
Alison Hampson
Richard Harding
Rhianwen Jones
Peter King
Audrey Kinsella
Anne Livingstone
Charles Lowe
Moira Mackenzie
Dave Miles
Alison Marshall
Billy Mulvihill
Patrick Mulvihill
Mike Orton
John Patterson
Janet Roberts
Manuela Schuette
Peter Smith
Chris Steele
Fran Taberner
Alison Thrower
Sue Tivey
John Webb
Anne Williams
Gareth Williams
Bob Woods

Foreword

This book is the result of over 30 years of work in the telecare and assistive technology industry. It is built on my personal experience of setting up and running a residential care home for older people in my home town of Caernarfon in North Wales (during the 1980s), and the discovery that most (if not all) the residents could (and perhaps should) have been in their own homes supported by home carers. Academic research at the University of Wales, Bangor, into the new subjects of telemedicine and telecare followed and laid the foundations for new products, systems and models.

There have been very few books published on telecare, as it applies to the UK, and as a successor to community alarm systems. This is hardly surprising because of the scale of the subject, and the speed of new developments. However, it is a long journey from identifying an application that might benefit one group of people, to designing, manufacturing and testing the product, and then introducing it commercially into a viable service. Many exciting new products fall by the wayside, and do not make the transition into the inventories of service providers. The reasons are complex but include a reluctance amongst front line support workers to introduce change, a fear of technology by many older people, and a lack of hard evidence of the benefits realisation.

Most of the barriers to implementation can be overcome if there is sufficient political will to support this change. It needs more than warm words; it also needs funding to pump prime the transformation process. With the notable exception of Scotland, no such funding has been made available to local authorities in England and Wales since 2007/2008, even though large sums have been invested in the NHS to achieve efficiencies through a greater use of modern technology. It may be evident to observers that much of this investment will be wasted if similar funding is not made available to social care through local authorities, as people may be left languishing in hospital as further examples of 'delayed transfers of care' unless they can be supported in the community.

As with most challenges, improved education is likely to be an essential component in the solution. It must begin with attempts to change the culture of the population; people need to understand that we must become partners in our own care, and not rely completely on the state to offer all aspects of support. Professionals then need to learn about the potential of technological applications in ways that give them the confidence to recommend (or perhaps prescribe) such solutions as preferred alternatives to traditional support mechanisms such as care homes, domiciliary care and day centres. This is not a trivial matter when the workforce is dominated by very caring people who may not be at ease in the world of technology, and where new care pathways are difficult to develop and introduce when some of the processes are not developed. All care and support staff need to learn quickly, especially if they are joining (or re-joining) care organisations after a number of years of absence bringing up a family or working in other roles.

This book can therefore be viewed as a resource manual for everyone from front-line care and support staff, through to managers, commissioners and families. Different groups can use it in different ways. All readers should regard it as an introduction to a future that will be described and predicted in Part 2 which takes on the challenge of digital transformation. They should learn to respect what has already been done in an analogue world, as described in Part 1, and take it forward to a digital future which is safer, more efficient, and more exciting.

List of Figures

List of Tables

Chapter 1

Introduction – care and support in the 21st Century

Changing healthcare

The establishment of the National Health Service in 1948 was perhaps the most important change in the culture and practice of care in the UK ever attempted. It created an environment of shared funding and peace of mind for millions of people who feared illness as much because of the cost of treatment as the possible outcome. The NHS has continued to evolve and must continue to do so if it is to survive and provide universal free care at the point of need for the next 70 years.

The shift of resources from secondary care (i.e. hospitals) to primary care and the community has been an aim of health services in all developed economies for most of this 21st century. Moving care 'closer to home' is an admirable concept as it has the potential to reduce the need for patients and their families to travel and allows the most valuable of health-industry resources (i.e. specialist staff, diagnostic equipment, operating theatres and intense care facilities) to be concentrated in larger centres. This approach is part of a medical model which assumed a simple 3 part strategy (Figure 1) in which people who are unwell are diagnosed, using high-tech medical devices such as scanners and imagers, treated, through surgery or pharmaceutical interventions , and then cared for by doctors, nurses and therapists until they have recovered sufficiently to be sent home (or to a local centre for a few days of recovery).

Figure 1: The proposed role of a modern (acute) hospital

The benefits of the above model include a reduction in the time a patient spends in the acute hospital, and a rapid flow through the system. This 'sausage-machine' approach would be efficient, but only if it is not overwhelmed by unplanned admissions, and by people turning up at an Accident and Emergency Unit without using the primary care system to triage their medical complaints. It also assumes that there are appropriate community facilities available to enable recovery to continue once a patient has been discharged. If people who are medically stable, but not well enough to self-care, are returned to their own homes too soon, they will return very quickly as a new hospital admission, often as an emergency case. The system needs to be balanced to yield good outcomes.

Table 1: A Comparison of Acute and Chronic Disease

Feature	*Acute Disease*	*Chronic Disease*
Example	Urinary tract infection	Arthritis
Onset	Abrupt	Gradual
Cause	Single	Multiple over time
Duration	Limited to a few days	Lengthy and indefinite
Outcome	Usually curable	Incurable
Effect of medication	Usually effective	Indecisive with adverse effects

A significant investment in community resources, staffing and technology enabled approaches may be required to achieve this end. In times of austerity, the money simply isn't forthcoming; commissioners, the NHS and service providers will struggle to achieve this type of transformation until there is more political support, and the public understands and accepts the need to change the model of care delivery. The health system needs to develop further if the 'closer to home' model is to work for those people who suffer major issues such as heart failure, cancers, acute disease, stroke, severe infections or sepsis. It is unlikely to be appropriate to people who have long term conditions such as diabetes, cardiovascular disease (CVD), and chronic obstructive pulmonary disease (COPD).

Table 2: Impact and Prevalence of the Three Most Common Chronic Diseases

Cardiovascular diseases (CVDs)	Type 2 Diabetes	Chronic Obstructive Pulmonary Disease (COPD)
The number one cause of death globally	Over 350 million people worldwide have the condition	Nearly 70 million people worldwide have the disease
30% of all global deaths	~4 million death annually as a result of high blood glucose	Generally, the result of long-term smoking of tobacco
7.3 million deaths from coronary heart disease in 2008	> 80% of deaths occur in low- and middle-income countries	It is responsible for about 5% of deaths globally
6.2 million deaths from stroke in 2008	7th most likely cause of death by 2030	90% of deaths occur in low and middle-income countries
Annual deaths > 23 million by 2030	Groups including those from Middle East and Indian sub-continent are at high risk	Total deaths are expected to increase by 30% by 2030 with women affected most
Most can be prevented by more exercise, less use of tobacco, and changes in diet, blood pressure and levels of lipids in the blood	The effects of poorly managed blood sugar levels are poor circulation, blindness, kidney damage and lower limb amputations	Sufferers have exacerbations that require antibiotics and hospitalisation frequently without monitoring and rapid intervention
>16% of deaths due to elevated blood pressure	Healthy diet, physical activity, maintaining a normal body weight and avoiding tobacco use can prevent/delay onset	Disease is incurable, but symptoms can be managed for some time using medication

These conditions can be managed medically, but sufferers can never be cured; people who suffer from one or more chronic diseases will therefore spend years in the community, sometimes enduring continued ill-health, but with management through physical and occupational therapies and medication treatments. Table 1 compares acute and chronic disease while Table 2 gives impact and prevalence comparison for the 3 most company chronic diseases described above.

It may be apparent that the impact of chronic disease will be profound both in the UK, and in other developed countries. It will require an acceleration of change in the way that healthcare services are configured, and a greater focus on prevention. This is the basis of this book; the impact of technology will be described, leading to the imperatives for changes service configuration.

An Ageing Population and the Need for Change

In the UK, the care of older people prior to the 20th century was based on families looking after their loved ones when they were unable to look after themselves. 24-hour care became routinely available in geriatric hospitals during Victorian times, though the quality is likely to have varied considerable between institutions. If older people had no family, they were consigned to the workhouse, an idea that scared many people long after they were closed and when local authorities assumed responsibility for providing support for vulnerable people. The replacement for the work-houses were residential care homes that ranged from private hotels for older people through to more institutional properties where residents shared a room with a stranger, and a bathroom with several others. Again, standards varied significantly, and there was little oversight to prevent poor practice. During the 1970s and 1980s, there was a rapid expansion of private care homes. Fees for residents (up to a certain maximum) were means-tested and could be met by the then Department of Health and Social Security. There were few eligibility criteria. Effectively, if a care home had a bed available, and a member of the public wanted it, then it was theirs!

But the cost to the public purse was unsustainable. There were already scares about a demographic time-bomb that would explode when there were more people needing care than available to deliver it. The population projections for 6 different age bands from 2014 over a period of 25 years are shown in Figure 2. They show relatively stable populations for ages up to 60, but with a 20% increase in the numbers aged between 60 and 74 by particularly over the next 15 years, but a monotonic increase in the older old population (those aged over 75) with a near doubling of numbers from 5 to 10 million by 2039. Because this group of people is more likely to require hospital attention than younger people, and because they are more likely to have several life limiting long term conditions, this will have a significant impact on the need for NHS support. However, it is likely to have a more profound effect on community care, including support with activities of daily living, because these are the people who are most likely to become frail and/or develop cognitive impairments which currently cannot be managed using pharmaceutical interventions. It follows that the potential and opportunities for supporting their independence and quality life using technology may be greatest. How this can be done effectively and efficiently is the subject of this book and has already been one of the drivers for developing Care in the Community as a replacement model for long term support.

The Care in the Community Act (1991) heralded major changes in long term care for people with disabilities and for older people generally. It encouraged a growth in private sector nursing homes and residential care homes and led to a rapid expansion of provision for a decade. However, it became increasingly clear that few people would choose to move from their own homes into residential care unless it became impossible to live safely without 24/7 access to care and support. The cost of this form of care also increased as minimum staffing levels were mandated and a regime of inspection was introduced; conditions improved and the expansion in the number of homes and beds had all but stopped by 1998, even though the number of older people continued to increase.

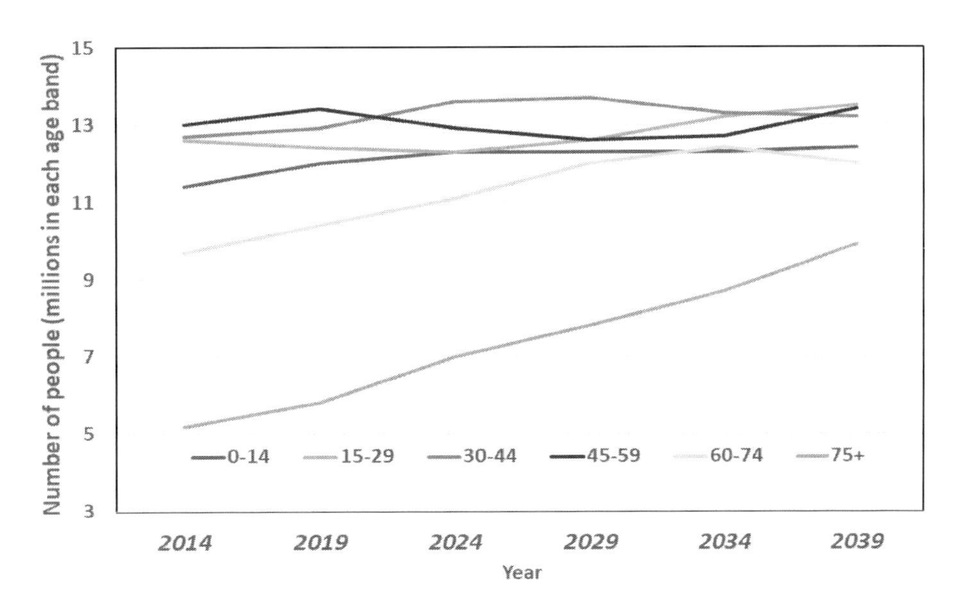

Figure 2: Population Trends in the UK based on ONS Principal Population Projections (2014)

Many more people began to appreciate the importance of independence in maintaining choice and Quality of Life and demanded more choice in where they lived out their final years. Home Help services offered by local authorities developed into Home Care propositions where generic help and support, such as assistance with shopping and cleaning, matured into personal care. These services replicating the support provided in care homes as improved training enabled such service elements to be delivered into people's own homes by a plethora of private agencies. This became the second era of long-term care (see Figure 3) which had a focus broadly on the assessed needs of individual, and was available through carers, trained personnel who could provide formal personal and/or nursing support for older or disabled people either in their own homes or in community care homes.

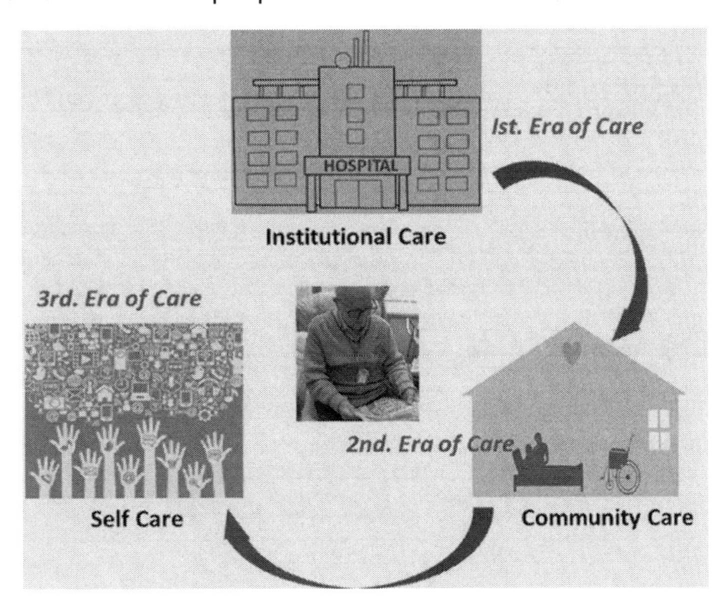

Figure 3: The Three Eras of Care

This era was dependent on a growing workforce of care assistants who needed to be trained within new domiciliary care organisations; the organisations could also, in principle, provide appropriate rostering and mechanisms to identify declining conditions and the need for increasing levels of support. The scale of operations was such that there were initially sufficient members of staff to provide cover for when their colleagues were off sick or on holiday.

One of the problems with the second era of care has been the need to manage multiple short care episodes, such as help with dressing or bathing, at times that suit people's needs. The issues become more profound when the population in need of this care is likely to be geographically dispersed. Carers may spend more than half their time travelling to and from the homes of people in need of services. More significantly, many service users will require similar sorts of help to perform Activities of Daily Living, and at the same times of day. For example, a great number of older people will need help to get up and out of bed (transfer) and then dressing in the morning and might require a service during a relatively narrow window of perhaps 7:30am to 9:30am. One carer might therefore be able to support perhaps a maximum of 5 service users.

A similar problem exists in the evening when they need to be helped to undress and get into bed. For safety and recruitment reasons, the latest that carers can be expected to be out working alone is 9 or 10pm. This means that some people have to be put to bed as early as 7:30pm and may then be expected to remain there for upwards of 12 hours if they are unable to get into and out of bed on their own during the night. Although some care agencies will provide scheduled visits during the night, on-demand support for personal care such as toileting is not generally available. Some of the issues associated with rural areas are considered in Chapter 11 of this book.

In more urban areas, there are issues relating to security and to anti-social behaviour to contend with. It is not surprising that some older people look for gated communities and the services of a concierge. Access to properties becomes important, especially if an individual lives alone, and if they are unable to answer the door without putting themselves at a high risk of falling. Sheltered housing schemes (see Chapter 2) offered some level of security through a scheme front door that can be monitored. Extra care housing extends many of the support functions of sheltered housing by ensuring that there are care and support staff available on a 24/7 basis. Such schemes have the potential to replace many of the old residential homes.

These options, despite recent expansion, are available to a relatively small proportion of the population; the majority will continue to live in properties that are dispersed in the community. If they need support with toileting or with transfers in the middle of the night, or at weekends outside the times when domiciliary carers are available, they would struggle – unless some form of 'uber-style' on-demand care service is available. This is why there continues to be a need for more self-care; individuals both perform some elements of DIY healthcare and organise others to support them in their everyday lives. A third era of care is then created. It will be dependent on information, communication and an increasing use of technologies that support self-care – all very consistent with personal budgets, direct payments and, unfortunately, a need for more people to pay for their care and support. These changes should become apparent in Figure 3. They will not be universally popular with many older people. This both because the expectation of many people that social care is free, and because they don't like change – especially if they believe that their service will be imperfect.

Book Contents and Style

This book is intended neither as a history manual of technology for supporting older people, nor as an academic review of how services based on technology have developed (and continue to develop). Rather it is designed as a story book, describing how services have evolved to such an extent that they are now routinely available to help people to live independently as they grow older or develop disabilities. That is very much the purpose of Part 1 – which covers the approximate 50-year period from 1968 to 2018. It is no coincidence that the end date corresponds to the 70th anniversary of the formation of the National Health Service in the UK.

Part 2 will describe how technologies themselves have improved, especially those based on telecommunications, digital processes, miniature sensors, and intelligent processes. It then offers suggestions for the direction of travel and offers a road map for commissioners of services and for service providers. This journey begins at the end of 2018 and looks at least 10 years into the future; it extends to equipment and software designers and includes some considerations for all stakeholders as they plan for future developments. It concludes with some blue sky thinking and some thoughts on how society much adapt and change in order that concepts such as social robots will be acceptable to recipients of care services.

Beyond this first chapter, there follows:

Chapter 2: **Community alarms** - the history of sheltered housing and the role of wardens in providing 24-hour support for tenants

Chapter 3: **Social alarms to telecare** – adding sensor devices and response protocols for risk management

Chapter 4: **Continual or continuous monitoring** – detecting and recording activities and vital signs for on-going assessment

Chapter 5: **Assistive and Smart Assistive Technologies** – understanding assistive technology and how they can help people with disabilities

Chapter 6: **Improving assessment, prescription and support options** - implementing a bottom-up, person centred care using technologies to improve outcomes

Chapter 7: **The journey to Technology Supported Living** - developing the 7 waves of technology enabled support at home to change the culture of provision

Chapter 8: **Technologies for managing frailty** – role of technology in helping to reduce the incidence and impact of falls, dehydration and poor nutrition

Chapter 9: **Supporting people with dementia, cognitive and intellectual issues** – exploring technology options for reminders and for keeping people safe

Chapter 10: **Caring for the carers** – discovering ways of supporting families and reducing the burden of stress through technology

Chapter 11: **Social isolation, loneliness and rurality** – ideas for helping people to feel valued connected, and happy using technology

These are followed by four appendices:

A: Vignettes - these may be used as exercises to test the reader's skills in assessment and in the possible prescription of technology support applications and equipment
B: Sources of equipment – a resource of companies and organisations that manufacture or distribute technology for support
C: Bibliography – ideas for further reading
D: Glossary of terms – definitions of expressions, terms and acronyms used in the text.

In Part 2, there are a further eleven chapters, though each will be subject to change as new inventions and technologies are mainstreamed:

Chapter 12: Connecting to the Internet – the importance of being able to access data and information through mobile and broadband provision, and personal devices, and of exploiting Big Data, machine learning and Artificial Intelligence, but in ways that are secure and avoid the scourges of cyber-attacks and hacking;

Chapter 13: Digital Transformation – opportunities for changing business practices in health, social and personal care using digital tools, electronic records, devices, apps and targeted applications of the Internet of Things to exploit data in beneficial ways to support independent living;

Chapter 14: Strategies for keeping people healthy – wearable sensors, the quantified self, new and emerging technologies, and applications for enabling people to remain safe out of hospital and institutional care, and to be able to remain in their own homes;

Chapter 15: Technologies for supporting mental health and well-being – management of stress, anxiety, depression and more complex neurological conditions using smartphones, apps, peripherals and digital platforms

Chapter 16: Supporting people with disabilities – smart assistive technology approaches to overcoming physical, sensory and communication challenges, and the role of 3D printing and emerging technologies to minimise the impact of disability;

Chapter 17: Volunteering in the digital world – the opportunities, knowledge and skills needed to support vulnerable people using digital services, so that prevention can be supported by society and a community base;

Chapter 18: Supporting care home residents – technologies for improving the support provided to people living in residential or nursing homes by empowering staff and families to offer improvements in care quality and better links;

Chapter 19: Improving properties, town environments and other support– improving design and opportunities for designing and retrofitting homes and shared spaces with technologies to keep people safe and connected;

Chapter 20: Digital and robotic assistants and chatbots – opportunities for providing intelligent voice interfaces and social robots for supporting independence by performing or helping people to perform domestic tasks and by giving ever-present reassurance

Chapter 21: Improving telecare service quality – specifying, measuring and implementing changes in service offerings to support improved outcomes for both the commissioners and for the individuals receiving support

Chapter 22: A vision for Total Technology Support – defining and creating a future of user empowerment, independence and choice based on technology, and the key developments that may be needed to make this a reality.

These are followed by 2 appendices:

E: Impact of Digital Transformation on support options – a comparison of the potential interventions to the vignettes presented in Appendix A with those that are possible in a digital world

F: Sources of digital equipment and platforms – a resource of companies and organisations that produce or license support services for care

The length of each chapter ranges from a few pages to more than 20 pages, depending on the volume of material that is available, and the depth that has been required to offer a comprehensive coverage. In particular, topics such as the support for people who have a cognitive issue (such as dementia) are covered extensively both because of the challenges posed to the care system, and because of the range of available applications for technology.

Readership and Style

The book has been written in a generally informal style; it is intended to become a handbook for the industry. Academic references have been deliberately omitted, but some mention is made of collaborators and of some individuals who have been influential in the development of the services that have been enabled through the introduction of new, or existing, technologies. Opinions are offered at all stages, generally without a level of evidence that might be expected in a text book. Indeed, some 'sweeping statements' are made that would have to be left out if subject to peer review. These are offered without apology!

The focus is on innovation and on applications of the technologies to enable people to live fuller and more interesting lives when challenged by disabilities or old age. The technologies will often be solutions to problems that have been identified through assessment. The ones that employ the most sophisticated technologies are not necessarily the most appropriate ways of providing support. Indeed, the simplest options are often the most acceptable, reliable and the ones that are easiest to implement because they don't rely on other products or services being in place. This comes as no surprise to many professionals, such as occupational therapists, who understand problem-solving from a practical perspective and who are expert at considering other approaches. The intention is not to make readers behave or think like occupational therapists but to keep an open mind on how keeping things simple can be a good option, especially when dealing with a wide range of issues.

The groups or organisations that might find the book interesting (and in no particular order) include:

- Local authority commissioners of support services
- Public health officials
- Clinical Commissioning Groups (in England)

- NHS service delivery organisations
- Telecare and similar service providers
- Registered social landlords
- Private housing providers and landlords
- Assistive technology equipment designers and providers
- Assessors across the whole health, social care and housing spectrum
- Allied professionals (especially occupational therapists)
- GPs and community nurses
- District nurses
- Mental health counsellors
- Housing associations and their architects
- Third sector organisations and social enterprises
- App and web developers and designers
- Anyone with any type of limiting disabilities
- Elected members of local authorities, national assemblies or parliaments; and
- The general public

The book has been written primarily for a UK and Ireland audience, but many of the technologies and services are available also in other European countries, Australasia and North America. The need for technology support applications is apparent across the developed world and in Japan, which has the highest percentage of older people. In China, there is increasing life expectancy. This, coupled with a policy of only one child per couple in most urban areas for over 20 years, is producing a population in which one grandchild is shared between 4 grandparents. Improved technology will become a major part of their future too. The digital transformation of care services, described in Part 2 of this book, will be relevant internationally.

Chapter 2

Community alarms

A Brief History

Sheltered housing is considered by many to be one of the UK's greatest social successes of the 20th. Century; it has been described as 'a key resource for frail older people'. In fact, their development is largely based on the Alms-houses model of housing and care for older people that had its origins over 1000 years ago. Indeed, the first recorded examples of Alms-houses were of those founded by King Athelstan in York in 937 AD; they were known as 'hospitals' because they provided hospitality and shelter for the poor and needy. Many of the early schemes were associated with the Church but were then passed on to land-owners, Livery Companies & municipal leaders following the dissolution of the monasteries.

During the Industrial Revolution, many new alms-house charities were founded by company owners; they provided retired workers and their widows with a form of 'retirement pension' including a home which was fundamental to keeping well in old age, and a key basic requirement for life as described in Maslow's Hierarchy of Need (see Chapter 6 for the role played in assessment processes and in Chapter 9 for applications for people with cognitive impairments such as dementia). By today, there are some 1750 such charities offering more than 30,000 homes up and down the country. Many of the older schemes have properties that have listed status and form an important part of British heritage. New schemes continue to be developed some of which offer resident wardens and levels of care up to full nursing home standards.

As a result of many homes being destroyed during World War 2, there was subsequently a need to demolish damaged properties as well as those that were no more than slums. There was then such a shortage of homes that radical housing policies were needed to enable the housing of families of ex-servicemen and a growing army of older people, many of whom had lost children in the war. Thus, while the establishment of the National Health Service in 1948 was seen by most people to be the greatest example of social engineering ever seen in the UK, it was but a single strand of a much wider strategy i.e. the development of a welfare state.

Central to the plan was an enhanced role for local authorities who were given new responsibilities and funding to build social housing at a previously unprecedented level. Enormous estates of council houses sprung up in every town and city throughout the country. Many of these estates included special provision for older people. An innovative step was the building of purpose-built bungalows within these new communities which was seen as a key requirement for the integration of older people into society. Sheltered housing schemes and new council-run 'Old Folks Homes' completed the provision in what was to be a New Social Model of Care.

Previously, a family home was often shared by grand-parents, parents and children. Few mothers worked outside the home and, thus, they were available to become de-facto carers to their relatives when the need arose. Older people who had no relatives or who became ill relied on charitable hospitals or the Workhouse. By 1926 there were nearly 600 of these institutions across the UK, which were home to nearly a quarter of a million people, mainly poor families. Conditions within the Workhouse were truly dreadful. The stigma attached to such an existence was so great that the threat of the Workhouse lay deeply embedded in generations of older people.

In 1929, the Local Government Act forced the closure of Workhouses giving responsibility for old and poor people's welfare to county and borough councils. Many of the old buildings were retained as geriatric hospitals or, in some cases, as mental asylums. Others were used as the first generation of local authority 'Old Folks Homes'. However, because they were often located on the outskirts of towns, and because of their previous use, they were generally unpopular. Thus, with the creation of the Welfare State and the new social agenda during the 1950s, it was only natural that new local care homes for older people (as well as sheltered housing schemes) would spring up within towns and cities and would form the basis of new inclusive communities. Conditions within these new care homes were considerably more comfortable than anything that people had been used to previously – even though shared rooms were the norm and communal bathrooms and toilets were also standard. Improvements that might create 'homes of the future' are described in Chapter 18.

Meanwhile, the facilities offered within the new sheltered housing schemes were a great improvement on the older housing stock in which most people lived in post-war Britain. Not only was accommodation in these schemes usually provided on a single level, but residents sometimes benefited from centralised heating schemes, communal meeting rooms and the sharing of gardens and other amenities. Their initial popularity ensured continued expansion and a recognition that the suitability of accommodation for people with specific needs depended both on the type of building (i.e. rooms, apartments or bungalows) and on the level of care required. In the 1960s, this was formalised by the introduction of the terms Category 1 and Category 2 to distinguish between two sets of minimum standards for funding purposes. In the former, communal facilities and a warden was optional and accommodation was in bungalows or in purpose-built flats (usually without a lift). The latter comprises self-contained apartments under one roof with a residents' lounge or laundry, a warden and a lift to serve those dwellings that are on upper floors. Clearly, Category 2 accommodation is intended for more dependent older people. It should be noted that Part 3 accommodation refers to residential care when provided by local authorities.

By 1997, there were 516,524 units of sheltered housing in England alone, with nearly 150,000 more in Scotland, Wales and Northern Ireland. Nearly 60% of these units were provided by local authorities (LAs); the remainder were shared between housing associations (HAs), charities and the private sector. Stock transfers from LAs to HAs have been changing these proportions on an annual basis and continue to do so. Table 3 describes the features of sheltered housing that make them attractive to both tenants and to their landlords and other stakeholders.

Table 3: Special Features of Sheltered Housing

Special Features of Sheltered Housing
Tenants have their own self-contained flats or bungalows
Properties are designed to be easy-to-manage
It provides a safe and secure environment
Help at hand from the scheme manager/warden
Tenant and householder status empower people
It provides a ready-made community
An alarm system provides access to support

However, the construction of new facilities has slumped over the past 25 years despite their being far more older people living alone . This implies a reduction in their popularity perhaps underlying both the increasing expectations of older people and a change in people's actual needs. It may not be accidental that these changes have occurred since the introduction of Care in the Community (1990) at which time it was suggested that much of the sheltered housing stock was obsolescent and in need of considerable remodelling. This was because many of the units were bed-sits, and few had been upgraded to comply with new housing standards.

By the mid-1990s, the difficulties in letting sheltered housing had become a topic for considerable concern for both local authorities and housing associations. Indeed, over a third of all Registered Social Landlords (RSLs) admitted that up to 10% of their stock could be correctly described in this way. It has long been the assumption that older people would be reluctant to move home under any circumstances, but it has been shown that one in ten was considering a move and the most likely reason was the need for increasing levels of care and support. It may be evident that many older people now believe that this old model of sheltered housing fails to provide these facilities and that a move to a home offering higher level of care, such as a residential home, may be inevitable as they grow older and especially if they live alone. Yet government figures in 2017 suggest there are 462,565 sheltered homes for older people for rent in Great Britain. An increasing number of these are Extracare, Very Sheltered or Housing with Care facilities. These are suitable for older people with more care and support needs and are increasingly being used to replace residential care homes. Their facilities and improvements are discussed below.

In practice, private sheltered housing continues to grow in the UK with companies such as McCarthy and Stone offering units described as 'Assisted Living' in prime locations up and down the country. It is thus suggested that the demise of sheltered housing provided by social landlords is a more complex function of location, size and changing patterns of ownership as well as the design and facilities available. If this is the case, then the trends towards higher specifications and facilities outlined above are likely to continue in the future as the proportion of owners to social renters increases.

Technology Features of Sheltered Housing

In the early days of sheltered housing, the fundamental requirement was for tenants to be able to summon and receive help quickly if an emergency situation arose. Most tenants could not afford their own telephone line. In general community situations, the task was to alert neighbours, friends, relatives or other dedicated wardens. A variety of alert mechanisms started to appear from the late 1940s. These have been described as 'bells and buzzers' systems, the earliest of which were introduced in rural Devon in 1948; residents could activate a bell to let their warden know when assistance was needed. Elsewhere, the problem of an unavailable (or non-existent) warden was overcome by using illuminated or drop-down messages in the front window. It became apparent that the larger the scheme, and the more dispersed the housing, then the more difficult it would be to ensure that the alert was received by the warden, and that help would be forthcoming. The notion of alarm systems received more attention as sheltered housing matured; this resulted in a 'standard configuration' proposed by the Ministry of Housing and Local Government.

Dedicated, hard-wired communication arrangements followed that, firstly, assured greater reliability and, secondly, paved the way for two-way speech through wall-mounted intercoms. There was considerable debate on the need for speech channels - some professionals suggested that speech might frustrate the role of the warden and cause delay and confusion. However, the consensus was that speech systems had some worthwhile advantages; this led ultimately to the demise of the simple call-bell system in sheltered housing, and the emergence of the social or community alarm system. Examples of speech units are shown in Figure 4. These were normally placed in the hall-way. They allowed wardens to check that a tenant was well without having to physically visit them every day. These calls were known as welfare checks and became a popular element of support for sheltered housing tenants and are still in use today in some schemes.

Figure 4: Example Speech Units in Sheltered Housing Units

The intercom approach become a useful a useful addition by the front door of apartments; it meant that tenants did not have to open their door to respond to callers who hadn't been invited. In many larger towns and cities, schemes were often large and sometimes in tower blocks. In the absence of a concierge (or doorman) entry for visitors was enabled through an electronic access panel in the doorway (see Figure 5 for examples). Callers were able to contact specific apartments and speak with tenants, who could admit them by pressing a button which activated a remote door release. Similar units are in use today in many sheltered housing schemes and in private retirement housing.

Figure 5: Examples of Door Intercom and Access Units

Over time, it was realised that in the event of illness, or a fall, a tenant might be unable to reach the intercom unit in the hallway. The supply industry responded by introducing pull-cords that were hard-wired into the intercom unit, enabling an alarm to be raised from elsewhere in the property, and opening up a voice channel. Pull cords were usually installed in the bedroom (by the bedside) and next to the toilet in the bathroom. Hard-wired and mains-powered smoke detectors could be connected to the unit in the same way, offering a linked alarm system that could both alert the fire and rescue service and the scheme warden.

Hard-wired pull cords were functional, but have become increasingly unpopular both with tenants and with those responsible for responding to them for a number of reasons:

- They can be pulled in error, especially by visitors (small children) and by pets
- Their length can be shortened (usually by tying them up) – then cannot be reached by someone if they are on the floor
- Their fixed position in the bedroom can prevent a tenant from moving their bed to a more suitable spot
- Bathroom modifications, such as the addition of a shower cubicle, can be thwarted by their position, or the wiring; and
- Adding additional pull-cords to match changing mobility levels is difficult.

Despite the short-comings of the early alarm technologies, the greatest threat to the original sheltered housing model lay in the issues of warden recruitment, retention and cost. By 1980, few schemes were able to continue to employ a resident warden who would effectively be available to respond to calls 24/7. The answer lay in 24-hour control centres.

Dispersed Alarms and Control Centres

Improvements in telecommunications equipment such as pagers and digital cordless telephones enabled community alarms to be relayed to on-site wardens when present, enabling messages to be sent directly to them irrespective of where they happened to be. This allowed wardens to become mobile for the first time and allowed the breaking of the chains that had previously kept them tied to their schemes. It also effectively accelerated the phasing out of resident wardens. At other times, and especially at night or at weekends, alarm calls could be rerouted to a24 hour control or monitoring centre which was originally local. Each town, city or county had their own centre.

These centres sprung up all over the country, supported by a new generation of call handlers who quickly learned how to deal with different types of situations ranging from genuine emergencies (including heart attacks, accidents, and security incidents) through to more mundane calls from wardens clocking on and off, and from lonely tenants who wanted a chat in the middle of the night. A new social alarm industry was born.

Meanwhile, it was suggested that the functionality of community alarms could be extended beyond sheltered housing schemes using the telephone as the easiest way to raise an alarm through to one of the new control centres. Ownership of telephones amongst older people had increased significantly since the 1960s and increasing numbers of older people were living alone in ordinary (general needs) houses, sometimes many miles away from their families. What was needed was a special telephone (and system) that would allow vulnerable people to raise an alarm in a simple way – a Personal Emergency Response System (PERS).

Andrew Dibner, a US physician, came up with the original Lifeline (or carephone) concept, a telephone with a big red button that could provide a fast dial through to a hospital or other response service. The idea was an immediate success both in the USA, Canada, Australia and in many European countries, but especially in the UK where more municipally-based approaches were available. However, a bigger breakthrough, perhaps, was his introduction of a wireless alarm trigger that could be detected (and uniquely identified) by a receiver in the carephone up to 50 yards away. The transmitters operated at low power in a narrow frequency band in the VHF or UHF bands that enabled them to cut through other signals and be uniquely identified by the carephone receiver which had a narrow passband filter (usually Class 1) to avoid interference. Different countries chose different frequencies, but standardisation by continent enabling manufacturers to export products more freely. European Norms were introduced to regulate the market, leading to UK transmitters moving from a frequency around 172MHz to new frequencies around 869MHz or 169MHz in 2004. EN50134 remains the key standard, from which the relevant components are:

- **EN 50134-1 2002 – System Requirements**
    - Specifies the minimum requirements for a social alarm system.
- **EN 50134-2 2000 – Trigger Devices**
    - Specifies the requirements for manually and automatically activated wireless trigger devices transmitting a triggering signal, and the requirements and tests for trigger devices forming part of a social alarm system.
- **EN 50134-3 2012 – Dispersed or Local Social Alarm and Controller**
    - Applies to social alarm base units and controllers that receive an alarm triggering signal from manually or automatically activated trigger devices and convert this into an alarm signal for transmission to the monitoring centre (or to an individual alarm recipient). It covers a range of functional requirements such as: the Power supply, 2-way speech, alarm processing, testing and fault reporting, documentation and controls.
- **EN 50134-5 2004 – Interconnections and Communication**
    - Specifies the minimum requirements for the interconnections and communications within a social alarm system including the radio frequency employed, the system availability, the quality of audio, transmission times, the reporting of faults and system availability.
- **EN 50134-7 2003 – Application Guidelines**
    - Provides recommendations to service providers (and their sub-contractors) for effective and efficient management policy and procedures for installing, testing, operating and maintaining a social alarm system, including technical facilities and organising assistance. It also requires risks management, enhanced record keeping, and documented procedures and training of staff employed to visit the homes of service users, and certain levels of staffing to ensure an adequate response.

Extensions to this European Norm continue to be added, but the social alarm systems covered by EN50134 are limited by one-way communication, which prevents confirmation that an alarm has been successfully received. For completeness, it should be noted that there are a number of British Standards that are also relevant to the operation of social alarms in the UK. They haven't been harmonised into European Norms or international standards but are important in defining signalling protocols and the properties of the alarm receiving centres.

BS 8521:2009 - Specification for Dual-Tone Multi-Frequency (DTMF) Signalling Protocol for Social Alarm Systems

• Specifies requirements for the transfer of information and controls within a social alarm system, by means of dual-tone multi-frequency (DTMF) tone signals via the public switched telephone network.

BS 5979:2007 - Remote Centres Receiving Signals from Fire and Security Systems. Code of Practice

• Makes recommendations for the planning, construction and facilities of manned and unmanned remote centres, and for the operation of alarm receiving centres

BS EN 50518- parts 1, 2 & 3 - Monitoring and Alarm Receiving Centre: Location and Construction Requirements

• Sets out location and construction, technical requirements, procedures and requirements for operation.

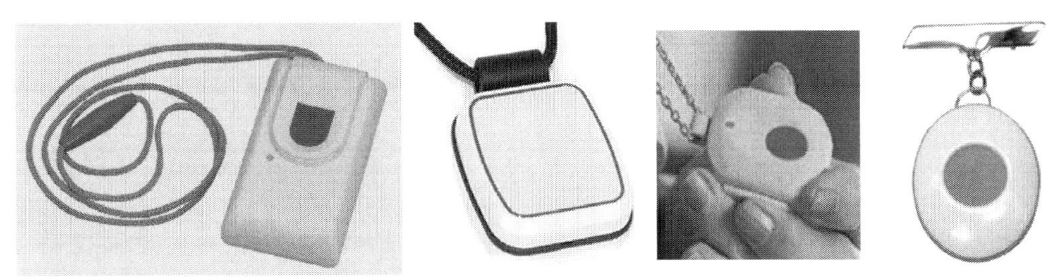

Figure 6: Emergency social alarm triggers designed for use with a lanyard

The introduction of the carephone (or dispersed alarm unit), together with the wireless trigger, overcame most of the problems previously associated with raising an alarm, enabling the community alarm (or PERS) to be used in any dispersed setting. This allowed the support of many more older and/or disabled people who wanted to continue to live independently in their own homes. The systems have been adopted across the world though, almost unique to the UK, control of the systems remained with local authority housing departments or with one of the rapidly growing housing associations. Figure 6 shows a variety of alarm triggers designed for wearing around the neck on a lanyard, or as a brooch on clothing.

Unfortunately, as more of these devices were deployed, many users complained that they were a badge of dependency, leading to poor wearing compliance, and a total reluctance to wear them at night when many emergencies arose. They were also likely to be taken off in the bathroom which was the place where people often felt unwell or suffered accidents such as a fall due to wet or slippery surfaces. Manufacturers responded both by making their triggers showerproof, and by offering wearing options that included the wrist. It was rightly felt that they could be worn in the same way as a wrist-watch, especially if the alarm button was slightly recessed to avoid the possibility of it being pressed in error. Most could be worn in bed at night without fear of false activation. Figure 7 shows some of the units available. They were particularly useful in dispersed settings and facilitated wider deployment in the community.

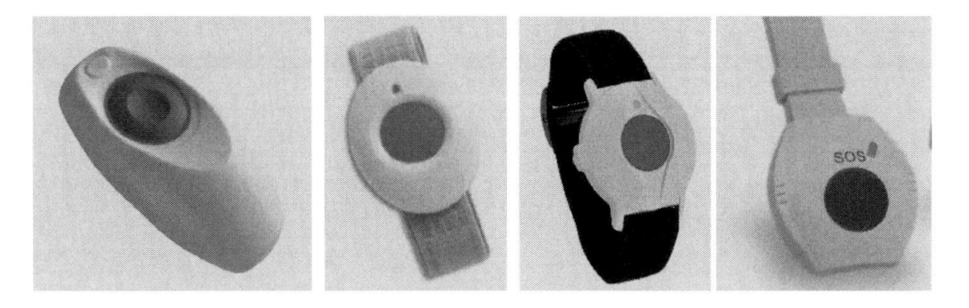

Figure 7: Wrist worn alarm trigger units for social alarms

Limitations of Social Alarm Systems and Services

The need for an alarm trigger to be worn at all times may be apparent, especially when accidents or illness can strike an older person down at any time of day or night. Some organisations produced statistics that showed the impact of a long lie following a fall, for example, demonstrating the importance of people wearing their alarm trigger continuously. With battery life extending to 5 years, in some cases, there was no practical reason to remove the device at any time. Nevertheless, surveys of users across the UK showed not only that that people didn't always wear their pendants, but also that in the event of an emergency, many were not likely to press their alarm button. Figure 8 shows the results of a review of emergency situations recorded by one London Borough's community alarm team in 2004 looking at why service users had not raised an alarm using their equipment. It may be observed that the biggest single factor (nearly 40%) was that their pendant was not with them at the time of the emergency. Over a quarter of the cases involved people hoping that the situation would resolve itself without intervention. Others did not want to disturb anyone (usually at night), while the remainder were either unaware that they were in a potentially dangerous situation or were reluctant to let on that they needed help, sometimes because they feared that the outcome might be hospitalisation or admission to a care home.

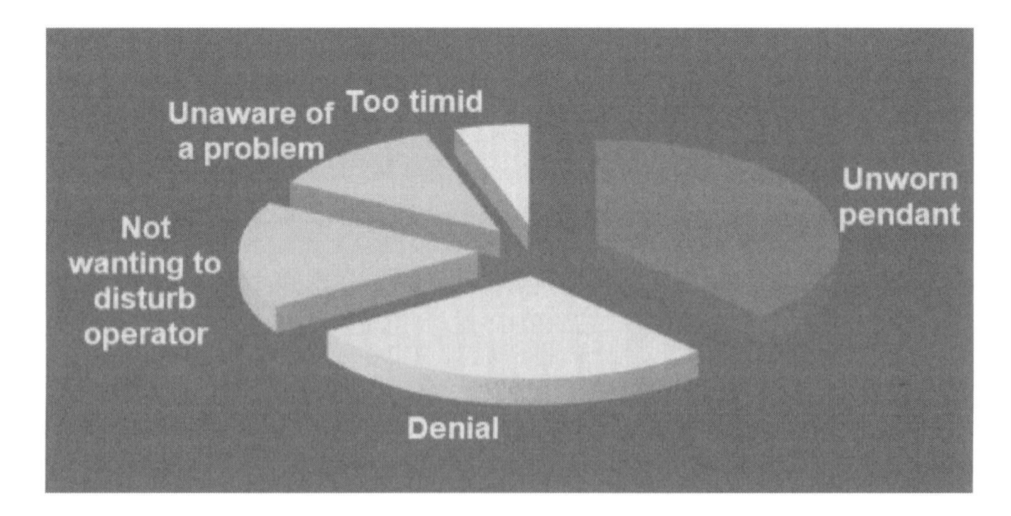

Figure 8: Why Social Alarm Users Failed to Raise an Alarm in an Emergency

The benefits of social alarms can only be realised if the users have enough confidence in the system to carry the trigger devices with them at all times. Several alarm services have surveyed their users to estimate their wearing compliance. The following is a synthesis of the results from such surveys:

- Fewer than a half of users wear (or carry) their alarm pendants at all times
- Two-thirds wear or carry them most day during the day-time
- Under 20% wear their device in bed at night
- One-third of users leave them on their bedside table, or in a drawer, at all times
- Nearly 10% don't understand that the devices have a limited range, and are generally not useful outside the home
- 5% don't know where their pendants are at any one time.

Much greater awareness is needed amongst members of the public. Perhaps the limitations need also be made clear to health and social care professionals. It is not uncommon for alarm pendants to be removed when a patient is admitted to hospital. On some wards, they were kept in a drawer without any indication of who owned each device. Patients might then be discharged with someone else's pendant, staff failing to appreciate that each device is programmed only into their own dispersed alarm system!

Despite the limitations of compliance, social alarms became an important component of Care in the Community, each local authority having their own service, including a call monitoring centre. In order to receive a service, customers needed to have 2 named individuals who could respond in the event of an emergency, and if a visit from the emergency services wasn't necessary. Few (if any) checks were made on these potential responders. In some cases, the responders lived many miles away, and had no means of transport. In other cases, older people named their friends or neighbours, who were sometimes frailer than they were, and either unable to respond, or perhaps unable to be roused from their sleep by a telephone call. In one audit performed by the management of a housing association, it was found that more than 6% of the named responders were either deceased or incapable of attending a property in the event of an emergency!

The original housing departments, and their successor housing associations, often utilised their former sheltered housing wardens as community responders. This overcame the problem of people having no family, neighbours or friends nearby who were willing to become a named responder. Over time, the roles of community responders changed, and this part of the service became an optional extra for which people had to pay a weekly charge. In order to make the response service successful, it was necessary for them to have access to a key. A key-holder element was then added to services but there was considerable debate about whether the keys should be held at a central location, or whether they should be carried in the vehicles used by the response team. The former approach, still used by a few local authorities, has the benefit of security, but can lead to a delay in a responder arriving at the property. The latter approach has all but been abandoned as the possibility of a vehicle containing keys to the properties of the most vulnerable people in an area being stolen is considered to be too great.

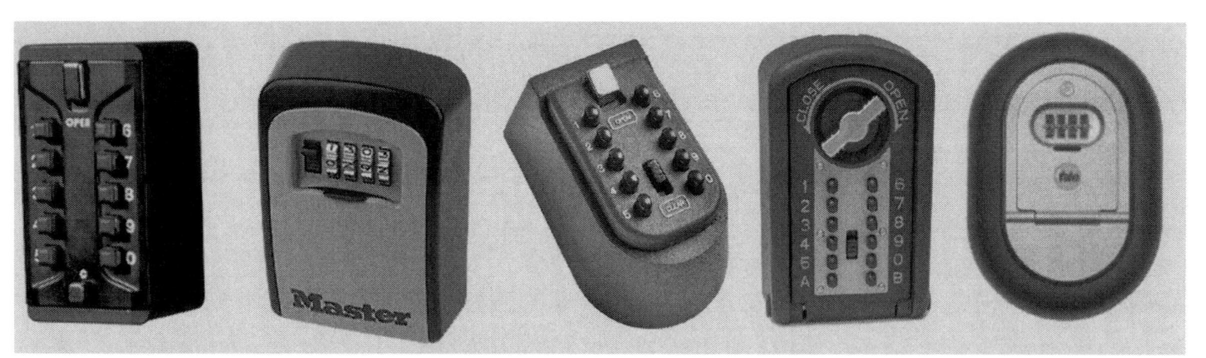

Figure 9: A range of mechanical key safes employed by social alarm services

Fortunately, the problem of access was solved by the introduction of key-safes, mechanical storage boxes that could be attached securely to an outside wall of a property, and which could be opened using a bespoke code. The code could be used by a responder (or by the police, paramedics or fire service) in the event of an emergency enabling them to quickly access a key. The code would be random and kept by the 24-hour call centre. This approach offers a robust approach to key security provided that the code is not shared with family and friends who might remove the key and fail to return it. The images shown in Figure 9 are from a trial of key safes conducted on behalf of Anglesey Council in North Wales.

All the ones shown were found to be relatively easy to install. They could all be broken into by a potential burglar, though the risk of them choosing to break into a key safe rather than the front or back door was quite low. In fact, the greatest potential problem lies in the selection of a random-access code; the number of digits used is normally 4, and in some models, no digit can be repeated. Although this restricts the number of available codes, this shouldn't be an issue – unless the year of birth is selected. In audits of alarm codes undertaken by the author in 2010, it was amazing how often 1920, 1923, 1924, 1925, 1926, 1927 and 1928 appeared; most would recognise that these are the year of birth of people who were 80, 83, 84, 85, 86, 87 and 88 respectively. Bearing in mind that the average age of service users was 84, would it not be surprising if people chose their year of birth as their key safe access code, rather than using a more robust approach? It should be added that the next generation of key safe devices are Home Office or Police-approved and are far more robust. However, there are many thousands of the original device out in the field.

There is little reliable information available on the number of connections held by each community alarm monitoring centre, partly because things change on a daily basis. However, in the last national survey performed by the TSA (the Telecare Services Association – the trade body for community alarms and telecare organisations) in 2011, and analysed by this author, 98 of 223-member organisations responded, yielding data on number of connections shown in Figure 10. The average number of connections per centre was 7600, but over a half of those responding to the survey had fewer than 5000 connections; at the time, fewer than 8000 connections was considered to be an unsustainably situation, and this has led to further consolidation of centres through take-overs, outsourcing and regionalisation initiatives.

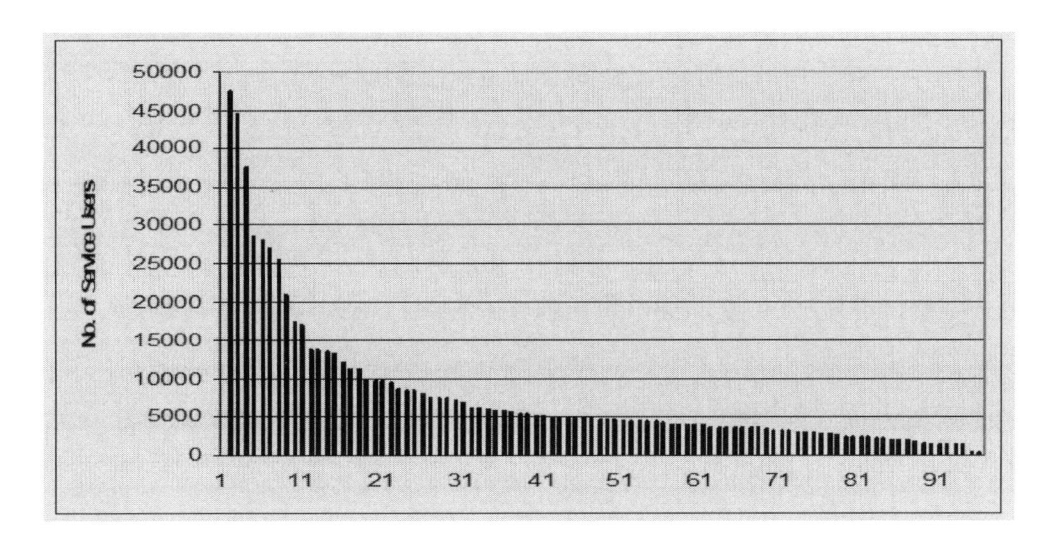

Figure 10: Number of Connections for 98 alarm monitoring centres in the UK

The number of calls fielded per day by different monitoring centres varies considerably, based on the day of the week, the number of connections, and whether or not the centre dealt primarily with private subscribers or with tenants in social housing. Most centres had a reasonable mix of connections, but about 10% had private subscribers representing fewer than one in ten of their connections. When the activity levels in 4 centres in North Wales were analysed in 2006/7, it was found that one some nights, the total calls per hour was less than 2 across the whole region. Yet, a minimum of 2 staff was usually in place at all times in order to ensure that they could have comfort breaks, and answer calls without delay. It follows that significant savings could be made by sharing a single centre. This approach also led to big improvements in service quality.

Concluding Remarks

Community alarms have a long history. They have been very successful, and acceptable to most vulnerable people – even though they choose not to carry their alarm buttons/triggers with them at all times. They are active devices; they need the user to press that button to raise the alarm. They are also reactive, so their relevance to prevention is limited to how an alarm monitoring service looks at trends, and their protocols for liaising with families and other supporters (or commissioners) to provide feedback.

Nevertheless, it is likely that social alarms remain the most popular and most used form of support used by older people both in the UK and across Europe, North America and Australasia. This is unlikely to change quickly because they have become a valuable component of support in many countries, especially those countries that have a community focus where people are prepared to go out of their way to respond to the needs of their neighbours at times of distress. The Senior Alert Scheme in Ireland provides funding for people aged over 65 to be offered a high-quality alarm system with a year's free monitoring, professional installation, and on-going maintenance. User satisfaction levels are high, and the number of referrals reached an all-time high during 2017-2018 though increasing numbers of service users need to use a GSM modem, because they no longer have a landline. This is one of the threats that will, sooner or later, lead to a transformation of services and a shift to digital technologies that will be described in Part 2 of this book.

Chapter 3

Social alarms to telecare

Introduction

It is apparent from the poor wearing compliance of social alarm pendants that many people will choose not to raise the alarm when in trouble, even if the process is made simple, and a direct channel to help is provided using the telephone network. In other cases, an individual may not be aware of an emergency until the situation may have become so severe that they are physically unable to respond. This might be the case when they are physically unwell, or when an environmental issue such as extreme heat, poisonous gases or smoke have overwhelmed them. There is a need for passive alarms that don't require the user to act.

Such sensors would need to take decisions automatically, so that the user is not then involved in deciding whether a situation is dangerous, and if help is needed. Such sensors may be described as SMART. The acronym stands for:

Specific to a given parameter
Modular – can be added or taken away according to circumstances
Automatic in operation
Rapid in making a decision; and
Tested to conform with quality standards (where applicable).

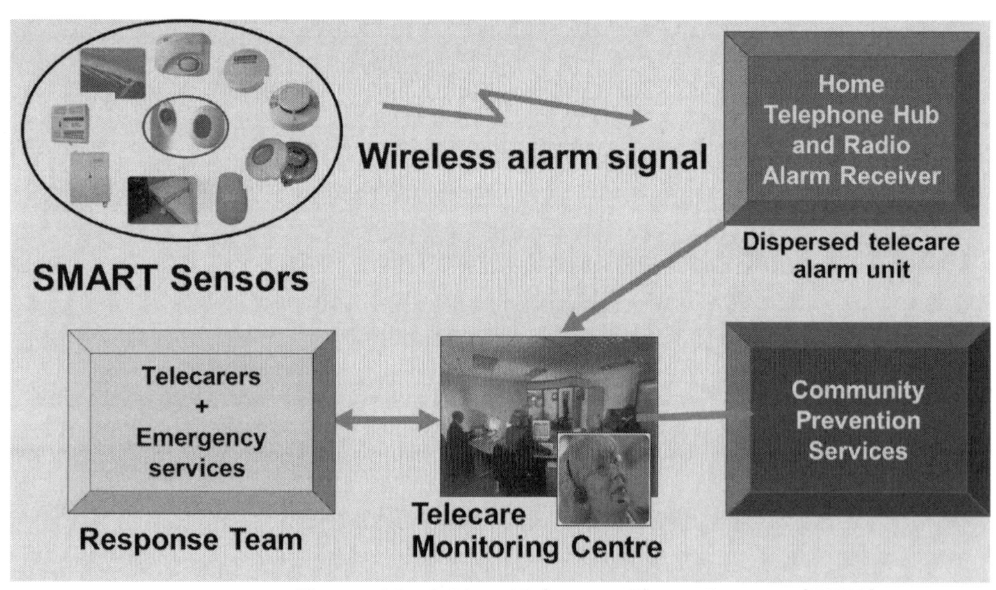

Figure 11: A New Telecare Alarm System (2006)

The SMART sensors would operate alongside existing portable alarm triggers in a telecare system as shown diagrammatically in Figure 11. In this model, the 24-hour telecare monitoring centre receives all incoming call traffic and is then able to deal with the situation appropriately. For some situations, this might involve giving reassurance over the telephone, but could also require the emergency services to be called. The system shown in Figure 11 should also comply with EN50134 in European countries, or their equivalent in other continents.

The model shown in Figure 11 became the basis for many telecare systems and services that were developed in the UK between 2006 and 2010. These were funded, initially, by grants from central government. In England, this was known as the Preventative Technology Grant which was allocated to local authorities with social care responsibilities; it was not ring-fenced, and was not, therefore, specifically to be used to pump prime telecare services. Indeed, some county councils (which had no social alarm services of their own) chose to spend their grants on more general assistive technology projects (see chapter 5 for more descriptions of AT). Similar grants were available in Scotland and Wales, where they were allocated with more direction from their respective devolved governments, leading to more direct benefits to the local services.

No further direct grants have been received from governments in London and Cardiff, but Scotland has continued to invest in telecare services and has established both a centre of excellence for a wider range of remote care services (the Scottish Centre for Telemedicine and Telecare in Aberdeen) and a network of local authorities to share learning. Neither housing departments of local authorities nor housing associations have benefited directly from grant funding, which many regarded as perverse because of their background in running telecare monitoring centres, and because the first smart telecare sensors to be employed in the UK were for detecting environmental emergencies.

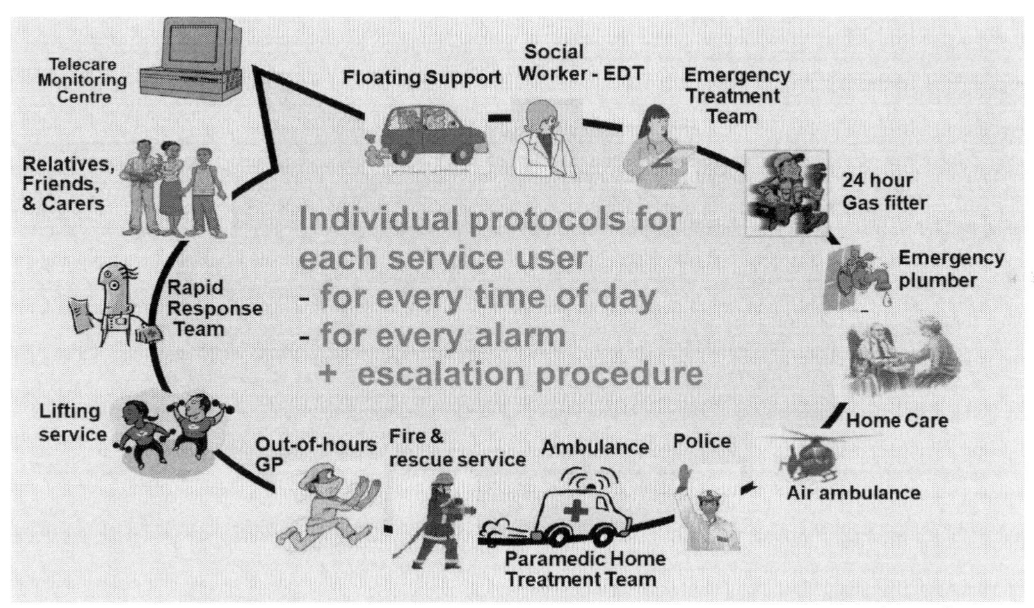

Figure 12: The 'Ring of Care' Community Response Team (2008)

One of the other elements of the telecare support mechanism shown in Figure 12 is the Emergency Response Team. This is an important change from the old social or community alarm model because it can be expanded to include many more services that are appropriate to deal with emergency situations. Figure 12 includes the author's 'Ring of Care' concept which embraces a much larger group of community organisations and services. This is particularly appropriate to the world of alarm telecare, because the use of smart sensors in the homes of service users means that the alarm receiving, or monitoring centre staff have a priori knowledge of the cause of an alarm call in many cases. This means that they won't call out paramedics to deal with a flood, and they won't call a lifting service when an individual needs the police to deal with antisocial behaviour. It should lead to more efficiencies resulting in faster and more appropriate responses. It is therefore essential for each service to have a full set of protocols in place to deal with all eventualities, as well as escalation procedures when the immediate response is not appropriate or possible. If call handlers are appropriately trained to operate such protocols, and to ask the correct questions of callers, they will often be able to reduce the number of occasions that a physical response will be required. This may be used as a Key Performance Indicator when comparing the quality of monitoring provision.

It may be apparent that there should be good reporting and feedback from all alarm incidents, both so that the service has a record of outcomes, and so that appropriate actions may be taken to avoid a similar accident or event occurring in the future. Figure 12 shows the inclusion of a community prevention service. Back in 2006, this was an idea rather than a representation of what was available at the time. Some examples of the services that might be included in a community prevention team are shown in Figure 13. The role of the alarm centre for coordinating these support services is apparent, and the coordination of such services can be used as another key performance indicator.

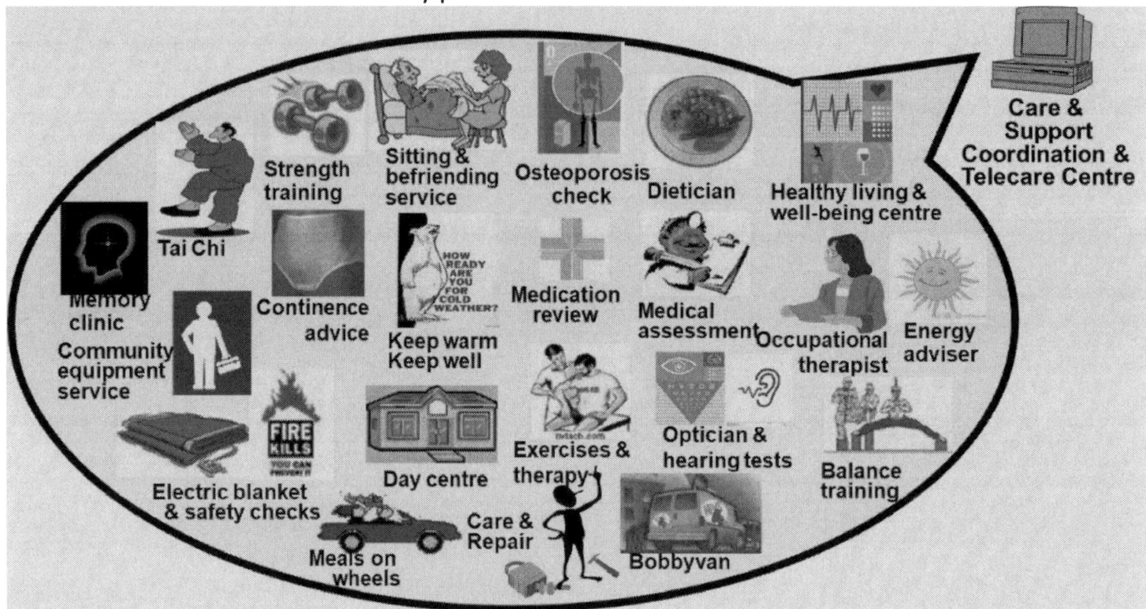

Figure 13: A Possible Community Prevention Team

Telecare Alarms in Risk Management

The aim of many telecare alarm services has been to manage (i.e. reduce) the risk of vulnerable people becoming ill or being involved in accidents that will ultimately impact on their quality of life, or that of others. More generally, such services can benefit all stakeholders in different ways as outlined in Table 4, though there is little robust evidence to confirm such outcomes. N.B. alarm telecare cannot support all these outcomes – they need other forms of intervention, either a wider range of assistive technologies, or an extended use of social prescriptions or services offered by third sector or statutory organisations. Nevertheless, it may be apparent that the use of more smart sensors can directly yield the outcome benefits sought be stakeholders.

Table 4: Benefits of Alarm Telecare to Major Stakeholders

Service Users	Carers	Social Services	Housing Providers	Health Economy
Increased choices	Improved peace of mind	Reduced or delayed care admissions	Reduced voids (sheltered housing	Reduced hospital (re)admissions
Support for self-care or prevention	Reduced stress burden	Improved assessments	Increased demand for special housing	Improved use of medication
Improved Quality of Life	Improved Quality of Life	Fewer delayed transfers of care	Fewer incidents of accidental damage	Fewer A&E presentations
Increased independence	Increased independence	Reduction of loneliness	New roles for wardens	Improved access to specialists
Improved access to services	More respite opportunity	Reduced demand for homecare	Improved security of properties	Shorter hospital stays – less DeToCs

It may be argued that many of these benefits are obvious, and do not need to be supported by major research studies. This is particularly the case when it comes to the management of risks. Generally:

The RISK associated with a potential
accident/dangerous incident or event
=
The LIKELIHOOD
of the event occurring
x
The DEGREE OF HARM
that occurs as a consequence

Both the LIKELIHOOD and the DEGREE OF HARM can be estimated as a score (1 to 5) based on actuarial evidence or on experience. In practice the harmful consequences relevant to many public body stakeholders need to be considered differently. Each may be relevant as they may fund telecare services for certain groups of people. Table 5 compares such outcomes from 5 different perspectives. N.B. ADLs are Activities of Daily Living such as bathing and transferring from bed to walking. In each case, the management of risk-taking action to reduce the risk score as much as possible, but certainly below 15, which is often used as the threshold of acceptable risk. Once a high level of risk has been identified, it follows that action may be needed to mitigate that risk, and this can be used to develop a risk action plan. This approach is not restricted to healthcare but may be used in any industry where managers have a duty of care over both their own staff and over the customers.

It may be evident that the overall risk in any particular set of circumstances may be managed (or controlled) by either reducing the likelihood of an accident or mishap occurring, or by reducing the level of harm possible as a consequence. Technology may place an important role in both elements; telecare alarms generally help to reduce the harmful consequences by providing rapid detection of the issue and ensuring that someone (or some organisation) is informed so that an early intervention is possible. Effectively, it plays the role of an electronic security or safety blanket using 'a stitch in time saves nine' principle.

Table 5: Harmful Consequences Scores for Different Stakeholders

Stake-holder Score	Social Services	Local authority	NHS	Mental health trust	Landlords
1	No threat to independence	No significant harm to public	No NHS requirement	No threat to self/ others	No injuries nor damage
2	Minor threat to independence	Potential for long term harm	Need for some NHS resources	Minor threat to self/others	Minor injury or need to redecorate
3	Unable to perform some ADLs	Some harm over short & long term	GPs, ambulance or hospital care	Threat of self-harm or violence	Injury or damage to property
4	Many ADLs not possible	Short- and long-term harm	Emergency & follow-up treatment	Significant	Serious injury or damage to property
5	Requires Long Term Care	Significant harm to individuals	Major trauma or death	Suicide, violence to staff/ public	Fatality or major property damage

The model for alarm-based telecare (and for two more generations of intelligent system) was published in 1996 jointly by a research group led by this author at Bangor University in North Wales, and British Telecom (BT Labs, Martlesham). This provided a blue-print for using the existing social alarm infrastructure described in Chapter 2 through three generations into a complete package for supporting older people remotely in their own homes. However, despite enthusiasm from many forward-looking social services and housing departments, there was a reluctance in the supply industry to invest in new designs beyond the alarm generation. The reasons may be complex, but at their core was a belief that health and care professionals, families and older people were yet prepared to accept more reliance on technology. This remains a back-drop to this book and may be a factor in the struggle to move from one type of technology to its successor, even if the new devices are smaller, neater, more powerful and less intrusive.

Types of Telecare Sensors

It is possible to design and fabricate an enormous number of different sensors to provide an alarm when an individual is unwell, has experienced an accident, or is exposed to danger in their own homes. Figure 14 shows a non-exhaustive list of possible applications. However, only a relatively few have been commercialised to date. This is either because the market for some devices is considered by manufacturers to be too small for commercial success, or because the sensors would look strange in the home environment, or because they would be too large to be worn unobtrusively. In all cases, battery life is an issue, and one that will limit the possibility of having wearable devices that link into this type of telecare system. Many of these issues can be solved in the digital world, and new solutions are described in Chapter 15.

Abuse	Catheter blockage or removal	Forgetting doors are open	Instability & giddiness	Orientation failure
Adverse drug reaction	Cooking accidents	Floods and leaks	Isolation & depression	Poor hygiene
Agitation	Explosion due to gas	Frozen pipes in kitchen	Lack of nutrition	Sequencing problems
Air quality and smells	Falls	Heat stress	Lack of hydration	Severe pain
Asthma attack	Faulty heating appliances	Hypoglycaemic incident	Medication non-compliance	Slip/illness in bath/shower
Blocked flues	Fever	Hypothermia	Mental health crisis	Sleep apnoea
Breathlessness	Fire	Immobility	Nocturnal epileptic seizure	Slump in chair/stroke
Cardiac arrhythmia	Forgetting cooker is on	Infection	Nocturnal enuresis	Stomach pain/blockage
Carer absence	Forgetting taps are on	Insecurity/fear of crime	Noise	Wandering

Figure 14: Risks and Applications for Reactive Telecare Sensors

The devices that can be used for risk management purposes of the type shown in Figure 14 can be separated into 4 major groups:
Environmental sensors – these are installed in the home, usually on the floor, walls or ceiling. They measure parameters such as levels of heat, smoke, gases, air quality, smell or moisture on a continuous basis, comparing them with defined thresholds that may be governed by official standards. These devices would be particularly relevant to a home owner or to a landlord. The addition of an appropriate radio transmitter to an existing smoke detector or carbon monoxide sensor is relatively straightforward. Such devices were the first to be deployed at scale in new telecare systems. The systems deployed in monitoring centres were already capable of receiving these alarms and of identifying them uniquely.
Security sensors – these devices are likely to be simple binary devices used in intruder alarm systems and are therefore relevant to landlords as well as home occupiers. They were also simple to introduce. Passive infra-red (PIR) movement detectors, and 'door open' detectors were also introduced into telecare systems but were more difficult to deploy at scale because

they required the system to differentiate between the activities of the inhabitants and those of intruders i.e. the householder must inform the system when they go out or go to bed.
Home safety sensors - these aim to identify situations or events that would threaten the well-being of the individual rather than the property. They are more difficult to design and to use as an alarm because they may contain more than one sensing device and are unlikely to be standards that govern the thresholds. There may be a need to have some programmable thresholds to enable them to be suitable for a wide range of users. Examples might include falls detection, a lack of nutrition, and social isolation. Some of these devices use security sensors but with local intelligence so that they were programmed to alarm for a lack of movement within defined time periods.
Medical or physiological sensors – these devices are intended to detect abnormal symptoms that might be indicative of an exacerbation of an illness. Many would be considered to be medical devices that could not be used in a conventional telecare system. Others measure more general parameters which, when combined with other factors could indicate a potential health problem such as hypoglycaemia and fever that may need to be investigated further.

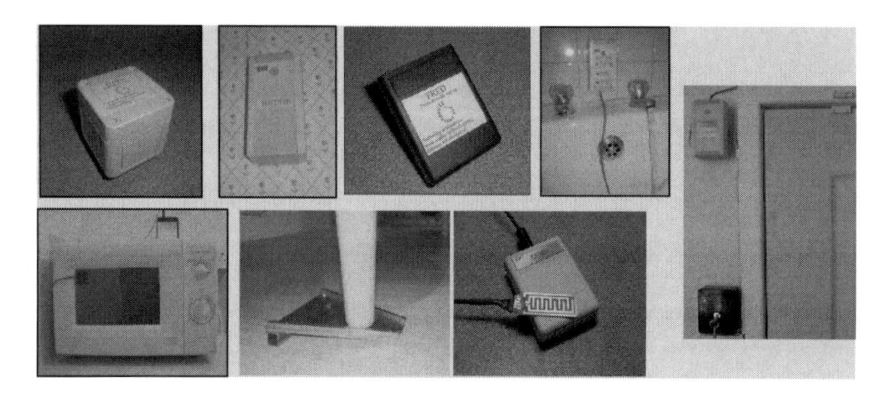

Figure 15: The Smart Sensors Introduced for Trials in 1998-2000

The first genuinely 'smart' sensors were introduced by a specialist company, Technology in Healthcare, that the author spun out of Bangor University in 1998. They gained Millennium Product status for their EXTRA (EXTended Remote Alarms) concept and for introducing a range of prototype devices including the ones shown in Figure 15. The first of these devices was known as FRED - Fall Response Emergency Detector (upper centre of Figure 15) - a worn device that was activated by an impact and a change in orientation that lasted 15 seconds or more. Other devices included FLORA - FLOod Response Alarm (top left), HETTIE – Heat Extremes Timed Temperature Intelligent Evaluator - (second top left), and ANWEN – Automatic Night Wandering Emergency Notifier – (extreme right).

These devices were fitted with Tunstall transmitters and, together with Tunstall linked smoke detectors, were deployed in a number of pilots and trials, the first of which was Tuag Adref (Towards home), a project run in conjunction with the local authority in Anglesey, the local District General Hospital (Ysbyty Gwynedd in Bangor) and Care and Repair Wales. This pilot was aimed at supporting older people with dementia to return home safely following a period in hospital. The greatest success was in developing ANWEN so that a family member, or a member of a domiciliary care team, was able to switch the device on when leaving a property at night. Particular issues relating to the support of people with cognitive impairments are discussed further in Chapter 9.

Other devices, including a range of bed occupancy monitors, enuresis alarms and other bespoke sensors were introduced in Cheshire and in County Durham where they enabled new response protocols to be developed at their monitoring sensors. Similar arrangements were also introduced into trials in West Lothian in Scotland and into housing associations including Fold in Northern Ireland and Hanover both in England and in Scotland. These developments accelerated in 2002 when Technology in Healthcare was taken over by the Tunstall Group. This led quickly to the redesign of the smart sensors and to their commercialisation over a period of 2 years. An entire telecare range was therefore in place by the time that preventative technology grants became available in England (and equivalents in Scotland and Wales). More significantly, it enabled processes to be put in place to ensure that a quality systems approach was possible.

Establishing and Evaluating Telecare Services

Over the course of a few years, it became apparent that telecare services would result in optimum outcomes only if the services were integrated with other community initiatives aimed at achieving similar results i.e. empowering people to remain independent and safe in their own homes for longer. It meant that the simple linear approach:
Referral → Information collection → Alarm Installation → Monitoring
would quickly become redundant. Perhaps the most significant change is that services could move quickly away from a 'one-size fits all' approach as it became clear that the devices appropriate for any individual would depend on their circumstances including the type of property in which they lived, and their attitude towards, and use of, equipment and technology. This led to the development of the process shown in Figure 16, which was first introduced in 2010 as an extension to the Telecare Services Association's R to R (Referral to Review) model.

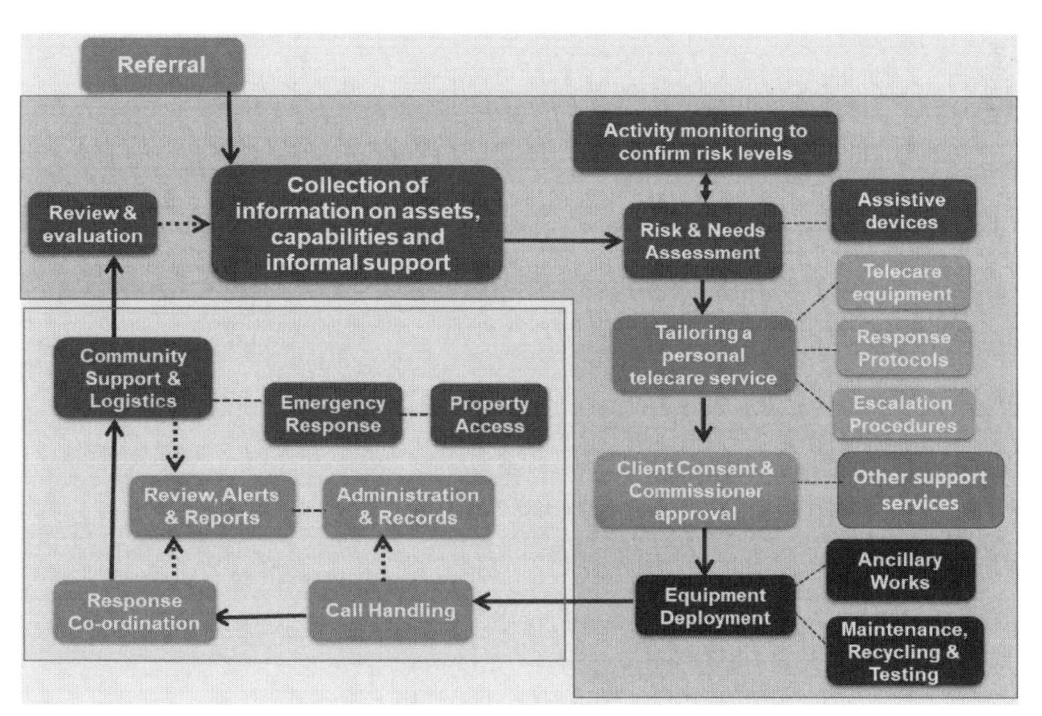

Figure 16: An Extended Referral to Review Process for Telecare Service Deployment

Figure 16 includes 2 major groups of steps. The first, from the collection of information through to the deployment of equipment, was usually seen as the responsibility of a local authority often performed in conjunction with local Community Equipment Services and organisations such as Care and Repair or Home Improvement Agencies. The second group, from call handling through to Community Support (including emergency response), has increasingly become outsourced by a commissioner to a neighbouring service or to a national provider. There are also a small number of managed service providers who take responsibility for all the steps in the process beyond Assessment. They are also expected to provide training to all stakeholders and to support the referral process so that more people can be engaged. The cost base has been reduced by the inevitable consolidation of service providers, but this process might also be responsible for stifling innovation because it doesn't reward the use of extended inventories of technology, and by the integration of different preventive measures into telecare service provision

Irrespective of whether service elements are outsourced or whether they are performed entirely by an inhouse provider, the telecare service itself remains the responsibility of the local authority. They are required to monitor performance and put in place arrangements to ensure that a high level of service quality is achieved. There are no agreed standards on the measurement of quality or performance for a full service, though there are various mechanical performance indicators that can be used to monitor call handling and response functions. Therefore, more subjective measures need to be undertaken. Figure 17 shows a 28 Domain Telecare Evaluation Matrix that has been used to compare services and to enable service providers to audit their performance to identify strengths and weaknesses. The domains cover 4 headings from Telecare Strategy to Measures for Service Improvement. A colour-coded arrangement can be employed to highlight areas where resources need to be focused in order to improve outcomes and outlook.

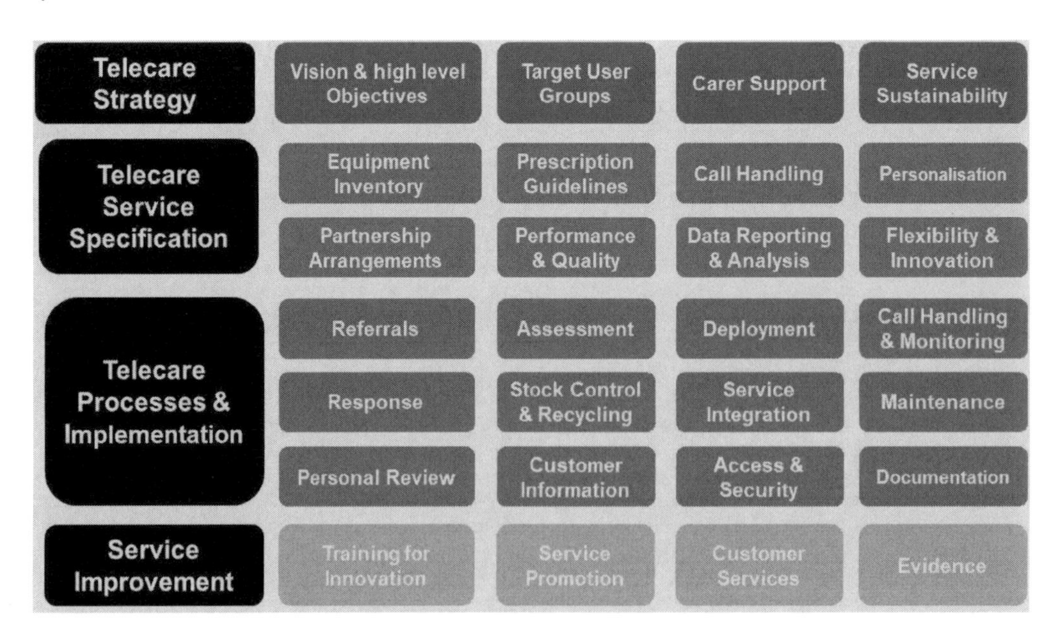

Figure 17: Domains for Telecare Service Evaluation

In general, services that have developed from a clear strategy are best placed to have well-defined service specifications that require an extended inventory of equipment in order to achieve the service objectives. The final elements of the matrix are aimed at improving the service; the requirements include gathering evidence of performance and success in achieving the required outcomes. Despite general agreement on the need for more robust evidence to support the deployment and expansion of telecare services, there are few standardised techniques available. Some results are discussed below but are often neglected in favour of simple cost avoidance (or savings) calculations.

Outcome Successes for Alarm Telecare

The Whole System Demonstrator projects performed in 2011 in Cornwall, parts of Kent, and in the London Borough of Newham, were intended to provide definitive evidence of the benefits that could be realised in an elderly population using standard alarm telecare equipment and services. Unfortunately, the results showed little reduction in social care resources could be achieved by using this form of intervention for the selected participants. This should not be considered to be a black mark against the cost-effectiveness of telecare as the projects were constrained by only including people who were already in receipt of a package of social care, and by excluding people who already received telecare. This approach prevented any investigation of telecare as a prevention measure. It also resulted in difficulties in identifying suitable participants because many of the most suitable people had previously been identified by the local teams and had already been provided with the most appropriate telecare service.
The result of the trial was therefore to conclude that telecare cannot cost-effectively be provided to people whose condition has already declined to such an extent that they are unable to perform at least one Activity of Daily Living. An important lesson was that technology was most likely to be beneficial if it was introduced early (and before conventional services such as domiciliary care) were introduced into a care package. It follows that telecare has a preventive role and should be introduced to people who aren't yet eligible for social care. This requires people to be responsible for funding their Other studies performed in Scotland, North Yorkshire, and the West Midlands, also concluded that alarm telecare cannot be used generally to replace existing services such as domiciliary care and day care centres. However, they found that the use of alarm telecare can increase the confidence of vulnerable individuals to continue to self-care, and more importantly, to feel more independent and in control. These groups might include:

- *People aged over 80 who live alone*
- *People recently discharged from hospital following accident, illness or surgery*
- *People who have suffered a bereavement or whose support circumstances have changed*
- *Those who have presented to a Memory Clinic*
- *People diagnosed with any form of neurological disorder including Parkinson's Disease*
- *Patients who have co-morbidities that require on-going polypharmacy*
- *Those who have poor gait or who are users of walking aids*
- *People who have a history of urinary tract infections or continence issues*
- *People aged over 80 years who have developed visual impairment.*

Some of these groups will be considered in later chapters of this book.

A change of circumstances relating to any (or several) of the above factors may be considered to be a trigger which could lead to a rapid decline into long term care unless interventions are provided. The most popular intervention in the UK is known as Reablement (or Enablement) which aims to help people to accommodate their condition (or illness) by learning or re-learning the skills necessary for daily living. Homecare re-ablement services are focused distinctively on supporting the instrumental activities required for independent living including the support elements shown in Table 6. They have been shown to have positive impacts on both health-related and social care-related quality of life up to 10 months after re-ablement, compared with users of conventional home care services. Services are extremely cost effective in reducing future social care costs.

Table 6: The 10Ps of Reablement

- **Personal care – e.g. assisting with washing, dressing**
- **Practical support – e.g. assisting with meals & laundry**
- **Priming people with information & signposting**
- **Psychological & emotional help – e.g. taking for a walk**
- **Proving their safety – e.g. attending social engagements**
- **Performing adaptations – making the home safer**
- **Providing equipment - simplifying domestic tasks**
- **Pushing best practice to reduce the risk of falls**
- **Prompting medication and nutrition**
- **Promoting confidence – 24/7 digital healthcare support**

Perhaps the best example of digital reablement currently in the UK is the Telecare service commissioned by West Sussex County Council and its Clinical Commissioning Group partners from 2011/12. The service is delivered through Welbeing, a private company originally run by and then part-owned by Eastbourne Borough Council. The service involves referred individuals receiving a personalised package of telecare, free of charge for a period of 13 weeks, during which time they experience the benefits of the telecare service that reduces both their levels of anxiety and those of their family members. At the end of this period, individuals may choose to have the equipment removed, but the vast majority (over 70%) elect to continue to receive the service as private payers, receiving no further funding contributions from the local authority. Many of those who discontinue the service following the trial feel sufficiently well and confident that they need no support moving forward. The result is that the number of people supported by an on-going telecare service in West Sussex over 5-years is over 12,000 with an approximately 20% annual increase in referrals. These come from a range of hospital-based staff as well as community-based assessment teams.

Each referral requires an indication of how telecare is likely to benefit the council and/or the NHS typically through delays to or avoidance of use of other services. They also estimate the period of time over which this would apply. All data is entered electronically through the referral portal, enabling rapid evaluation of the number of service users, and the saving to be made continuously.

The cost avoidance (rather than bankable savings) to the local authority budget can be calculated using a figure of £244.10 per week (net of bed and breakfast costs) for the avoidance of admission to care homes, while a figure of £15.34 per hour is used for the avoidance of domiciliary care. In the same way, timely discharge from hospital (i.e. avoidance of delayed transfers of care) saves the NHS £240 a day. The avoidance of hospital admission (or readmission for a similar issue) results in an average saving of £1992 per incident using tariff figures. The council currently estimates that for every £1 invested in their telecare service, the avoidance of cost (i.e. savings) is nearly £10. It follows that the total cost avoidance per annum seen in Table 7 is over £2.3 million by 2015/16 with further savings to the Fire and Rescue Service, which is also funded by the local authority. Overall, about £2 out of every £3 avoided through the telecare service benefits the NHS, who are increasingly providing funding through the Better Care Fund. This helps to ensure that the number of service users continues to increase; this drives future cost avoidance.

Table 7: Increasing Levels of Savings to Health and Social Care Due to Telecare Service

Health savings	13/14	15/16
Timely transfer	£629,520	£839,360
A&E Admission	£468,932	£625,243
Ambulance Trust	£69,300	£92,400
Total potential savings	£1,167,752	£1,557,003

Social care savings	13/14	15/16
Dom Care	£447,606	£596,808
Care Home	£115,475	£153,967
Total potential savings	£563,081	£750,775

The London Hillingdon Council reported that the mainstreaming of its basic TeleCareLine service and its reablement service led to cost savings of £4.7m after 12 months and a 50% reduction in residential care home admissions, though the evaluation failed to disaggregate the savings. Unfortunately, the baseline for such intervention was not well established, nor any comparison made with comparable boroughs prior to the mainstreaming of their reablement telecare. However, the initiative greatly increased the number of suitable service users recruited into the service and will without doubt yield long term benefits to the local authority, especially if telecare is introduced as a preventive measure following a period of illness or a change in circumstances.

Generally, there are a number of problems to solve to increase the use of telecare services:

- A lack of awareness of telecare as a mature approach to supporting vulnerable people;
- Little knowledge of the latest sensor technologies (and associated assistive device) and applications amongst professional care and support staff;
- Limited assessment processes that fail to match applications to the unmet needs and ambitions of individuals;
- A poor knowledge of the potential outcome benefits of using more advanced telecare services rather than a basic social alarm;
- A fear of technology by many older people, and
- A failure to differentiate between high quality, forward looking telecare service providers and those that only offer older, more traditional alarm-based services.

These problems were recognised in Hampshire, a large county council whose citizens, like those of many county councils, had used a number of different social alarm services for many years, based on a history of services provided by district councils (or their successor housing associations). This had resulted in a wide range of approaches and quality propositions. The council recognised the need for a more standardised approach using a defined range of equipment which was not limited to basic telecare devices, and which included some options on devices. The local authority chose to procure a fully managed service from Argenti, a consortium led by PA Consultants, who ensured that all relevant staff received formal training in assessment and prescription of telecare. At the end of year four of their five-year contract, over 8,600 people were receiving telecare as part of their mainstream care package. The consultants claimed evidence of net savings of over £7 million.

The significance of training is therefore evident. The Welsh Government, through Care Council Wales, procured both basic and advanced assessment training course packages (in both English and Welsh) for free use by all local authorities, housing associations, domiciliary care organisations and the NHS. However, the uptake was low though many of the resources have been used by service providers outside the Principality. The reasons for the poor deployment are many but include a lack of strategy (and a road map for telecare deployment), a lack of investment and resources, and a significant gap in many local authorities between the range of available technologies and their applications, and the capabilities of service delivery organisations and partners, most of whom remain in-house.

It is sadly the case that some local authorities, especially those that run relatively small operations, don't charge their service users an economic subscription for telecare. Many charge users the same for a basic social alarm (around £3 a week) as they do for a more complex arrangement that includes several smart sensors. This results in significant annual losses. It follows that there is little incentive to perform full telecare assessments and, indeed, a financial imperative to reduce the number of telecare service users in order to contain the service losses. This problem can be overcome by increasing efforts to recruit more self-payers, and a 2-tier service, but this may not be politically acceptable in some areas.

Although housing associations were amongst the pioneers of alarm-based telecare, especially the use of environmental sensors to provide early warnings of potential fire, gas leaks and unlimited water flow, they were generally unable to tap into additional funding. However, those social landlords who did invest in sensors saw a return on investment within five years, due to reduced damage and insurance claims. The sensors that had a significant impact were:

<u>Temperature extreme alarms</u> - these provide rapid warning of fire in kitchens and of risk of freezing water pipes (and the potential for advising people to increase their heating in order to avoid the possibility of hypothermia)

<u>Flood detectors</u> - these offer an early warning of plumbing leaks, blocked toilets and overflowing sinks and baths (which can cause damage to floors and ceilings of other properties as well as to electricity and communications systems); and

<u>Gas alarms</u> – methane and carbon monoxide detectors can be used to prevent explosions and to save lives due to poisoning or suffocation

There remains a good argument for fitting these sensors in all properties with permanent connection to the electricity supply, but wireless devices can be installed or removed quickly for people with particular needs. Their role can become crucial for people who have short term memory problems. Applications of telecare for people with cognitive impairments, including dementia and/or learning disabilities are discussed in Chapter 9.

Quality Measures and Improvements

There has been significant debate on the relevance of measures of quality that might be used to identify best practice. The focus has previously been on providing key performance indicators that have meaning to corporate clients i.e. commissioners such as local authorities or housing associations. This is changing as the view of the individual service user, or consumer, becomes more important. Consumers rightly demand quality products and services. There are many ways of describing the quality of manufactured products through the use of minimum technical standards, some of which are international while others are country or continent-specific. They lead to descriptions of quality as measures of excellence that imply a lack of defects and meeting a uniform set of characteristics. For example, ISO 9000 defines quality as the degree to which a commodity meets the requirements of the customer at the start of its life. Quality can be assured through strict and consistent commitment to the standards that are applicable to meeting the user requirements, leading directly to improved reliability, often a key factor from a consumer perspective.

But service quality is more complex when applied to those services that are based on interactions with the public i.e. consumer services. For the consumer (or customer) quality is an experience. Thus, product quality becomes a perception which is governed by design specifications and manner of service provision as well as the manufacture standards achieved. Service quality perception comes from the service process design and the customer contact impressions. This may be helped by using a quality management system (such as ISO 9001) though this is actually a framework rather than a set of instructions on how to design and build processes and products that will satisfy customers and their needs. Service providers need to develop their business systems and processes to deliver all the quality and reliability intentions wanted by the consumers – the determinants of quality which may vary significantly between different service types and products. Leading providers of telecare services adopt these quality measures rather than industry-specific standards.

Table 8: Attributes or Principles Defining a Quality Support Service

Attribute	Description
Safe	Aims to do no harm to the consumer or customer
Person-centred	Focuses on the consumer's individual care or support needs
Timely	Reduces delays in providing attention and treatment to a minimum
Effective	Uses good practice and evidence-based principles wherever possible
Efficient	Reduces interaction times, consumables and waste as much as possible
Accessible	Available whenever and wherever the customer requires service
Choice	Offers the consumer different options when appropriate
Equitable	Care provision should be equal to all consumables with similar needs
Dignified	Treats consumers and their personal information with respect

Quality care and support services are no different; they may be defined simplistically as doing the right thing (getting the support services needed and expected by the consumer or service user), at the right time (i.e. when they need it), and in the right way (i.e. using the appropriate products, tests or procedures), to achieve the best possible or meaningful outcomes. This implies a number of attributes for such a quality service as those described in Table 8. Others might also be applicable and specific to some types of service.

Some consumers believe that getting quality support from a telecare service means achieving a rapid access to a monitoring centre at all times, being treated courteously by all members of the service provider's staff or having individual responsive support in their home whenever the need it. Others make their judgement on their own subjective feelings about how the service makes them feel better, safe and free from anxiety. For hospitals, there are quality indicators such as measures of how many patients die during or shortly after surgery, how many are readmitted within a month of discharge, and how many achieve full independence following a hip replacement, for example. These measures have been developed over several decades and can therefore be used for comparison between different hospitals on a national or international level. However, for remote support services, such as telecare, that rely on assistive devices and connected technologies, many, and probably most, of these measures are inappropriate. This is both because the services are novel, and because the consumers are looking to improve or maintain their quality of life and well-being rather than overcome an acute medical condition. The differences between services being offered in a densely populated urban multi-cultural environment, and a large rural area are also significant (see Chapter 11), as should be measures of quality.

The principles that define service quality, as described in Table 6, remain applicable for virtual and remotely delivered services in the same way as they are relevant to hospital services or conventional healthcare delivery using physical face-to-face procedures. They need to be articulated in a different way in order that they make sense in technology supported service propositions.

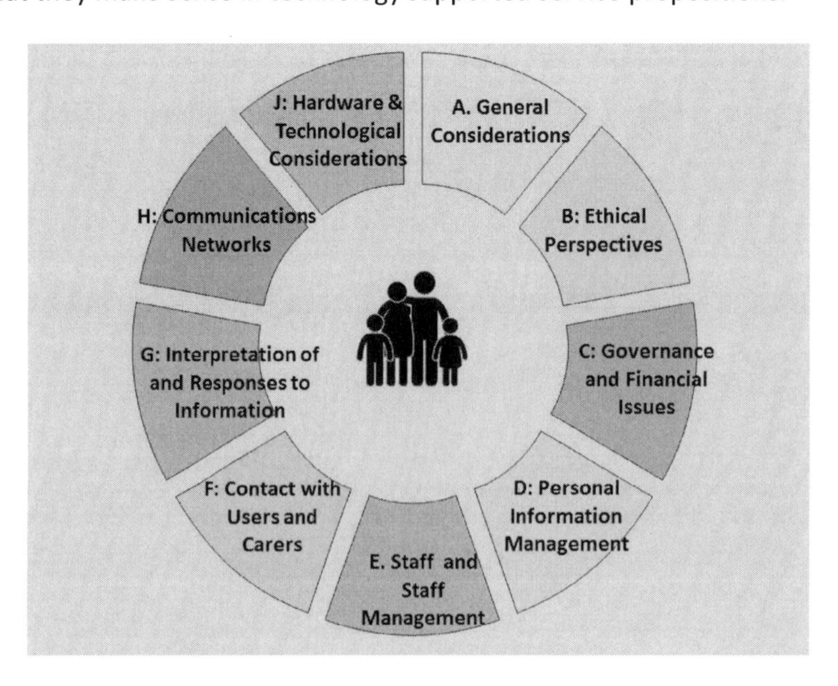

Figure 18: The Nine Domains of the International Code of Practice for Telehealth Services

Such principles are embedded in the International Code of Practice for Telehealth and related services. This provides a framework for all services that relate to consumer support using Information and Communication Technologies. It has sufficient breadth to embrace all current services and an agility that will allow future developments also to be included. Figure 18 shows the 9 domains. 54 different clauses (or standards) are introduced of which 47 are considered to be relevant to all types of service, including those described in Table 8. This ensures an agile approach which will remain appropriate to new service propositions as they develop using the latest technological applications. For example, they allow expansion of Section D in line with new initiatives on data security including GDPR (General Data Protection Regulation) which was introduced in May 2018.

The framework approach allows other operational codes to be produced which can be used to define performance standards that are specific to single service types, such as alarm systems. These may build on existing codes of practice such as those used to accredit social alarms or telecare systems which use a remote monitoring centre as a service hub. However, their relevance to services that use mobile or web-based approaches to define personalised alarm conditions and which can generate alarms automatically and then transmit alerts using either email or SMS, may be questioned. Indeed, the significance of the time taken to answer emergency calls may, at an individual level, be unimportant to someone who is more concerned about the appropriateness of outcomes, and who may wish to see actual performance data in real-time rather than rely on averaged data used in an inspection or audit undertaken perhaps before 2016. Systems based more on mobile and web services are introduced in Chapter 7.

The governance and financial issues section of the International Code may have a different significance to consumers compared with commissioners or service funders. The latter will have concerns over the financial viability of a service provider because they recognise the risk of a service collapsing at some stage - leaving potentially hundreds of service users without support and leaving responsibility for an alternative service with the funding agency. Consumers who have purchased a service privately may not appreciate that their service provider is no longer trading until they are in crisis and unable to access support. This is analogous to the situation in the travel industry where travellers can be stranded in foreign countries if their holiday tour operator becomes insolvent. This problem can be remedied by requiring service providers to deposit a bond with an industry group which would provide financial support for consumers affected by such circumstances, but also by insurance policies that cover these eventualities. It may be apparent that the support industry must learn lessons from other sectors.

Final Comments

It is easy to see how the existing telecare alarm service can be improved through using more technology. However, the service remains, thankfully, more about people than about the technology deployed. A number of points need to be considered before considering a full transformation of service delivery:

1. Existing services are generally understood and liked by the end users;
2. There are many improvements in outcomes that could be enabled by using newer and more sophisticated technologies;
3. New devices are unlikely to have undergone years of rigorous testing to identify any shortcomings and to prove the business case;

4. There needs to be a balance between risk management approaches, and the use of technology that actually supports quality of life more directly;
5. Interoperability issues exist in the home between peripheral sensor devices supplied by one company, and the hub if supplied by another company;
6. Manufacturers continue to use their own signalling protocols between dispersed alarm units and monitoring centres, resulting in potential failures;
7. Social workers and other non-expert assessors are unlikely to be able to keep an up-to-date knowledge of all products and services that are available; and
8. Telecare call handler and members of the telecare team new to learn all new products and procedures and to be confident in their use before service users are exposed to them.

It is not therefore surprising that many service commissioners and providers are content to continue with the status quo for many months (and probably years) after products have been superseded by newer and better devices and solutions. Innovation therefore needs to be applied to all parts of the telecare service from procurement through to maintenance and emergency response. Transformation of services may begin with new technology, but it can't stop there. Referrers and assessors need to understand the benefits and limitations of all products and need considerable training in how to introduce new technologies to potential users. Most service providers seem to believe that their staff are well-trained, but fail to appreciate the limitations of provision, especially the need to think 'outside the box' in order to offer optimum solutions.

Commissioners are more realistic about training limitations and are therefore keen to transfer responsibility to service providers. Providing that the contract holders are agnostic about technology providers, this can lead to significant improvements in knowledge, especially if training is offered through academic institutions that ensure quality through testing and through reference to quality frameworks. They also need to offer some grounding in basic occupational therapy principles, so that assessors look for the simplest and most appropriate solutions rather than offering sophisticated technologies that may not be appropriate to the individual.

A lack of interoperability between products remains a limitation, but manufacturers should accept their role in not agreeing communication standards. Ultimately, they are protecting their business rather than trying to drive up quality. This must be recognised, and service commissioners need to be prepared to exclude them from their inventories until they open up their communication protocols to competition.

Service propositions can be improved by engaging with consultants who offer to coproduce better processes and systems with stakeholders by embedding themselves within existing staff teams. There remains a risk that financial savings could be spent on consultancy fees and training costs rather than on benefiting the funding authority.

Chapter 4

Activity and health monitoring

Introduction

Alarm-based telecare service provision, as described in Chapter 3, involves identifying the risks to the independence of vulnerable people and managing those risks by using dedicated alarm sensors. These automatically detect emergency situations at an early stage and signal the alarm locally, if required, but also through the telephone network to the monitoring centre where the origin and location of the cause of the problem are immediately recognised. Thereafter, an agreed, measured and rapid response can be initiated, minimising the potential for harm to the service user, his or her neighbours and their properties. This form of telecare has been particularly useful in protecting people who live alone and those who may suffer from physical, sensory, intellectual or cognitive deficiencies or disabilities. Sensors which detect environmental problems such as flood, fire, heating failure, blocked flues and gas leaks have proved both popular and reliable in this reactive role.

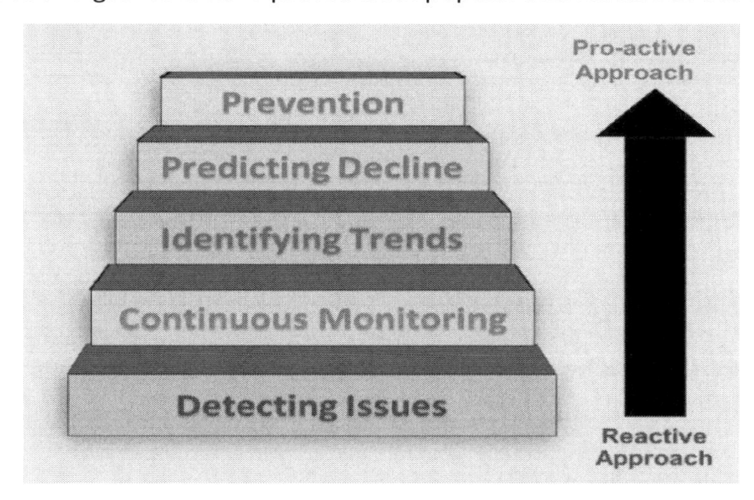

Figure 19: The 5 Steps from Detection to Prevention

Incidents such as falls, and kitchen accidents are often the events that indicate that an individual's independence is under threat, and that interventions are needed to maintain safety and wellbeing. Unfortunately, a single accident can destroy an individual's confidence to live alone or with family or friends in the community. Therefore, an ambition of intelligent monitoring systems should be to predict such events on the basis of dynamic changes in risk, and then to provide strategies for preventing such issues arising. These form the steps to the pro-active approach shown in Figure 19 that will form the basis of Digital Telecare described in Part 2 of this book.

A second generation of telecare systems was proposed in 1996; it would be capable of continuously monitoring lifestyle or domestic activity levels so that the ability of an individual to live independently in their own home could be gauged. This would be achieved initially by adding more sensors, or by adding more intelligence in the sensors, the home interface units, and at the monitoring centres. The first such system was MIDAS – Modular Intelligent Domiciliary Activity System. The MIDAS activity monitoring system – was demonstrated live at the County Durham telecare conference in 2000.

Figure 20: Bespoke sensors providing input into MIDAS

MIDAS included standard wireless security sensors that monitored door and window status, PIR movement sensors for detecting movement (or activity) in each room, and other binary sensors such as KATIE, which could monitor the use of the kettle or other electrical appliances and BORIS, which indicated if the bed (or a chair) was occupied – see Figure 20. Perhaps the most important development was MAVIS, which combined a movement detector with sensors for temperature, humidity, sound and light levels. MIDAS was first installed for Fold Housing Association in Londonderry in 2002.

It had many potential applications; simply by adding up the number of actions, and without providing weighting, it was possible to identify particular types of behaviour and actions that could be considered risky or indicative for an individual who was unable to cope with performing everyday activities. This could be extended to monitor the role of a carer, for example, as shown in Figure 21 which compares patterns between successive days. The opportunities to use this technology for observing trends or for assessing capabilities is evident. By splitting the activities into those requiring physical and cognitive inputs, it was considered possible to look separately at how the contributions to a well-being index were tracking. However, the cost of providing specialist multi-function sensors in every room of a property would prove to be prohibitive, both on the grounds of cost, and because ethical issues such as intrusion and privacy had not been highlighted and discussed at that time.

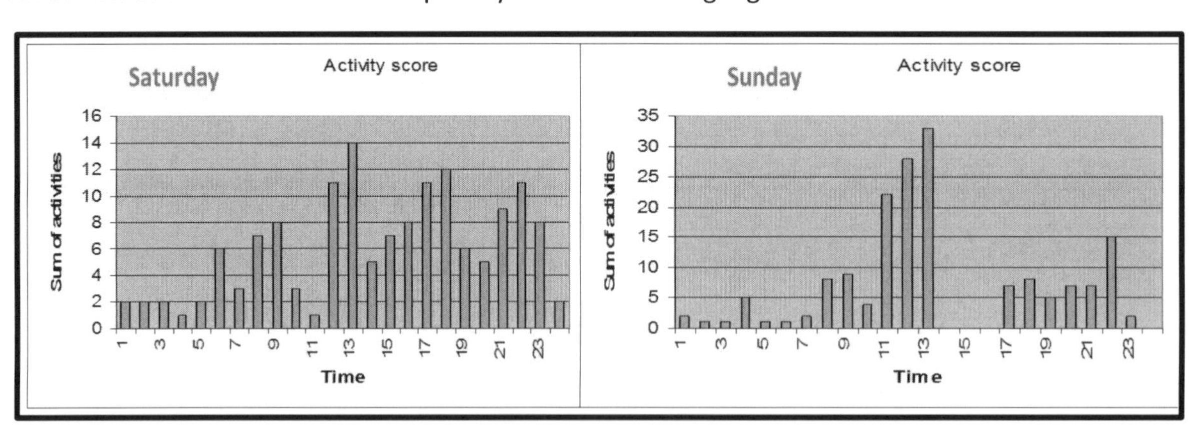

Figure 21: Basic Activity Monitoring Plots from MIDAS

But the wealth of information being generated by the sensors could be used for real-time monitoring and for the detection of problems as they happened if sufficient intelligence could be provided in a processing unit in the home, or if all data could be relayed almost instantly to a central remote system. It would be another few years before developments in telecare hubs and in telecommunications would make this a viable proposition, and a concept that would become core to the digital systems described in Part 2 of this book, Nevertheless, the idea of continuous local monitoring and alerts was the basis of the **V**irtual **I**ntegrated **C**are **E**nvironment (VICE) concept studied by Gareth Williams for his Ph.D. published in 2002. It was seen as a way of keeping safe the many vulnerable people who lived in an assisted living facility or a specialist home for dementia sufferers. It would enable care staff to be alerted as soon as a resident exhibited unusual or challenging behaviour, enable them to intervene before everyone became distressed.

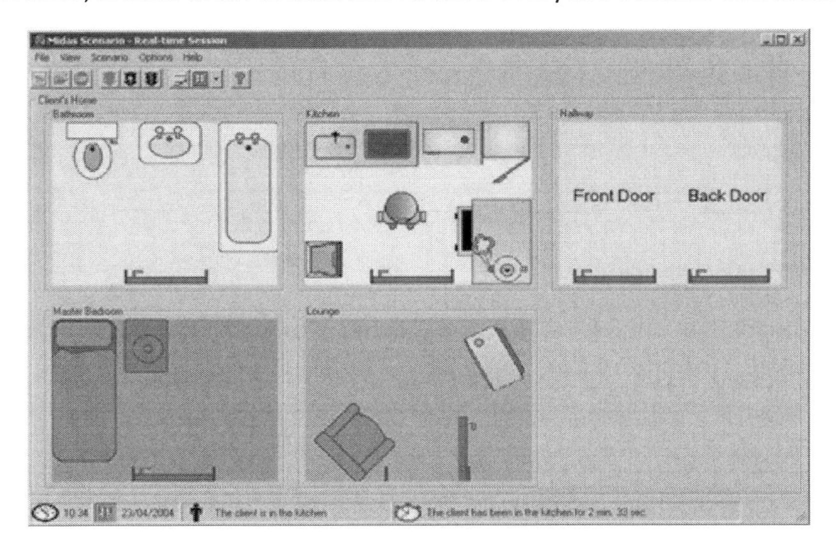

Figure 22: A Virtual Care Environment Screenshot Providing Realtime Information

This approach is often known as Ambient Assisted Living (AAL) by Gerontechnologists, especially in mainland Europe and in North America. However, the term hasn't become a part of the industrial or commercial lexicon in the UK due to the slow introduction and take-up of the relevant technologies, and by the dominance of alarm systems for both commissioners and the alarm receiving centres. The VICE principles might also become the basis for a genuinely smart home for older people. Figure 22 shows a screenshot from Gareth's tool based on a real-time MIDAS scenario where the resident is alone in the kitchen boiling a kettle for a cup of tea. The system would run through a set of rules every few minutes to establish if the individual was safe, enabling rapid reporting of dangerous situations, as well as potentially offering support, advice, reminders and feedback to the individual.

Advanced telecare systems, using new sensors and combinations to monitor household actions on a continuous basis are likely to become the dominant form of support technology in future years. However, they need a major investment in new technologies, and a greater acceptance of all technology by older people and their families, in order to make such a significant impact. Digital Transformation may become the vehicle for these changes - see Part 2 of this book.

However, variations of MIDAS would be introduced within 2 years by Tunstall in their ADLife System (see below), and by Tynetec, a few years later in their i-Care systems, though neither proved commercially viable for large scale deployment in the UK. Meanwhile, QuietCare in the USA used similar sensors to produce a traffic light analysis of activity patterns as shown in Figure 23. This proved to be successful in many Assisted Living facilities but failed to offer the commercial scale that was expected, though its principles have subsequently been replicated in many other systems.

System Status		Low Meal Prep Battery
Room Temperature		75F
Bedroom Exit		Jane Smith left the bedroom at 07.37am
Bathroom Fall		No suspected bathroom falls
Medication		4 Medication events yesterday
Activity		Activity yesterday was in normal range
Night Bathroom Visits		5 Night bathroom visits yesterday
Meal Preparation		8 Meal prep events yesterday
Bathroom Visits		10 Bathroom visits yesterday
Motion Through Door		12 motion through door events yesterday
Wander/Door Exit		No door exit alert triggered
Night Motion		No night motion alert triggered
More Information and System Settings		Click here for more information

Night Bathroom Visits ● Too Many Events

Date	Start Time	Duration
Wed Nov 15, 2006	12:08am	2mins
Wed Nov 15, 2006	01:18am	5mins
Wed Nov 15, 2006	02:12am	3mins
Wed Nov 15, 2006	02:58am	7mins
Wed Nov 15, 2006	04:08am	2mins

Night Motion Through Door

Date	Start Time	Duration
Wed Nov 15, 2006	12:08am	2mins
Wed Nov 15, 2006	01:18am	5mins
Wed Nov 15, 2006	02:12am	3mins
Wed Nov 15, 2006	02:58am	7mins
Wed Nov 15, 2006	04:08am	2mins

Figure 23: QuietCare's Living Independently 7 Day Activity Summary

Continuous Monitoring for Assessment

Tunstall's ADLife system used existing telecare alarm sensor and communication frequencies to link with the advanced telecare dispersed alarm unit to be used as the home monitoring hub. This limited the amount of information that could be transmitted to the hub, and also the level of data and the frequency of transmission to the central server and computer. The market, and many service providers, may not have been sufficiently knowledgeable and mature to make the leap from alarms to continuous monitoring in 2007/8. Furthermore, the systems had limited use in the homes of people who didn't live alone, and in larger properties, including quite small terraced houses where there were several rooms and spaces outdoors that were used on a daily basis.

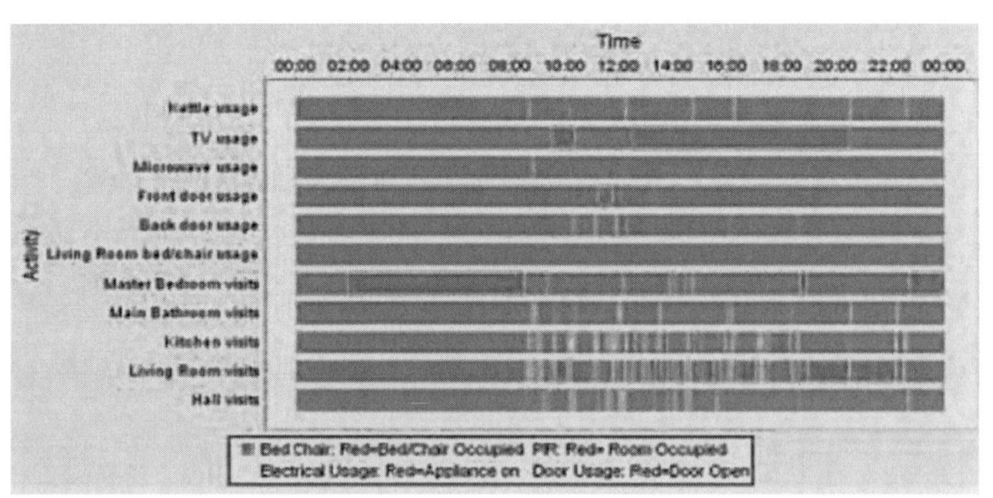

Figure 24: Web-based View of a Day's Activities in an Older Person's Home

Although improvement in the dispersed alarm unit followed on an almost annual basis, the product failed to make the breakthrough in terms of new applications and sales even when the system was extended to enable the use of different electrical appliances (such as microwave oven, TV, kettle, toaster, reading lamp) to be added. Figure 24 shows the complex patterns for a day in the home of an older person living alone.

A company called *Just Checking* then appeared with a rather different telecare proposition (Figure 25); it was aimed at local authorities that were finding it increasingly difficult to identify the needs of, and provide effective support to, people who might be suffering from a cognitive impairment. They found that accounts of the 'strange' behaviour and lifestyle of some people were being reported by neighbours, and sometimes by family members, who either had genuine concerns about the ability of their elderly relative to live independently or were trying to pass on responsibility for their support to the local authority. A typical claim made by neighbours was that they had observed a person leaving home in the middle of the night, and not returning for several hours, and that this had occurred increasingly often in recent weeks. The stories were often exaggerated, and some of the incidents were clearly in the imaginations of the neighbours but had led to some people being admitted into residential or nursing homes against their wishes, and certainly when they might have been supported more appropriately (and at much lower cost) in their own homes.

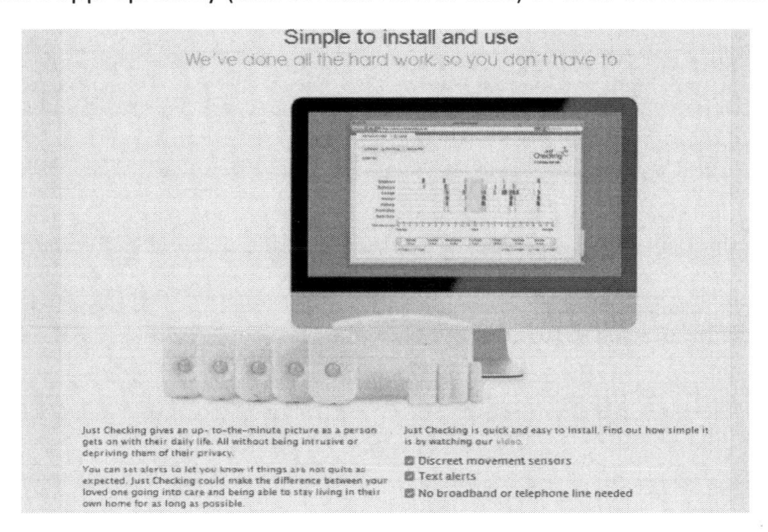

Figure 25: A Just Checking Advertisement Showing the Hardware and an Assessment Screen

The Just Checking approach combined a redesigned set of wireless movement and door status sensors (which had much more aesthetically appealing form factors) with an online assessment tool that showed movement activity in each room on an hour by hour basis (see Figure 25). The sensors were connected to a wireless hub in which a mobile phone communication system had been installed. This meant that the system could be installed and set up very quickly in virtually any property irrespective of whether they had a telephone line. Many older people were giving up their fixed lines (with a quarterly line charge) in favour of a pay-as-you-go mobile phone. In areas where there was a variable quality mobile signal, a roaming SIM card was subsequently deployed that greatly reduced the number of properties in which the system would not operate correctly.

The Just Checking system was quickly found to be ideal for identifying abnormal behaviour patterns and could be used to review and compare daily activities over periods of several weeks. It was found to be particularly easy for social care staff to use and interpret the results. This simplicity was an important factor in its success, as was the use of software at the data centre which could interpret results to show periods of when the house was empty. Figure 26 shows an example of absence during the daytime when the individual is actually out at a day centre. Movement in the lounge during the night was due to the service user's cat! Alongside this plot, is another activity pattern (and for a different individual) which shows severe agitation. Changes to the medication regime of the latter enabled better sleep at night, something which could also be confirmed using the same system.

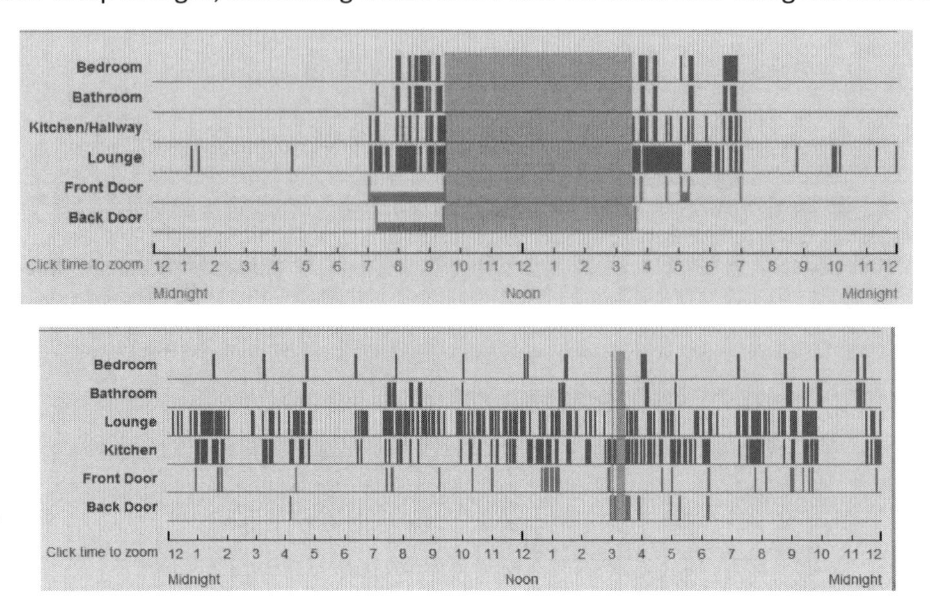

Figure 26: Movement profiles for 2 individuals using the Just Checking System

A number of local authorities purchased several Just Checking systems to be used for assessment purposes. Typically, they would be installed in the homes of individuals being assessed for a period of 4 to 6 weeks before being withdrawn and installed elsewhere. The systems enabled initial opinions on capabilities and risks, obtained by interviews with the individual and their family members, and a visit to their home, to be supplemented by objective information before considering the elements of a support package. This overcame many of the issues associated with relying on a 'snapshot' approach; it enabled many people to continue to live in their own homes for an extended period of time. The outcomes were considered to be so favourable, that some councils required everyone who was referred to them for assessment to undergo a period of several weeks of activity monitoring prior to admission to a residential care home. This process led to significant cost avoidance, through delays in residential care home admission and through providing domiciliary care earlier.

The limitation of the original Just Checking system was that it was designed for use in properties where an individual lived alone for most of the time. This is because it was not possible to differentiate between two or more people in order to assign specific activities to one or the other. This had little bearing on most applications involving people with suspected cognitive impairments, because the focus was on people who lived alone and who didn't therefore have the close attention or support of a partner. Multi-user systems have subsequently been developed.

There are now more than a dozen commercial systems available in the UK that can also provide activity monitoring on a continuous basis. These include Canarycare, TextCare, Acticheck, Kraydel, Mimocare, Howz, Alcove, Tec Angel and 3rings. However, these systems offer increased levels of machine learning and artificial intelligence through their use of all-digital approaches to data collection, analysis and alert generation. They are likely to become the mainstays of telecare monitoring in a digitally transformed world and will be considered in depth in Part 2 of this book.

Support for People with Learning Disabilities

A second major area of application for Activity or Lifestyle Monitoring was in remodelling the level of support, especially at night, needed by people with intellectual or learning disabilities. Since the closure of most long-term hospitals and large residential care homes for people who lacked the ability to live independently, many younger people with specific needs were moved into supported housing. These properties could typically house 2, 3 or sometimes 4 service users, and usually had a spare bedroom where a support worker could sleep overnight. The cost of providing overnight support increased rapidly when a review ruled that a 'sleep-in' rate could no longer be paid if a member of staff if he or she was required to sleep in a bed other than their own. This forced many local authority commissioners, and supported housing providers (which included many charitable organisations), to consider whether overnight support was needed.

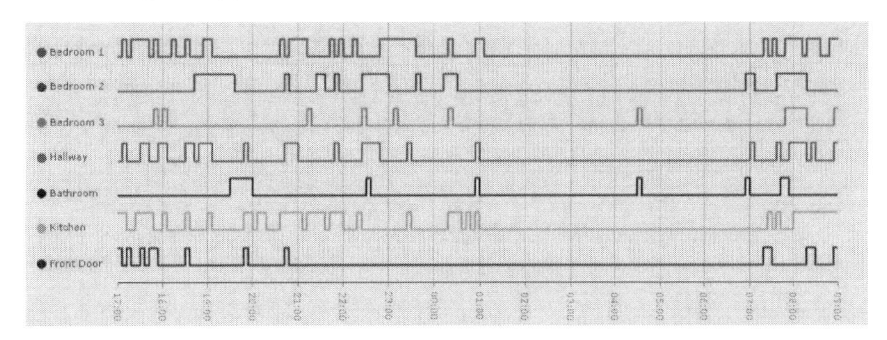

Figure 27: iCare Activity Information from a Supported Housing Property

Multi-room activity monitoring systems were introduced and were able to identify those people who needed no support overnight for most of the time. Essentially, information of the type shown in Figure 28, from a supported housing provision managed by a charity in Ireland, could be used to answer a number of questions, such as:

(a) Do service users sleep well during the night, most of the time or all of the time?

(b) Do they use the kitchen or lounge at night with or without supervision?

(c) Do they need help with personal care tasks such as bathing and toileting during the night?

(d) Are members of the support staff able to sleep well at night, or are they frequently disturbed by the activities of service users?

(e) Do support staff have to get up during the night *to* deal with problem presented by the service users, or by environmental or safety issues never or infrequently?

(f) Might service users, with appropriate technology support and/or remote monitoring, be able to deal effectively with potential emergency situations without having to rely on the interventions of support staff?

In the example shown in Figure 27, the overnight support member of staff slept in Bedroom 1 while the two service users slept in Bedrooms 2 and 3 respectively. It may be observed that both service users had undisturbed night's though one was able to use the bathroom without help from the carer. This approach enabled the service provider to withdraw the carer at 10pm saving approximately 25% of the cost of support. In other properties, the system was able to identify service users who needed wakeful night provision and others who needed other support elements such a security system, or kitchen appliances with built-in safety features.

This approach has enabled service providers to implement major changes to their strategies for support. This has included purchasing properties in groups where those service users with 24 support needs are housed together, and where wakeful night staff are available to respond to emergency support needs in any of the properties when summoned using conventional telecare sensors and alerting systems to inform them of potential issues. The longer-term use of activity monitoring has also helped to identify individuals who could benefit from living on their own. Consequently, many service providers and social landlords have developed houses or apartments that are suitable for individual tenancies, using conventional telecare alarm systems to link them to a 24-hour monitoring centre (or directly to a support worker) in the event of an emergency.

Table 9: Comparison of Health of People with Learning Disabilities with Others

Condition or Outcome	People with LD	Entire Population
Chances of dying before age 50	6%	0.1%
Death from respiratory diseases	49%	16%
Lifetime incidence of epilepsy	22%	1%
Avoidable death rate per annum	1.2%	0.3%
Dementia in people aged 65+	21.6%	5.7%
Women having cervical smear test	19%	77%
In-hospital patients per annum	26%	14%

The strategies have also involved more attention being given to the medical support needs of service users. Table 9 compares the incidence of some diseases and the use made of healthcare resources by people with learning disabilities (LD) with the general population. It may be observed that the former is challenged by much higher levels of ill-health. They are often unable to identify incidents of exacerbation at an early stage, and therefore suffer unwelcome consequences. Some incidents, such as epileptic seizures can be detected automatically by telecare alarms systems, enabling a rapid intervention to be made when necessary, and preventing carers from having to sleep in the same room as them during the night. Monitoring of vital signs can also help achieve improved outcomes.

Remote Vital Signs Monitoring

During the 20th century, infectious diseases such as Tuberculosis, Pneumonia and Diphtheria that caused up to 25% of all deaths in the developed world, were successfully treated with antibiotics. This produced prevalence by orders of magnitude. They have been replaced as the major causes of death by non-communicable (or chronic) diseases. Chronic diseases, such as heart disease, stroke, cancer, and diabetes, are today the leading cause of mortality in the world, representing 63% of all deaths – see Figure 28 for UK statistics. Out of 36 million people who died from chronic disease in 2008, 9 million were under 60 and ninety per cent of these premature deaths occurred in low- and middle-income countries. This number will increase rapidly as obesity levels increase and as people in developing nations replicate the lifestyles of North Americans, Europeans and Australasians.

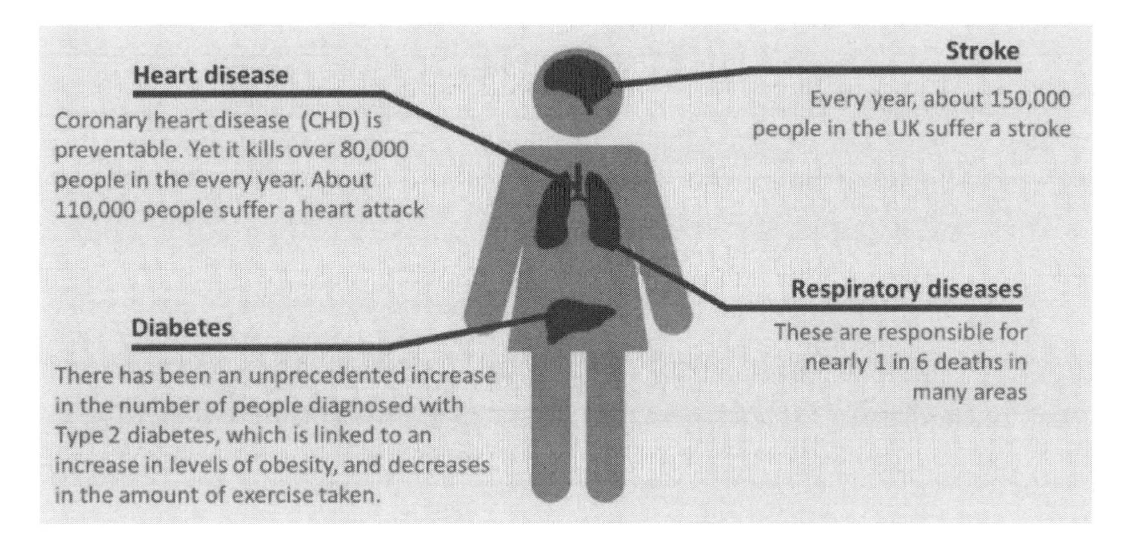

Figure 28: Impact of Chronic Diseases in the UK

Healthcare systems that focus on treatments are unlikely to be able to deal with the impact of these changes. In those countries which have privately funded healthcare, insurance costs may become prohibitively expensive for a large proportion of the population. On the other hand, countries that have health services that are funded mainly by the public purse will need to commit an ever-increasing percentage of their Gross Domestic Product on providing healthcare services. Otherwise, they may need to accept that rationing will be essential if the quality of care is to be maintained and the outcomes improved.

A greater emphasis on a prevention agenda would benefit all healthcare systems; this is based on an improved understanding of the causes of chronic disease as shown in Figure 29. This requires public health campaigns to encourage healthier behaviour including:

- improved diet,
- increased levels of physical activity, and
- the avoidance of risky habits such as smoking and the taking of lifestyle drugs.

This approach shifts responsibility from expensive surgery, therapy and medication prescribed by healthcare professionals to the actions of the individual and their support network (mainly family and friends). It encourages self-care which combines personal responsibility with some elements of DIY (do it yourself) healthcare, thus avoiding overuse of statutory services including Accident and Emergency units and helplines such as NHS Direct and 111 services. This approach is also consistent with Integrated Care ambitions as it can lead to more efficient delivery of services.

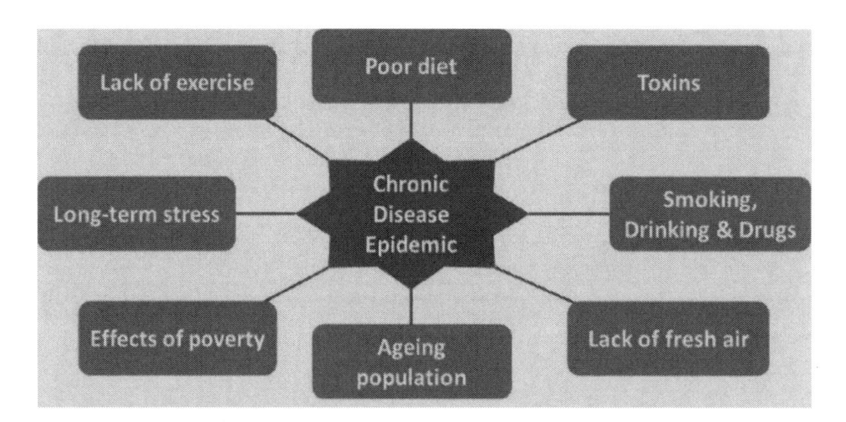

Figure 29: Causes of Many Chronic Diseases

One of the most difficult aspects of chronic disease management, especially its prevention, is the time delay between the causes and the effects. In some cases, symptoms only become evident 40 or 50 years after poor lifestyle decisions were made. It is not surprising therefore that chronic diseases are most likely to impact people in middle-age when it can be more difficult to promote healthier options. Fundamental to the above principles are the need to monitor relevant physiological or medical parameters in a way that informs the individual of their success in controlling their disease. Invasive techniques which involve taking a blood sample can be very unpopular; but people might also find it difficult to take measurements with a portable device every day if they are feeling well.

Similarly, the compliance of people with chronic disease with their medication prescriptions can also vary significantly both as a function of time and with respect to the actual medications required. This demonstrates that the people will exercise choice in adhering (or not) to their treatment regimens for a number of different reasons depending on what they plan to do on a particular day and how they feel. Monitoring is therefore a far from trivial activity which is likely to be less than perfect in many people with chronic disease. This should not be a great surprise as it might often be their resistance to following public health advice (such as not smoking) which was a major factor in the development of their disease.

The success of managing chronic disease must therefore consider:

- making the process as simple and non-invasive as possible for individuals, carers and NHS staff using devices that are part of normal lives
- allowing measurements to be taken at any location (including the homes of family and friends) so that individuals do not need to stay at home
- providing regular and independent feedback to people to allow them to take control, so that they can become co-carers, and
- not referring to those with chronic disease as patients but as individuals who are in control and partners in supporting their own health.

These factors may be particularly relevant when their impact on quality of life is considered, especially if the disease limits an individual's ability to perform various domestic activities without help. It follows that activity monitoring, as described above, might also provide important indicators of an individual's condition. Indeed, changing trends offer an early warning of exacerbations. However, physicians continue to use traditional measurements of vital signs as indicators of condition. Parameters may be measured in the home on a regular (and sometimes frequent) basis. This offers an approach to medical telecare (sometimes erroneously called telehealth by vendors) that can provide useful information on changing conditions. Telehealth is now seen as a term to describe a wider range of modalities that can allow remote interaction between patients and their medical support team, and which can be more interactive.

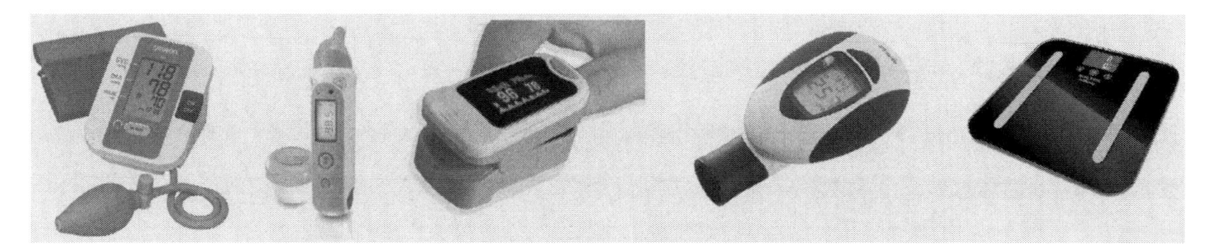

Figure 30: Electronic devices for Measuring Vital Signs

Examples of the devices employed in systems for chronic disease management are shown in Figure 30. They include, from left to right, an electronic sphygmomanometer (for measuring blood pressure), an electronic thermometer, a pulse plesythmograph (for measuring the saturation of oxygen in the blood, a spirometer (for measuring the flow of air during breathing in and out), and a weighing scales and body composition analyser (for measuring fat levels).

Daily or twice daily measurements can be taken and compared with target values to detect a possible exacerbation. In the original systems, the sensors were connected to the home hub by wire. More recently, the connections have been wireless, making the systems easier to use and less conspicuous in the home. The results are normally uploaded automatically through a control box connected to the telephone line; Internet-based software converts readings into a traffic light arrangement enabling a remote call-handler to ring back if there is a technical issue such as missing data, and to refer to a clinician if there is a need for intervention if there is a decline in conditions. More advanced systems asked the patient to answer some subjective questions such as "Do you feel worse today than you did yesterday?" or "Are you in pain today?"

Outcomes of Physiological Monitoring of People with Chronic Diseases

A number of projects and pilot studies in the USA and in Europe had reported major financial benefits to health systems by adopting the use of remote vital signs monitoring to support people with chronic diseases in their own homes. The research was generally sub-optimum either because they failed to fully consider the possibility of selection or reporting bias, or because their scale was too small to offer statistical certainty in their outcomes. Furthermore, they were in some cases commissioned by, or were part-funded by, suppliers of equipment or of services that had a vested interested in demonstrating positive outcomes for the use of technology in this way.

In England, some local authorities had been persuaded to invest in remote health monitoring equipment using their Preventative Technology Grants, in the hope that this would lead to their telecare services being funded by local NHS groups. Again, the outcomes were positive, but clinicians were reluctant to support this approach in many areas for a number of reasons, including the lack of robust evidence, the high apparent costs, and the loss of control in cases where services were being outsourced to national providers. This led to calls for a major trial, and the setting up in 2010 of the Whole System Demonstrator System in Cornwall, parts of Kent and the London Borough of Newham. These three areas were chosen following a competition process; their combined populations of over 2 million people, offered sufficient scale and diversity to ensure that the results could be seen as representative of the entire country. Their geography and population details are shown in Figure 31.

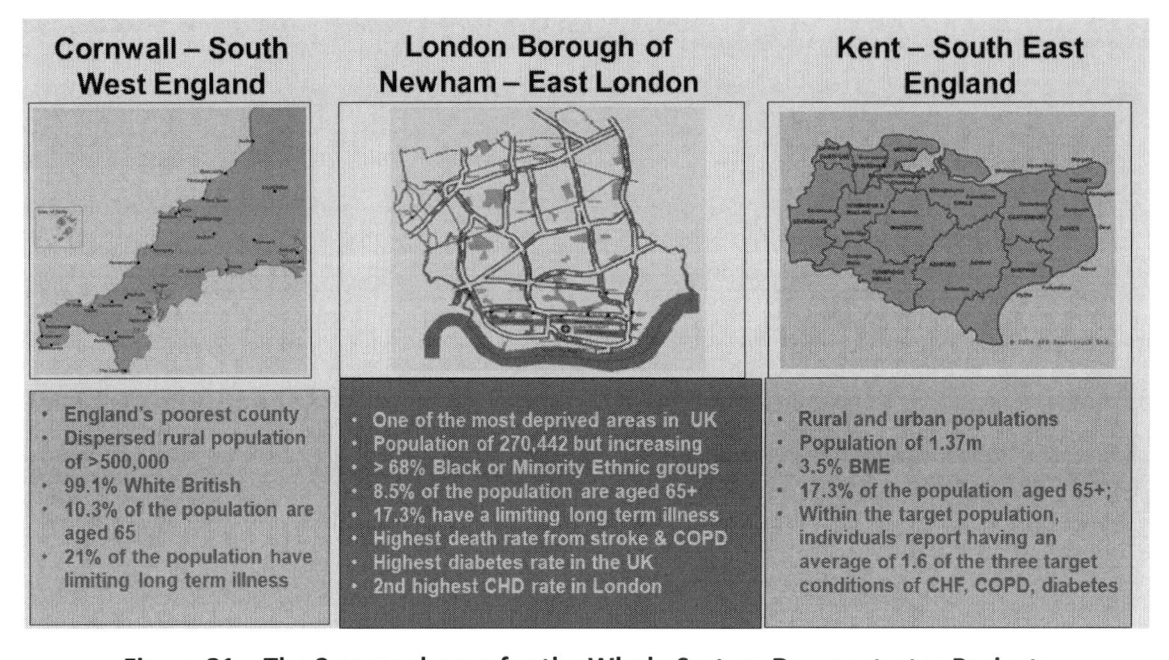

Figure 31 – The 3 areas chosen for the Whole System Demonstrator Projects

Each of the areas was able to select its own technology, but each used systems that were basically similar in that they offered complete sensor and hub arrangements that communicated through telephone lines to the centres operated by their service providers. The inclusion criteria, relating to confirmed diagnosis of the target diseases (diabetes, COPD and coronary heart disease), meant that it was easier to identify eligible people for the trial (than was experienced for the alarm telecare element of the project). However, the process was not simple especially when nearly 240 GP practices needed to be involved, and a total of 6191 patients recruited.

The academic study looked at the cost-effectiveness, clinical effectiveness, and organisational issues, as well as the impact on family carers and on the workforce. The findings were peer-reviewed and published in a number of academic papers. The headline figures for the use of remote monitoring of vital signs in the homes of people with long term conditions were released several months before the papers were published, as their conclusions were awaited by all sectors of the industry.

The results were extremely positive:

If used correctly telehealth can deliver a...

- **45% reduction in mortality rates**
- **20% reduction in emergency admissions**
- **15% reduction in A&E visits**
- **14% reduction in elective admissions**
- **14% reduction in bed days**
- **8% reduction in tariff costs**

However, the results also showed that there were significant costs associated with these outcomes. In a sub-group the incremental cost per quality adjusted life year (QALY) of using the remote monitoring technology when added to usual care was £92,000, and over £30,000 more than would have been required to gain NICE approval for cost-effectiveness. The QALY gain by patients using the technology in addition to usual care **(including social care)** was similar to that by patients receiving usual care only, and total costs associated with the telehealth intervention were higher. The technology approach <u>did not therefore seem to be a cost-effective addition to standard support and treatment</u>. The published papers analysed the reason and concluded that this would remain the case unless there was a significant reduction in equipment costs and its use increased to most who might benefit, and at an earlier stage during disease progression. In other words, value for money can only be achieved if the services are scaled up AND if equipment costs are reduced by at least 80%.

In practice, this might only be possible through

- early interventions,
- major reductions in the cost of fully-connected (Bluetooth) peripheral devices such as pulse oximeters and weighing scales,
- the use of more standard equipment as hub devices to link with clinical systems
- the provision of individual feedback to patients; and
- changing the delivery model so that NHS organisations took charge of viewing data and of organising appropriate interventions.

These changes have already started to take place. Organisations that had focused on providing full managed solutions to the NHS, had to accept that their pricing model was wrong. It became increasingly apparent that clinicians within hospitals and community services wanted to maintain control of both the data and the interventions, but not necessarily the procurement, installation and maintenance of the equipment.

Subsequently, even more impressive outcome benefits have been highlighted in projects undertaken in Northern Ireland and in Portsdown respectively, the former claiming to have reduced the unplanned hospital admissions rate for people with COPD supported by the technology from 42% to only 9%; the latter study, also for COPD patients, led to a 68% reduction in GP appointments and an 85% reduction in home visits.

In future, vital signs monitoring is unlikely to remain a 'black art' as more people are encouraged to buy and use their own low-cost devices such as pulse oximeters, blood pressure monitors and blood glucose monitors. This has led to many simple vital signs monitoring systems, of which the most established is Florence (or Flo for short) which was set up in Stoke in 2010. Flo is a text messaging system that sends patients reminders and health tips tailored to their individual needs. Clinicians (doctors or other health professionals) can adjust the settings on Flo for each patient, defining when messages should be sent, what information they are asking for, and how the system should respond. Flo then sends regular text messages to patients helping them to monitor their health, sharing any information sent back by the patient with the person managing their care.

For example, a patient with hypertension could be asked to check their own blood pressure each morning and then to text the results back to Flo. If the results are outside agreed limits, Flo will pass this information on to their clinician, whilst also suggesting that the patient speak to someone at a primary care centre on the phone. As with other more complex (and expensive systems, clinicians are also able to view real time information about their patients a via a simple web interface.

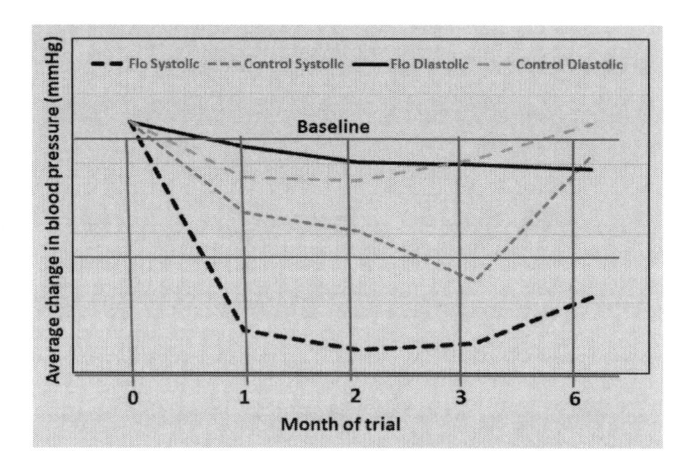

Figure 32: Reductions in Systolic and Diastolic Blood Pressure Readings with Flo Support

An early clinical trial showed the impact using Flo had for patients with hypertension and chronic kidney disease. The trial showed better and faster clinical results in managing people's blood pressure compared to a group of patients who didn't use Flo. Figure 32 shows that the impact of using Flo on blood pressure readings not only shows a greater reduction in both systolic and diastolic readings than those of a control group, but also helps to maintain the benefits at 6 months while early benefits for the control group decline over time. 'nhssimple' and NHS Stoke-on-Trent CCG have since rolled Flo out to more than 70 health and social care organisations across the UK. Flo is being used to support an extended list of conditions, including asthma, heart failure, chronic obstructive pulmonary disease (COPD), and stress incontinence. It's also being used to make sure patients take their medicines, to aid recovery after surgery, and to support people who want to stop smoking or lose weight. An adapted version of Flo, named Annie (after Annie Fox, the first woman to receive the Purple Heart for combat, is also now being used to support veterans in the US.

The use of technology alongside other early warning systems to treat people with long term conditions sooner for illnesses such as flu and chest infections has led to thousands of hospital days being saved in the West Midlands. The iCare system has produced savings of £7 million for the local care economy. This field is developing rapidly as new sensor configurations emerge. These are likely to result in further improvements in assessment and in the automatic identification of emergency situations.

These will be considered again in Chapter 6, but more extensively in Part 2 of this book when Internet of Things opportunities will be considered.

Concluding Remarks

Monitoring people and their lifestyle through their activities and their vital signs is a significant step away from basic alarms because their direct output is likely to be an early warning dashboard rather than confirmation of an acute event or emergency. The challenge is to collect sufficient, relevant data continuously, or actually continually, over an extended period of time to populate the dashboard in a way that isn't intrusive even though many more sensor inputs and data processing are required. The sensors that were used for activity monitoring sensors were initially those simple binary sensors used in the security industry. They were, and are, low cost and reliable. Information can also be transmitted from the sensor to the telecare hub using slightly different reserved frequencies to those use for telecare and social alarms. Indeed, they can use frequencies that are reserved for security and automotive applications. This enables low cost transmission modules to be employed, reducing overall costs. It should be noted that the systems do not need to comply with EN 50134, thus enabling the use of more novel, bi-directional communications and the replacement of a monitoring centre with an on-line approach .

Sensors for the measurement of vital signs are medical devices and are therefore very expensive compared with other sensors. Measurements are taken only once or twice a day, with options for transmission of data to a home hub ranging from wires through to new wireless approaches such as Bluetooth. The quantitative data, along with answers to subjective questions are sent over telephone lines (fixed or mobile) to servers which can be interrogated by physicians or nurses. Such systems have been very expensive; service providers have tried to provide fully managed solutions. This trend has declined as equipment costs dropped rapidly and the roles of service providers have changed to those relevant to data storage and protection, and in offering clinicians training in tools that enable them to make better use of the data.

Digital transformation will involve the development of new sensing techniques, including those based on wearable devices, and on non-contacting techniques. Better data processing and advanced analytics, including the combination of activity data with physiological data will improve prediction and allow conditions to be managed more efficiently. These techniques will be introduced in Chapter 6, and their further use will be described throughout Part 2 of this book.

Chapter 5

Assistive Technologies and Telecommunications

Introduction

The World Health Organisation in 2011 defined Assistive Technology as
"any piece of equipment, or product, whether it is acquired commercially, modified, or customized, that is used to increase, maintain, or improve the functional capabilities of individuals with disabilities".

The vast array of possible Assistive Technology (AT) products and solutions reflects the diversity of the needs of people with disabilities – ranging from new form of communication technologies that can support social engagement, employment, learning, memory, planning and safe guarding services through to products and devices that support mobility and personal care requirements. As the sophistication and complexity of solutions increase, so do the potential cost of provision, unless more universal design principles are employed that could lead to greater uptake, utilisation and competition. It follows that the potential risks to users also increase if not appropriately set up or maintained, and if the technology is not provided within a support system.

However, a more widely used definition of AT may be the one proposed by the International Standards Organisation (ISO):
"Any product (including devices, equipment, instruments and software), especially produced or generally available, used by or for persons with disability: for participation; to protect, support, train, measure or substitute for body functions/structures and activities; or to prevent impairments, activity limitations or participation restrictions."

This is a more inclusive approach as it recognises users as both people with a defined and recognised 'disability' and older people, or those with limiting conditions, who might experience functional or sensory challenges under certain circumstances. It includes:

- Assistive products for personal medical treatment
- Assistive products for training in skills
- Orthoses and prostheses
- Assistive products for personal care and protection
- Assistive products for personal mobility
- Assistive products for housekeeping
- Furnishings and adaptations to homes and other premises
- Assistive products for communication and information
- Assistive products for handling objects and devices
- Assistive products for environmental improvement and assessment
- Assistive products for employment and vocational training
- Assistive products for recreation.

This directly supports the role of AT to help support students in education, but also its role in helping people to gain or maintain employment. Increasingly, it is also being used in supporting personal care and in making domiciliary tasks easier, either for the individual, or for their carers. It may be apparent that AT can be complementary to telecare services.

AT solutions have traditionally been separated into 3 domains:

- Adaptations to the home environment;
- Portable devices that are aids to daily living; and
- Functional aids.

This approach enables products that are mobility aids to be described as portable devices, so that walking sticks and frames may be described as portable devices, while pavement scooters which are larger, more complex (and more expensive) may be described as functional aids. Similarly, a hearing aid or a pair of spectacles are portable but are also functional. The approach is limited in its recognition of electronic devices, and especially the classification of telecare devices. Whilst environmental control systems are clearly functional aids, the addition of sensors and intelligence can make the home "smart", and might be considered to be an adaptation to the home.

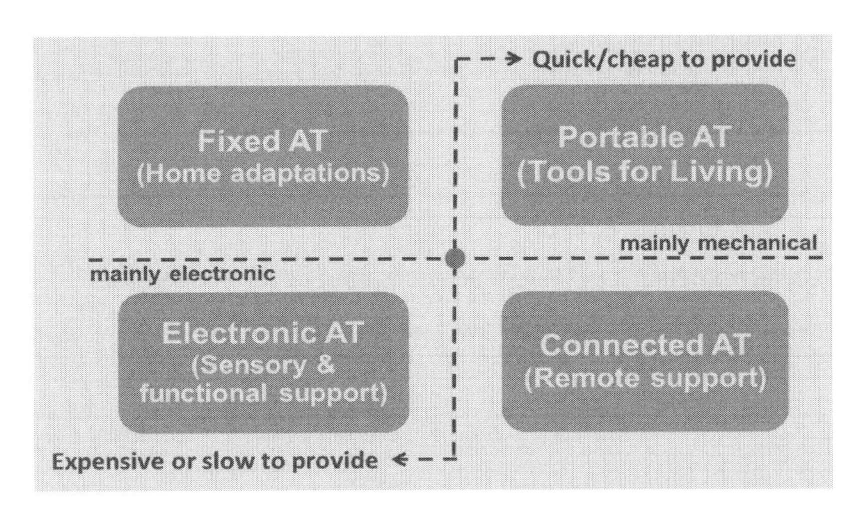

Figure 33: The Four Blocks of Assistive Technology Model

To overcome the above deficiencies, a 4 blocks of AT model is proposed as shown in Figure 34. It allows the separation of devices into those that are mainly mechanical in nature (upper blocks) and those that are electronic, smart or intelligent (lower 2 blocks). The blocks can also be split left and right; those that are on the left are generally more expensive and take longer to install or specify, while those on the right are more standard devices, and are consequently lower cost and readily available for installation. Consequently, someone who is in hospital and is medically stable and able to go home safely as soon as they are provided with AT, may become a delayed transfer of care if their needs are on the left of Figure 33 but not if they are on the right.

The fourth element of AT is effectively telecare, but in its widest sense. It is not restricted to alarm services that link a vulnerable person to a monitoring centre but includes a much wider range of products and services that are enabled by advances in electronics, processing and telecommunications. These advances including the telecare alarm services described in Chapter 3, but go beyond this, and provide the context for the remaining chapters of this book. However, the need for service integration is becoming more evident as both end-users and professionals understand the need for single assessment processes and 'one stop shops'. It may be apparent that people need a balanced approach that mixes mechanical equipment, electronic equipment and a range of support services if they are to overcome life's challenges and have a high quality of life.

Home Adaptations (and DFGs)

The relevance of the home environment to successful ageing in place, as well as to better healthcare and Quality of Life, has been shown in several studies spanning over 20 years. Furthermore, surveys of housing in England have shown that:

- Almost half of people aged 75-84, and over 3 in 5 of those aged 85 or over, live alone;
- 3 million households with an adult aged 65+ report a long-term illness or disability;
- Most older people live in housing not built to modern accessibility standards - a quarter of those aged 75-84 and a third of those aged 85+ live in homes built before 1945;
- One in five homes occupied by older people in England (2 million households) failed the Decent Homes Standard in 2014;
- Close to half a million households with at least one adult aged 65+, with a long-term illness or disability, reported the need for at least one adaptation to make living easier;
- Very few people needing adaptations to their homes have an ambition to move; and
- The desire to move declines with increasing age.

It follows that there should be a high demand for adaptations to the homes of older people that will help them to stay put. Many (or perhaps most) of these people could also benefits from aids to daily living, other assistive technologies for functional and sensory support, and a range of telecare devices and systems that will relieve their anxiety and give them 24/7 support in the event of an accident or health incident.

Table 10: Examples of large and small adaptations to the home and their cost

Large adaptations	Av. Cost	Small adaptations	Av. Cost
Installing a downstairs extension with bathroom	**£30,000**	Fitting a second bannister on stairs	**£150**
Level access shower to replace bath	**£5,000**	Installing grab rails in bathroom and by doors	**£200**
Fitting a through floor lift	**£12,000**	Adding electric bath hoist	**£400**
Widening doorways	**£3,000**	Installing external security lights	**£300**
Kitchen redesign with lowered worktops	**£9,000**	Improving level of internal lighting	**£200**
Fabricating outdoor ramps	**£4,000**	Providing security doors/locks	**£250**
Fitting a stair lift	**£5,000**	External rail for steps	**£500**
Adding an extra toilet	**£3,000**	Shower/bath seat	**£500**

Some adaptations are relatively quick and easy to install and are also quite low cost. These are known as small adaptations and cost less than £1000 to provide. These are generally provided free of charge following an assessment by an Occupational Therapist that indicated that such an adaptation is necessary in order to maintain an individual's safety and independence in their own homes. Examples are provided in Table 10. The most popular are grab rails either by doorways or in the bathroom, followed by special seats for attaching to the bath or the shower wall.

There is strong evidence that minor adaptations are particularly effective at improving outcomes and reducing risk, especially when they are combined with other necessary repairs and improvements in lighting and removal of hazards, and reduction of risk (through telecare strategies). They are accepted quickly if they help an individual to achieve their own objectives but are far less successful when proposed contrary to the wishes of the individual and/or their family. The best outcomes are achieved when the adaptations are provided in a timely manner, preferably before an incident occurs. Sometimes, older people reject quite small adaptations, such as a grab rail on the wall by their front door, because they perceive it to be stigmatising, in much the same way as some might choose not to wear an alarm pendant.

Major adaptations cost more than £1000 to provide. The most popular are the replacement of a bath with a shower (preferably level access), and the provision of a stair lift. The latter can make the difference between an individual being able to stay in their own home, especially if their bathroom and bedrooms are upstairs, and moving into an apartment or a care home. Some people are eligible for a means-tested Disabled Facilities Grant (DFG) from the local authority if an adaptation can make it easier to enter and leave their home. Support in making applications, performing surveys and in selecting and managing contractors is often provided by Home Improvement Agencies or by Care and Repair organisations. Grants can be for up to £30,000 though income and savings of over £6000 are considered in assessing the claim. It should be noted that once an adaptation has been installed, it becomes the property of the householder, whose also must accept responsibility for its safety, repair and maintenance, for a period of at least 5 years. This can be expensive. In rented accommodation, a landlord must be consulted before adaptations are made because they represent a modification of their property. Some adaptations need to be removed, or changes reversed before the property can be relet.

In Wales, there is a Care & Repair organisation operating in all counties. The parent body, Care & Repair Cymru manages the Rapid Response Adaptation Programme (RAAP) on behalf of the Welsh Government. This pays for small works (up to £350 in value) to be performed to support timely hospital discharge. In Bridgend (a relatively small county to the west of Cardiff in South Wales), Care and Repair's Hospital to Home Service in 2016/17 helped over a thousand patients to achieve a safe discharge from hospital to their own homes with a cost avoidance potential of nearly £3 million. Based on modest assumptions on bed days saved, this can be equated to providing additional hospital capacity of a 16-bed ward. The social return on investment for bed days saved was calculated to be £5.50 for every pound invested. There were other cost avoidance savings too, including the prevention of falls (which will be discussed further in Chapter 10.

An extended role for Care and Repair and Home Improvement Agencies may be evident in the operation and management of older people services. They are well-positioned to introduce telecare services to the people who might benefit most from them and they could provide a service to a wide range of house-holders in a commercially viable manner.

Community Equipment

The *Integrated Community Equipment Services* (ICES) initiative was introduced in 2001 in England (and replicated elsewhere in the UK) in order to:

- Provide more, and better, equipment;
- Provide responsive services to people of all ages;
- Reach larger numbers of individuals; and
- Facilitate intermediate and community care

Community equipment enables children and adults, who require assistance, to perform essential activities of daily living to maintain their health and autonomy and to live as full a life as possible. It includes most portable (i.e. not fixed) equipment used for home nursing, and for performing tasks that are fundamental to their independence. Their integration into a new service, was designed to lead directly to benefits for users, for professionals responsible for their care, and for social services, and the NHS in general. The approach enabled the procurement, provision, maintenance, repair and recycling of AT that had previously been prescribed separately by healthcare and social care personnel, to be offered together enabling an avoidance of duplicated assessments, home visits and maintenance arrangements. This equipment includes, but is not limited to:

- Prevention devices, such as pressure relief mattresses and commodes.
- Furniture and equipment for daily living, such as children's special seating, shower chairs, raised toilet seats, teapot tippers and liquid level indicators.
- Taps turners and special handles for door
- Ancillary equipment for people with sensory impairments, such as flashing doorbells, low vision optical aids, textphones and assistive listening devices.
- Communication aids for people who are speech-impaired; and
- A limited amount of alarm monitoring equipment that can provide a warning to carers in the same property

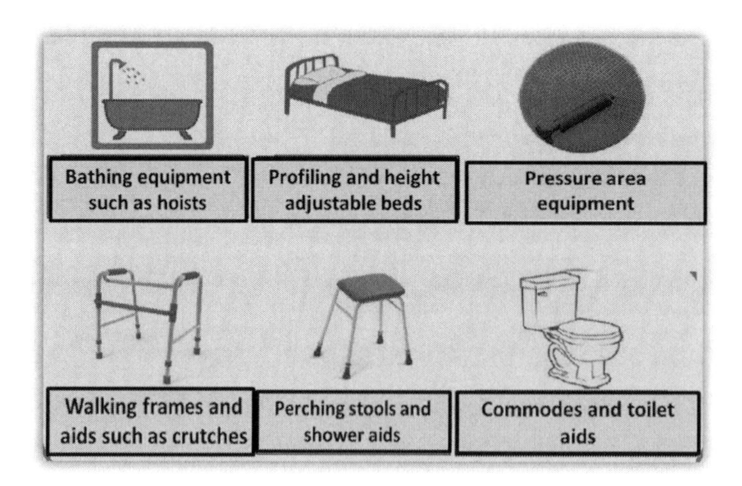

Figure 34: Most highly prescribed items of community equipment

The most popular items of Community equipment are shown in Figure 34. They are generally prescribed by Occupational therapists who assess both their needs and the suitability of their home environment. Other community staff, including nurses and social workers are playing an increasing role in assessments so that delays are avoided enabling patients to return home safely from hospital as soon as they are medically fit enough to do so. The equipment is provided on a short-term loan basis. Small items are taken home by the patient whereas bigger items (especially beds and bathing equipment) are delivered to the home by dedicated personnel. These same members of staff also collect the equipment for cleansing and recycling when no longer needed. The recycling rate (by value) is in excess of 80% though many smaller items such as walking sticks and frames are no longer included in the returns because the cost of collection, cleansing and testing exceeds the cost of replacement. For this reason, many community equipment services are restricting their inventories to products that are deemed to be more complex, and which need regular servicing and maintenance. This includes the more expensive items such as electric beds and hoists (top half of Figure 34). The expectation is that people who need simpler products (such as those in the lower half of Figure 34) will either be issued with a prescription that can be exchanged at an approved and accredited retailer, or they will choose to buy their own equipment. This is part of a new retail model that aims to broaden the sources of supply of community equipment and encourage a network of accredited retailers with staff trained to a minimum competency level for this purpose. It is envisaged that both private sector providers and third sector organisations that are currently involved in providing this equipment would play enhanced roles in improving access and choice for the consumer.

In order to operate efficiently, and to minimise the risk of inappropriate provision (or overselling by retailers) the model also assumes the creation and training of new independent, or trusted, assessors. They would decide what equipment was needed and most appropriate and make summary recommendations. Existing professionals, especially Occupational Therapists and Physiotherapists would support them and provide specialist assessments for people with more complex needs. (appropriately qualified professionals such as occupational therapists and physiotherapists) to assess equipment needs and make recommendations. More people are now being offered a Personal Budget, rather than an arranged service, following an assessment. This enables them to purchase their own support package, which should increasingly include equipment and community services.

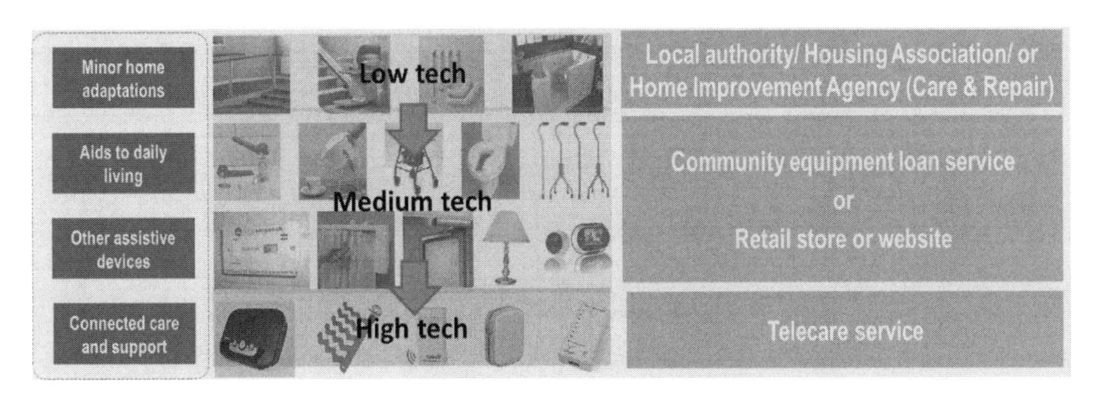

Figure 35: A Mixed AT Approach to Providing Equipment for Support

There are now dozens of web-based providers of assistive technologies, offering an extended range of products, styles and prices, and they are usually capable of offering next day delivery. However, Disability Supermarkets have also sprung up in most large towns, where customers can see and sometimes test equipment before purchase. These organisations often specialise in mobility scooters and similar devices that would not generally be available through loan stores. The emergence of digital assistive technologies opens up another opportunity for private retailers as well as not for profit organisations. These will be considered in greater detail in Part 2 of this book.

An ageing population and a corresponding increase in the number of people living with long term conditions and disabilities suggests strongly that there will be a significant increase in the number of people who need one or more assistive devices. Indeed, a significant proportion will need a combination of assistive technologies to address all their needs and to help them to achieve as much independence as possible. Currently, different types of equipment are provided by different organisations, as shown in Figure 35. This shows that minor adaptations to the home are provided by the landlord, or by the local authority, or one of its agents. Aids to daily living and other assistive devices that can reduce the likelihood of an accident occurring, can be provided free of charge through the Community Equipment Service (on loan) or purchased privately.

Finally, connected support devices (such as telecare alarms) are provided by telecare services that are increasingly privately run or through local authorities. Most will pay a monthly subscription charge. However, as shown by the arrows in Figure 35, the priority should be to provide the simplest approaches first – no technology. Only for more complex cases should high technology be necessary. Medium level technology may play a significant role in prevention. It remains a matter of contention how technology should be funded. The NHS continues to provide community equipment free at the point of care, whereas social services (and private equipment providers) are able to charge when deemed appropriate. This inevitably results in Occupational Therapists and other NHS staff being restricted in the equipment and devices that they can recommend and supply. This has an impact on successful outcomes following assessment (see Chapter 6).

In an ideal world, all of the above elements could be provided by one organisation following a single assessment; they could be installed at the same time for maximum efficiency and customer benefit. Local authorities such as Leeds have recognised the potential of merging community equipment with telecare services for a number of years. Conwy council in North Wales has successfully merged their services. Sunderland is considering a fully integrated service that would be capable of offering all elements to its citizens. As public service austerity continues to bite, more people are likely to have to buy their own equipment and services, leading to more demands for assessment, sign-posting and independent support. It is also likely that many people with more complex needs will become more dependent on equipment for support. Fortunately, technology improvements may provide them with major benefits, as described below.

These examples are separated into 3 groups:

- Sensory disability technologies
- Technologies to support people with communication issues; and
- Functional support technologies, including those larger items that can help people with mobility issues to transfer and to move around both in their homes and in the wider world.

Assistive technologies are also used increasingly in rehabilitation, as components of systems that can help to restore function for people following accident or traumatic medical events such as a stroke. These are likely to use advanced digital technologies and will be considered at length within Part 2 of this book.

Technology for People with Visual Impairments

Vision Impairment (VI) is the consequence of a functional loss of vision rather than the eye disorder itself. It is not necessary therefore to define VI in terms of the catalogue of permanent conditions or chronic diseases that result in a loss of vision. However, the impact may be described through 3 terms:

- **Low Vision:** a person can see, but not well, and full vision can't be enabled by surgery.
- **Legally Blind:** someone lacking visual perception due to physiological or neurological factors so that they can't see well or have minimal vision e.g. they can't see at 6 metres what others can see at 60 m.
- **Totally Blind:** someone who can't see at all (less than 20/200 vision in their better eye), and who must therefore depend on their other senses.

The World Health Organisation believes that there are 285 million visually impaired people in the world, including 39 million who are blind. 90% of them live in "low income" countries while a similar percentage are impaired as a result of having a chronic condition such as diabetes. They are three times more likely to be unemployed or be involved in a motor vehicle accident, or to suffer from anxiety or anxiety disorder. 85% are aged 50+ years, and this percentage will grow rapidly as life expectancy increases. It is not surprising that people with VI are twice as likely to suffer a fall compared with fully sighted people.

Table 11: The Major Causes of Vision Impairment

Cause of VI	Explanation
Loss of Central Vision	makes it difficult to read, recognize faces and distinguish distant detail
Loss of Peripheral Vision	loss of peripheral vision ('tunnel vision ') affects mobility and reading.
Blurred Vision	both near & far vision are out of focus, even with good correction
Generalized Haze	a sensation of a film that can extend over the entire viewing field.
Extreme Sensitivity to Light	ordinary light levels overwhelm the visual system, producing a washed-out image and/or a glare and pain.
Night Blindness	inability to see outside at night under moonlight, nor in dimly lighted interiors
Colour Blindness	decreased ability to see *colour*, or *colour* differences, with normal lighting conditions

Table 11 lists seven common causes of Low Vision some of which affect acuity, while others affect the visual field and ability to process colours. Most of these causes make vision especially difficult when the available level of lighting is poor, demonstrating the importance of ensuring adequate lighting both inside and outside the home. The exact number of people who fall into each category is difficult to estimate, especially as some will suffer from more than one cause.

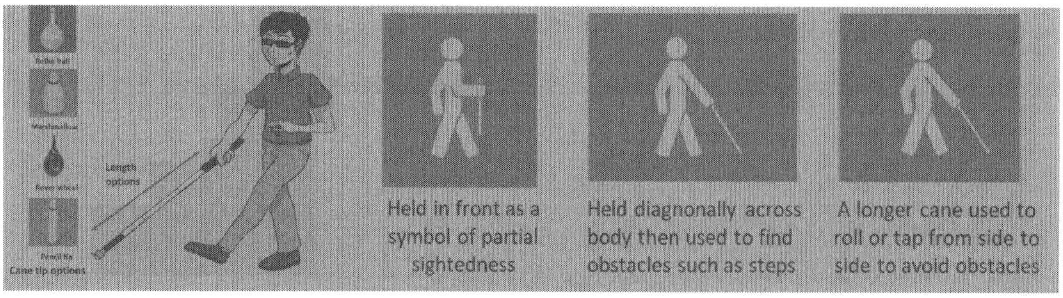

Figure 36: Styles and Techniques of Canes Used by People with VI

The most common form of assistive technology used by people with VI is a cane. These come in different lengths and formats as shown in Figure 36. Each requires a different technique for use, and using a longer cane is quite difficult without proper mobility training. As technology improves, new forms of AT are emerging to support people with VI.

Night vision enhancers are available commercially on various web stores and are able to amplify available ambient light levels in order to transform a very dark night time scene into a brighter picture. This can enable the user to recognise features and objects that would otherwise remain masked by darkness. Similar technology could be used by people with Visual Impairment to make the most of any vision that remains, much in the same way as a hearing aid might boost a nearly deaf person's ability to hear sounds.

One leading example is the eSight Vision system which involves a pair of glasses that can enable someone who has got very little sight to allow them to walk around unfamiliar places, to recognize obstacles, and to get a greater independence. They are constructed using transparent OLED displays, 2 small cameras, a gyroscope, a GPS locator, and a headphone. Data collected by the cameras can be processed and used in various ways including the use of brightness to show depth so that wearers can detect shapes including people and large objects. Machine learning can then be applied to enable the identification and classification of different shapes. This could be used to read signs and provide spoken directions. Another example of image enhancement and processing comes from NuEyes, who use a combination of custom software and smart glasses, to restore some sight to people with serious vision loss. A camera on the front of the blacked-out glasses acts as eyes. Captured images are projected on the lenses. A speech-recognition feature lets wearers use voice commands to improve the image projected through the lenses to best remedy their particular condition.

One of the most common visual problems is Colour Vision Deficiency (CVD) sometimes referred to as colour-blindness. It can affect one in 12 men but far fewer women. Most people with deficient colour vision have inherited a gene from their mother, but some have developed the condition over their lifetime either as a result of a long-term condition, such as diabetes, or as a result of ageing of a side effect of medication. Most people with CVD can see things as clearly as other people but they unable to fully appreciate or differentiate between red, green or blue light. The most common form of colour is known as red/green colour deficit which means that they mix up all colours which have some red or green as part of the whole colour. For example, they will confuse a blue with a purple because they can't 'see' the red element of the colour purple. CVD can lead to many problems in daily life ranging from clothes selection, to driving and food preparation. For example, they may not be able to tell whether a beef steak has been cooked to be well done or rage. Similarly, they may struggle to identify ripe bananas and tomatoes, and to differentiate between tomato ketchup and brown sauce.

These problems can increase safety concerns both regarding the appearance of people who are "off-colour" and in industries and the military where the correct identification of colours is essential. Different coloured lights are used to control access and movement on the railways, in shipping lanes, in aircraft flight controls, and at sea. People who have CVD are often denied employment in safety critical positions. Manufacturers of set-top-boxes for digital TV can now include Eyeteq chipset as an option for customers with CVD. They use image processing algorithms to boost some colours in order to reduce the impact of the condition on the viewing experience. Electronic contact lenses that use 3D printing to produce multi-colour LEDs on contact lenses have found that they can improve and augment native vision. Similarly, EnChroma Cx glasses provide high colour contrast, producing an image in which the primary colours of red, blue, and green "pop" and are perceived correctly by the wearer allowing the effects of CVD to be overcome.

Lazy eye, or amblyopia, is likely to affect between 2 and 3% of children and can lead to long term vision loss. Current methods of treatment include eye patches and drops, which require a disciplined approach for success. The Amblyz electronic glasses are programmed to automatically build the brain's reliance on the weaker eye by blocking out vision in either eye for programmable periods of time using LCD lenses. They are available in the USA for ~$450. On the other hand, many older people develop peripheral vision problems often as a result of glaucoma, concussion or stroke. They result in a reduced field of view, sometimes creating the sensation of seeing through a narrow tube, a condition referred to as "tunnel vision" which leads to problems seeing in dim light and a decreased ability to navigate while walking.

Table 12: Example of Hardware Used for Reading and Writing in Braille

Device name	Braillenote Touch 32 Notetaker/Tablet	Brailliant BI 32 Braille Display	Embraille Desktop Embosser
Picture			
Price	**£3995**	**£1985**	**£1595**
Description & features	Combines benefits of traditional note taker with power of tablet Uses TouchBraille – a natural way to type Allows easy access to apps and cloud storage	BI display is designed for extensive reading Offers full comfort and intuitive display navigation of HumanWare's signature thumb keys	Lightweight & compact design Polymer platen creates smoothed rounded dots USB connection

The Braille tactile writing system was developed in 1824 by a Frenchman, Louis Braille, who lost his sight in a childhood accident. In 1829, he published detail of the system which was subsequently revised and improved to include musical notation. Braille characters are small rectangular cells which contain bumps known as raised dots. The number of dots, and their spatial arrangement, are used to distinguish one character from another in its most basic form but can also be used to represent numbers, punctuation marks and some complete words in arrangements that are specific to each language. Braille education has been an important element of support for blind and visually impaired children for a number of years for developing reading skills; it continues to be used today by specialist teachers who visit primary schools on a peripatetic basis. Braille literacy correlates well with higher employment rates. However, braille usage has been declining throughout the 21st century. This is, in part, in recognition of the fact that the vast majority of people with VI lose their sight later in life when out of full-time education (and when their ability to learn new skills has diminished often along with their tactile sensitivity).

Braille is traditionally written with embossed paper using the original slate and stylus method in which each dot is created from the back of the page, writing in mirror image, or it may be produced on a braille typewriter or Perkins Brailler. Unfortunately, braille letters produced in this way cannot effectively be erased and written over if an error is made. Braille-users can read computer screens and other electronic supports thanks to refreshable braille displays or they can type it on a braille writer, such as a portable braille note-taker, or on a computer that prints with a braille embosser once they are happy with what has been written. Table 12 shows a number of items of computerised hardware that are available to support braille users. It should be noted that they are very specialised and therefore quite expensive. It is likely that further advances in digital technologies may offer alternative and lower cost approaches to reading and writing for people with VI. These will be discussed in Part 2 of this book, along with the impact of such transformation on the skills and training agendas.

Technology for People with Hearing Deficits

Hearing impairment is the most common sensory disorder in the human population; progressive hearing loss is particularly common. It poses several challenges for people to communicate with friends, family and work colleagues, for appreciating music, entertainment and nature, and in keeping safe and secure in a world where many warning cues take the form of audio alerts. Overcoming these issues and being able to enjoy a full life is a goal that can be supported increasingly through appropriate Assistive Technologies.

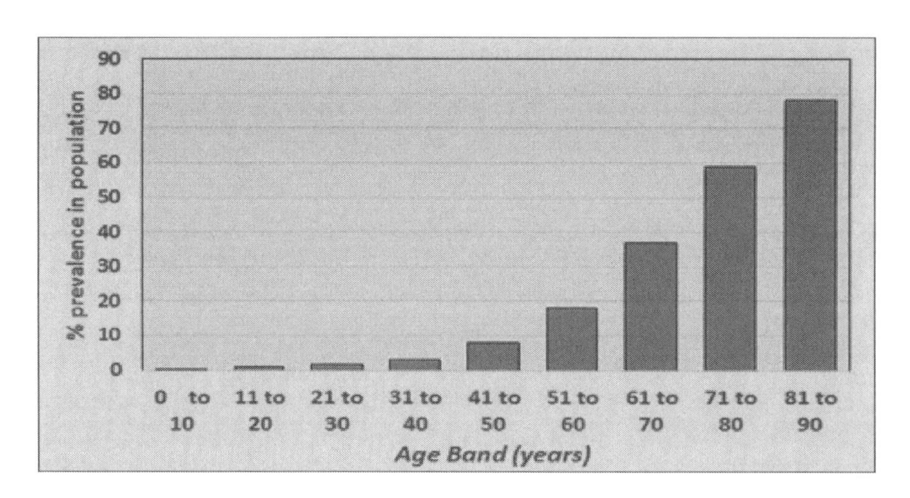

Figure 37: Hearing Loss as a Function of Age in the UK Population

A partial or total inability to hear is known as hearing loss or hearing impairment. It can be temporary or permanent in one or both ears and caused by a number of different factors including noise exposure, genetics, some infections and medications, and trauma. However, age-related progressive hearing loss is the most common cause and, because people are living longer, has been the dominant factor for the past century. Figure 37 shows the age dependence of mild to moderate hearing loss in the UK as a function of age; it shows that the vast majority of people aged over 80 will have a noticeable loss of hearing sensitivity. Hearing loss in the over 80s is likely to be under-reported – mainly because the level of professional testing of this group tends to be lower, as they also may have reduced mobility, which makes it difficult to attend testing sessions.

Different strategies may be required to support these populations to enable them to enjoy a good Quality of Life, and to play a full and meaningful role in society. Younger people may be able to learn to read lips or to use sign language. Indeed, there are 25,000 or more hearing impaired people in the UK who use sign language as their main language. Lipreading (sometimes called speech reading) is a difficult skill to acquire. Even the best lipreaders can catch only 25% - 30% of what is being said. The percentage may be a little higher if the deaf person knows the speaker well and the vocabulary that they tend to use. Deaf people look at the context of the discernible words in the sentence to fill in the words that they did not understand; this is because many mouth movements appear similar on the lips and may look the same to a deaf person.

Tinnitus is the term for hearing sounds that come from inside the body, rather than from an outside source. It's often described as "ringing in the ears", although several sounds can be heard, including: buzzing, humming, grinding, hissing, and whistling. Some people may hear sounds similar to music or singing, and others hear noises that beat in time with their pulse. Around one in every 10 UK adults has tinnitus. This increases to 25-30% of over 70 years-old and can be very distressing and a reason why they feel stressed or unable to get to sleep.

There are essentially two types of assistive technology that can help with hearing loss issues:

1. *A cochlear implant - this is an electronic medical device that replaces the function of the damaged inner ear (cochlea) to provide sound signals to the brain.*
2. *A hearing aid – this uses a microphone and amplification to make sounds louder.*

Cochlear implants use a sound processor usually worn behind the ear, which captures sound and turns it into digital code for transmission through a coil on the outside of the head to the implant. This converts the digitally-coded sound into electrical impulses, sending them along the electrode array placed in the cochlea. The implant's electrodes stimulate the cochlea's hearing nerve, sending the impulses to the brain where they are interpreted as sound. They can help people who:

- have moderate to profound hearing loss in both ears
- receive little or no benefit from hearing aids
- score <50% on sentence recognition tests by a hearing professional

Many people have cochlear implants in both ears (bilateral) - listening with two ears can improve the separation of wanted sounds from background sources. Many adults with profound hearing loss who have had cochlear implants:

- Hear better with a cochlear implant than with a hearing aid achieve an average of 80% sentence understanding, compared with 10% sentence understanding for hearing aids
- Can focus better when in noisy environments enabling them to have conversations with people across meeting tables, in restaurants and in other crowded places.
- Feel safer in the world as they can hear alarms, people calling out and vehicles; and
- Take part in meaningful telephone conversations , enjoy music and sounds of nature.

However, the success of implants depends on a range of other factors including the length of time that an individual has previously lived with hearing loss, the severity of the condition, physiological factors, and the amount of practice that they can have in using their implant system in normal life. About 1000 people in the UK receive implants every year. However, while the estimated uptake of cochlear implants in children aged under 18 is about 94%, this drops to 5% for adults, even though

60% of people with severe hearing loss are aged 70 or over. It is possible that the referral rates are low and older people are more reluctant to accept surgery if they consider hearing loss to be an inevitable consequence of ageing.

In the UK, nearly 7 million people could benefit from hearing aids. On average, it takes ten years for people to address their hearing loss Hearing aid use statistics suggest that fewer than 20% of adults who could be helped with a hearing aid actually wear one – this includes a quarter of hearing aid owners who don't regularly wear them! This figure suggests a significant aural rehabilitation challenge. If people do not acknowledge that they have a hearing loss, there is no way to help them. All the recent developments in assistive listening devices are useless if people will not wear them. Table 13 describes some of factors which have been studied by psychologists as possible reasons to explain these problems.

Table 13: Explanations for Poor Utility in the Use and Wearing of Hearing Aids

Factors Explaining Poor Utility	Explanation
Lack of clarity	Inappropriate design for individual requirements
Uncomfortable	Difficulties wearing device, putting it on and taking it off
Maintenance	Dexterity issues for changing settings and batteries
Denial	Failure to accept the need for aid with hearing
Poor design	Acoustic feedback, low battery life and performance
Few opportunities	Use not compatible with lifestyle and work requirements
Cost	Private purchases, maintenance and batteries expensive
Nuisance	Difficulties in remembering to use it appropriately
Medical issues	Risk of infection or problems with ear wax

In the USA, where people generally have to fund the purchase of their own devices, the cost is often cited as the major barrier. However, as digital hearing aids are available free of charge through the NHS in the UK, it may be apparent that the reasons for non-utility must be similar for hearing aids as for other devices such as walking frames and other assistive aids. Research has concluded that the dominant factor was one of pride, many people believing that a hearing aid makes them look old even when the devices are now so small that they are barely noticeable. As increasing number of younger people use earphones and special headsets to listen to music or to make telephone calls, it is possible that soon wearable devices to improve hearing or other sensitivity will not be seen as stigmatising and age-related but will be thought of "cool". This could become a major 'digital dividend' as transformation occurs.

Technology for People with Communication or Speech Issues

People can have difficulty with face to face communication for a range of different reasons including physical disabilities and motor co-ordination problems that can make the production of speech difficult or impossible. Some with learning disabilities can find it hard to produce or manage speech. Others struggle with speech following a medical incident such as a stroke. The term AAC (Alternative and Augmentative Communication) is used to describe the different methods and assistive technologies that can be used to help people with disabilities communicate as an alternative or supplement to speech. We all use augmentative communication methods some of the time, whether it's through gestures with our hands or by our facial expressions. It is part of our articulation that can often add emphasis or context.

Manual signing systems have been in use for some time, especially in the deaf community. Different systems have been developed to meet the needs of individuals with learning and motor disabilities. Signing systems have the major advantage of not requiring any additional equipment or materials but can be harder to learn. On the other hand, written or computer-generated text can often be used to convey messages with more description and meaning. It can be produced easily using desk-top computers and then used as required or spoken out either using text to voice systems or using pre-stored messages.

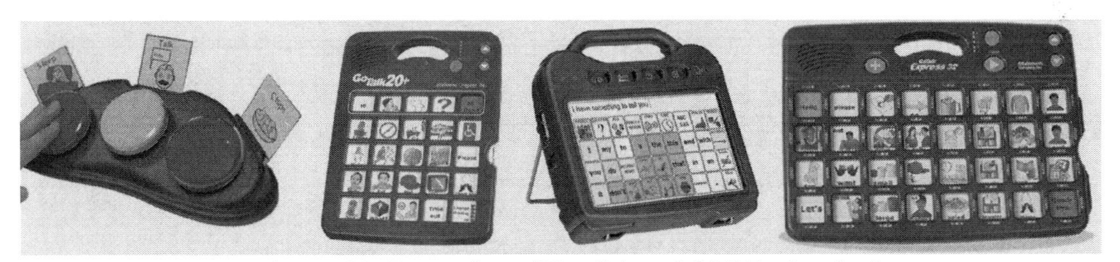

Figure 38: Examples of Traditional AAC Technologies

Most AAC methods today use integrated electronic systems that combine message selection with speech output. They range from simple buttons that correspond to pictures or short messages, through to more sophisticated methods of selection and far wider vocabularies. Most are portable (see Figure 38) but are relatively easy to damage. Therefore, their use needs to be supported with appropriate maintenance and repair arrangements and must be restricted to situations where they will not easily be damaged. They remain relatively expensive due to the limited and extremely fragmented market. Opportunities for improvement using consumer electronics and digital communication and display methods will be significant in the future. Most will be based on the use of digital consumer electronics, the thrust of Part 2 of this book.

Technologies for Functional Support

There are many types of physical disability that can be supported through assistive technologies that enable people who cannot control or use all their limbs to perform tasks associated with daily living. These used to be referred to as electronic assistive technology systems, manufactured and supplied by companies such as Possum, and championed by Biomedical Scientists within the NHS. Systems known as environmental controllers offered the means to remotely open doors or windows, to switch on lights and appliances such as a television, radio or music player, and control intercoms and telephones using jelly switches, eye movement controllers and other sophisticated systems often on a bespoke basis. Capital costs were high, but so were the maintenance contracts needed to ensure that life critical support arrangements were available at all times. Improvements in wireless communications and in local intelligence have compromised the viability of the older systems. The support options are likely to be transformed in 'smarter' homes that already have flexible communications infrastructures in place, and which will benefit from Internet of Things arrangements when Digital Transformation takes places. These opportunities will be a focus for housing redesign which will be covered in Part 2 of this book.

Prosthesis

There are many people who require an artificial limb or limbs to be fitted at some stage of their lives. Although some limbs may be amputated during the final years of limb as a result of diabetes, for example, most people who lose limbs are younger and, at the time, relatively fit. They may be the victims of road traffic accidents, industrial injuries, or soldiers who lost limbs (usually legs) in explosions or terror incidents. Originally, prosthetic hands were no more than hooks (see left of Figure 39). Improvements achieved through 3D printing, in different colours for children, and by micro-circuitry, embedded intelligence, flexible sensors, artificial skins and compliant, composite construction materials, have revolutionised. The other pictures in Figure 39 show a personalised prosthetic for a little girl, the Raphael smart glove, and an artificial hand made by i-Limb, from the world's leading developer and manufacturer of such technology, who are based in Scotland.

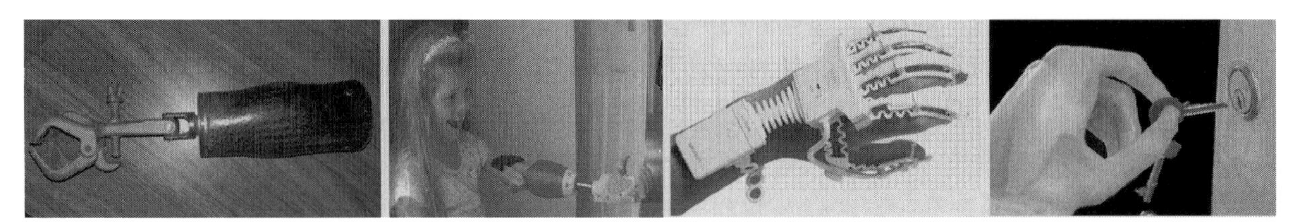

Figure 39: Improvements in hand prosthetics due to 3D printing and smart electronic design

Mobility

Walking can become difficult or dangerous, without support, for many people suffering from a wide range of diseases and disabilities. Wheelchairs have been used both as assistive devices that allow carers to take people outside rather than being confined to bed or to the inside of their homes, and as vehicles that enable users to propel themselves if they have sufficient strength in their arms.

Figure 40: Examples of advanced electronic wheel-chair designs

Electronic wheelchairs have benefited from advances in battery power, and in more responsive control units. Figure 40 shows some more advanced designs that allow height adjustment so that the user can, for example, take their place at a kitchen breakfast bar, or reach to a high shelf, or with adapted treads, climb steps.

Many older people are able to use portable assistive devices such as walking sticks, crutches or rollators to move around the house, but struggle to get up and down the stairs safely. Quite often, properties have only one toilet, which is upstairs, requiring them to climb the stair many times a day, or somehow restrict visits to the 3 or 4 times when they have a carer present. For this reason, others bring the bed downstairs and rely on a commode for toilet functions. The stair lift was introduced to help with this problem and proved to be especially popular in the UK because so much of the housing stock is on two or more levels. It was a modification to the home described earlier in this chapter.

Figure 41: An example of a stair lift, a StairSteady system and a through-floor lift

An example is shown in Figure 41, alongside a StairSteady device which has a rachet mechanism that provides support while the user climbs or descends the stairs. The advantage of the latter over the former is both that it is easy to install as it requires no mains power, and that the gripping back can be raised and parked in a vertical position when not in use, thus ensuring that the stairs aren't blocked for other people. Stairlifts can be cumbersome and are usually removed as soon as the user no longer utilises it. Some stairlift manufacturers offer telecare alarm trigger as integrated components for people to report either a system failure or that they are feeling unwell whilst using the system. Also shown in Figure 41 is an example of a through-floor lift. These cost about twice the price of a comparable stair-lift but are only suitable for larger properties. However, they can accommodate a person in a wheelchair.

Muscle Augmentation Systems

The alternative to the mobility technologies shown above is to use muscle support systems (or partial exoskeletons) of the type shown in Figure 42. These have been under development by companies such as Lockheed Martin and Ekso for the military to help soldiers carry more equipment on the battlefield but are also been used in factories to enhance the performance of workers by up to 300%.

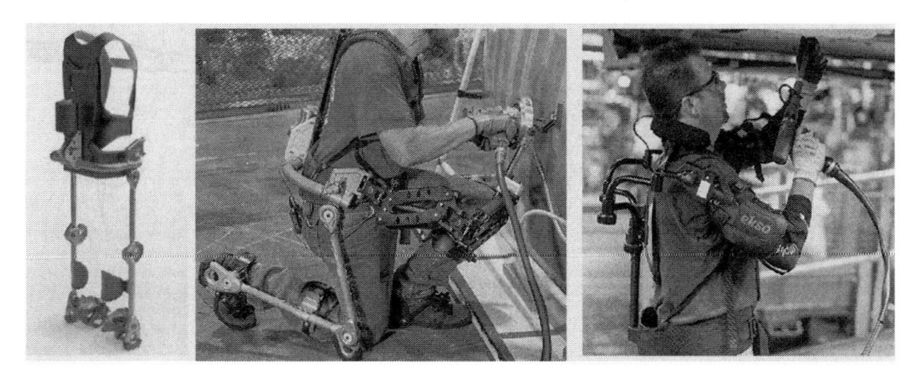

Figure 42: The Fortis and Ekso Augmentive Exoskeleton Systems

More advanced systems can be used by paraplegics (especially those with spinal cord injuries) to rehabilitate after accidents and to enable them to move away from wheelchairs for mobility, improving their outcomes, especially with respect to cardiovascular function, breathing and general health and well-being. Full wearable exoskeletons for people with spinal damage are shown in Figure 43. Rex P (on the left) enables users to walk and stand with their hands free, helping them to be able to perform more activities. There is a dedicated rehabilitation centre in Lincolnshire to support people to use the Rex system, which is often funded (typically £70k) through insurance following a motor vehicle accident. The ReWalk system (centre and right of Figure 43) is lighter and enables use with crutches. It was used by Clair Lomas, a paraplegic from Leicestershire, to walk a complete marathon. She was the first person to use an exoskeleton for daily and independent living about the house, even walking up and down stairs.

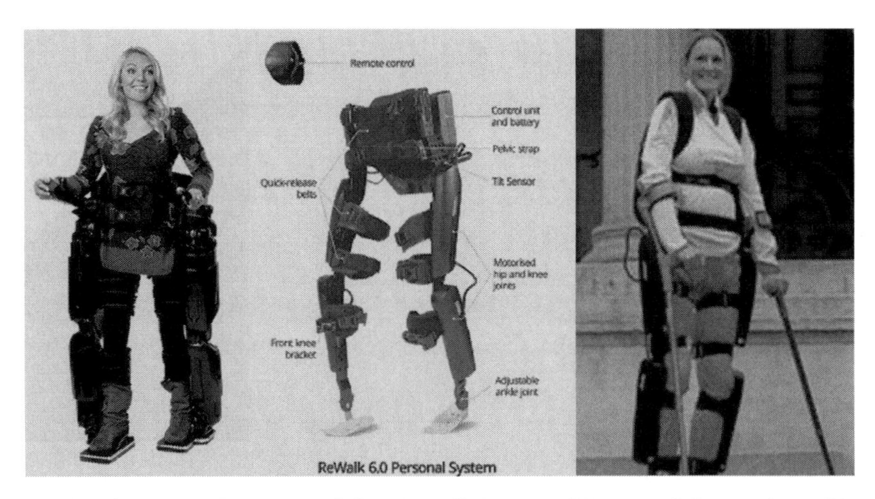

Figure 43: The Rex P System and the Rewalk Personal System Schematic and in use

Figure 44 shows developments in practical and domestic exoskeleton support devices that have emerged after years of research and development by Japanese companies which have significant expertise in motor vehicle design. Companies such as Cyberdyne aim for the rehabilitation market, targeting people who have reduced mobility following a stroke. The versions on the far right are produced by Honda and might be seen as viable replacements for stair-lifts and would enable the user to climb stairs anywhere rather than only in their own homes.

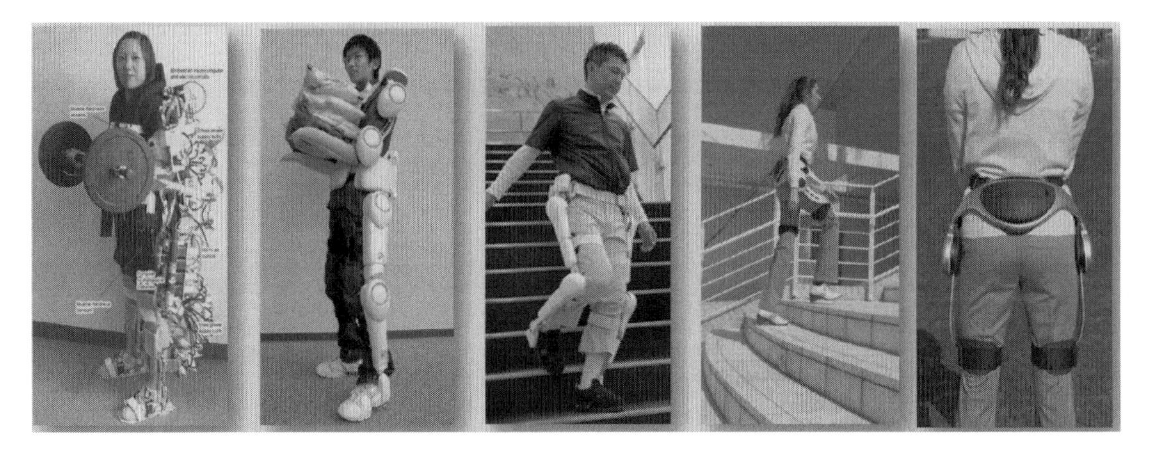

Figure 44: Wearable exoskeletons to overcome a lack of body strength

Overcoming the Effects of Paralysis

When the spinal cord is severed, an individual may lose the ability to move his or her limbs; they may become a quadriplegic who may be limited to head movements. Neurostimulations systems (sometimes known as Neurobridge) use an electronic interface embedded in the brain to bypass the damaged spinal cord and reconnect the brain directly the muscles controlling fingers and hands. Future systems may offer the ability to control devices that allow breathing, and bladder and bowel movements. These will rely on though alone and may benefit from headsets such as those shown in Figure 45 that pick up and decode electrical activity from the brain. The EEG skull caps have been used to control games, simulators and exoskeletons, while the Emotiv headset (extreme right) is a $300 headset that connects to a laptop through Bluetooth to allow the operation of games and apps using mind control.

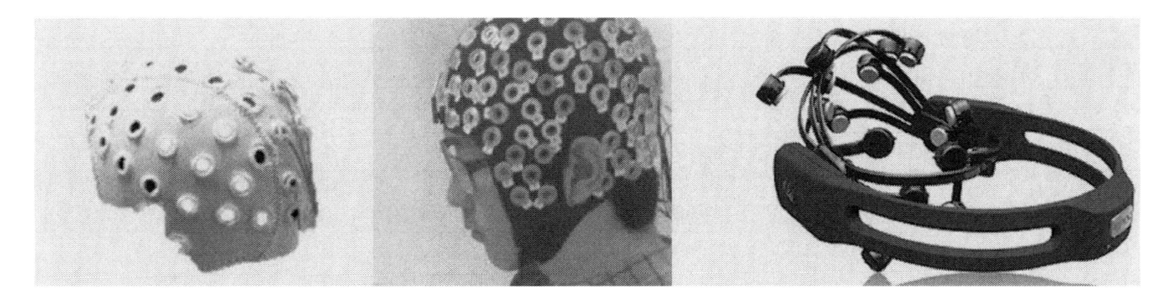

Figure 45: EEG Skullcaps and the Emotive mind control headset for gaming

People are unable to move their hands and arms, are also unable to perform activities of daily living such as feeding themselves. Figure 46 (left) shows the new Obi device from the USA, a robotic system that learns the user's feeding requirements, and is designed for use in the home. In the centre of this figure is the Besta feeding arm from Scandinavia, which is an earlier design with fewer joints, On the right is a Korean development designed for eating food from the Far East. It has two arms – a spoon arm for transferring food, and a grab arm for picking up and releasing food.

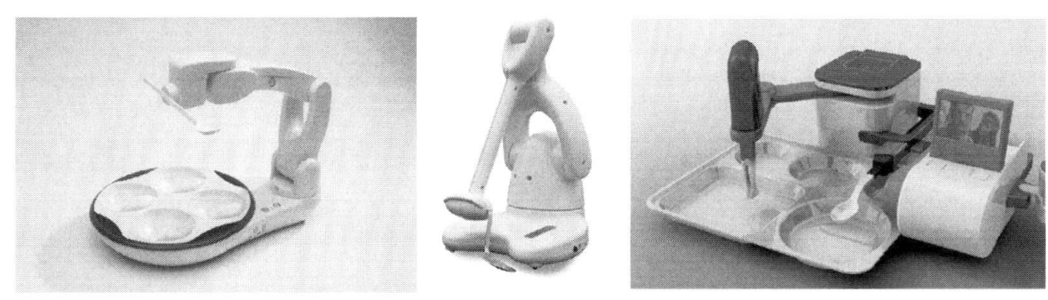

Figure 46: Robotic feeding devices – the Obi, the Besta and a novel South Korean system

<u>Inability to Transfer Without Help</u>

Independent living involves an ability to perform many domestic tasks including transferring from sitting to standing several times a day (see chapter 6). This is a fundamental requirement for people to get out of bed safely, as well as get up from a chair and mobilise. Domiciliary care staff are often employed to support people, especially older ones, to get up in the morning, and to safely get back into bed at night. Unfortunately, this gives people little flexibility because the schedules of staff have to be worked out for maximum efficiency. Figure 47 shows two different approaches to technology that helps an individual to get up from bed and, in the case on the left, into a standing position, and in the case on the right, to transfer into a wheelchair. The mechanical lifting device (left) has been used successfully with many younger disabled people. The bed system (right) is manufactured by Panasonic but has not be exported for the European markets.

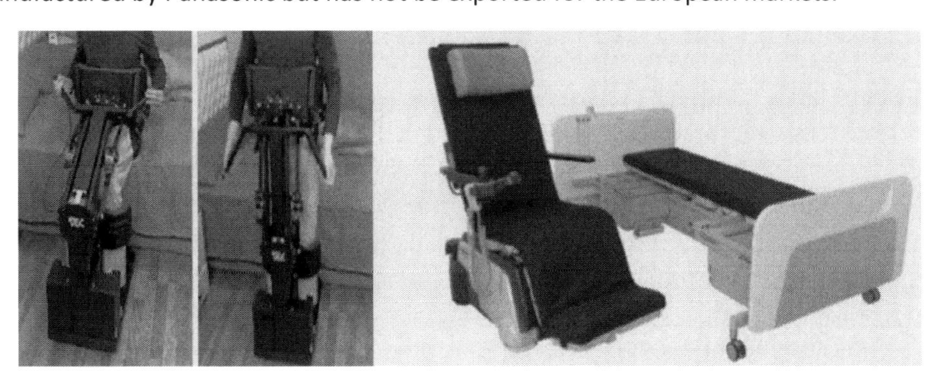

Figure 47: Individual lifting device and a Panasonic bed-wheelchair transfer system

Alternative approaches involve the use of rising or profiling beds. The former can be lowered to minimise the risk of injury if the user falls out but can also be raised to help carers to affect a transfer. The sophisticated system shown in Figure 48 enables the user (or their carer) to control the raising of different parts of the bed using a control pad, and also its rotation and elevation to make it easier for the user to be got up. This approach the advantage that it often needs only one carer, rather than two, to help the user, and can therefore be used in initiatives based on the 'right-sizing' of care. Double handed carer visits can be very expensive but remain a requirement for patients who are heavy or who have a complex condition.

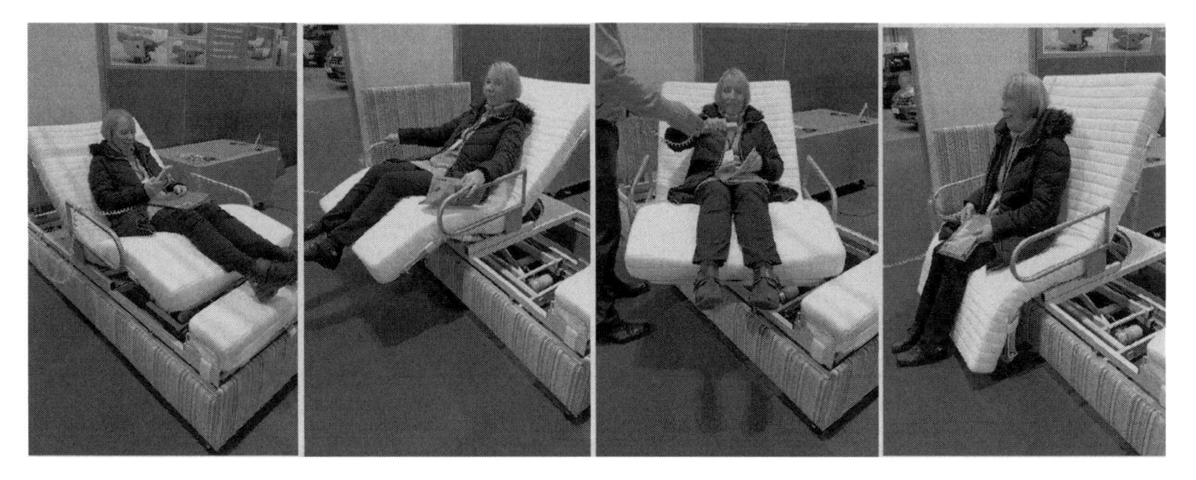

Figure 48: A Profiling Bed to Simplify the Process of Getting out of Bed

Concluding Remarks

Assistive technologies have benefited significantly from developments in electronics, signal processing, local intelligence and new manufacturing techniques. However, it is the ability of devices to communicate with each other and with remote support staff and family members that have the potential to increase uptake and utility that can give users the means to live more independently and to perform more activities on their own.

By including new connected and enabling technologies, such as alarm systems and telecare, under the umbrella of assistive technologies, the potential of an integrated approach to equipment provision is greatly increased. The challenges include making the entire spectrum of AT available through one service so that there are any gaps in provision. It is evident that telecare services can be integrated with community equipment services, or the other way around. But neither currently deals with specialist equipment for people with disabilities, which are currently the responsibility of specialist OTs in sensory, physical or cognitive issues. The 'one stop shop' approach is clearly a worthwhile ambition but is compromised unless all staff are suitable trained in ALL the available technologies and their characteristics.

In the same way, there are significant funding issues that need to be resolved. Adaptations to the home can be expensive and generally funded by individuals though sometime supported by Disabled Facilities Grants. Functional and sensory products may be provided free of charge to eligible people, but often without access to the latest and most useful products. Telecare service are generally not funded free of charge, unless for those who haven't the highest eligibility (and who pass the financial tests). The solution lies with an improved funding model for social care, which is outside the scope of this book.

Chapter 6

Improving assessment, prescription and support options

Introduction

For 20 years after the introduction of Care in the Community, the need for support was assessed using 3 different groups of activities or requirements that were deemed to be necessary for successful and independent living. Figure 49 shows a number of factors that could be used to identify gaps in capability that need to be filled by a social service. Those on the left are known as Activities of Daily Living and involve personal tasks such as bathing and using the toilet. An inability to perform at least one of these tasks could lead to a need for long term care, every day. This can be provided, of course, in a 24-hour care facility, but otherwise through visits by domiciliary carers, typically up to 4 times a day.

On the other hand, instrumental activities (towards the left and centre of Figure 39) are practical tasks which are necessary for independent living for most people, but which don't need to be performed every day (or several times a day). They are desirable but not absolutely essential. People who aren't able to manage these tasks often learn coping strategies, even though some of these may pose additional risks of them having accidents or of being socially isolated. Quality of life activities or options, shown on the right Figure 49, are also desirable to improve well-being and to help in achieving a level of fulfilment. But their importance is generally subjective; certainly, their relative significance is a matter of personal choice.

It follows that support available through local authorities, or through commissioners has tended to begin with those tasks or requirements on the left of Figure 49. As demand for support grew, through an ageing population, and available funding reduced, the priorities have moved sharply to the left, leaving many people with no support at all.

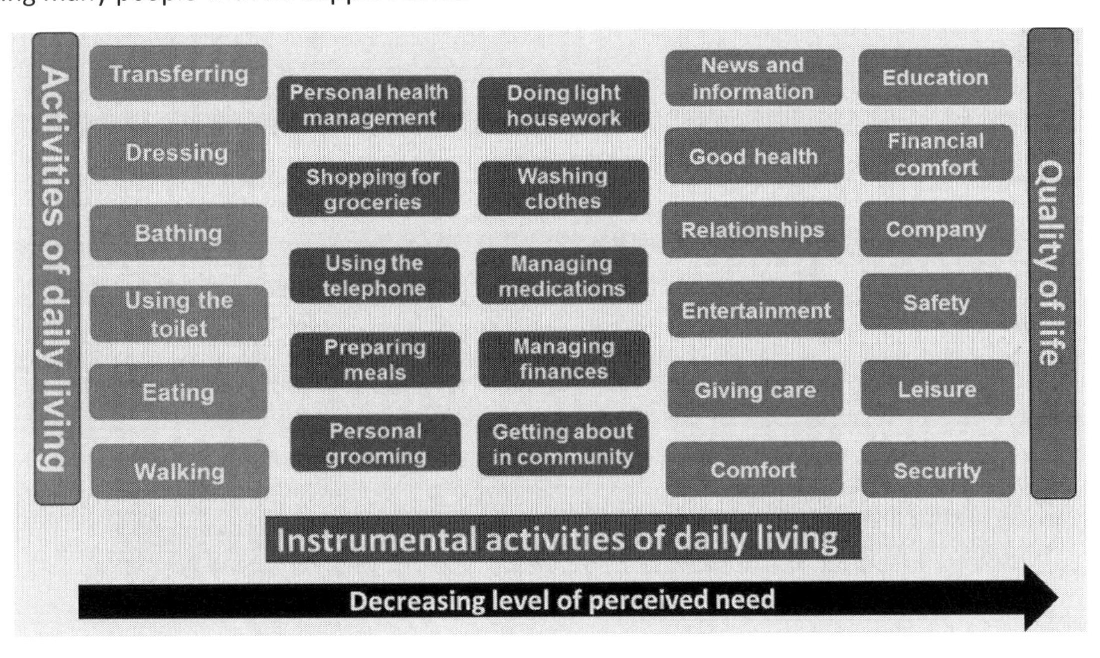

Figure 49: Three levels of need for individuals requiring support

Modern Assessments of Need

The English Care Act (2014), and equivalent legislation in Wales and Scotland, are designed to put people and their carers in charge of their own care and support arrangements. The differences between the new approach and the old one, and shown in Table 14, mark the change from an assessment system that was driven by professionals who looked for what an individual couldn't do, to one in which there is an increasing interest in where the strengths of an individual lie, and the support that they receive from their family and friends, their community and from their personal rewards due to their interests and leisure activities .e. an assets-based approach. This means a change in what services are offered, where they are offered, and how they impact on an individual's perceived Quality of Life – and, indeed, a questioning of whether services can provide the answers in some cases.

Table 14: A Comparison of Old and New Approaches to Assessment

Old Approach		New Approach	
Feature	**Result**	**Feature**	**Result**
Practitioner-based	Professional take all the decisions	Person-centred	Focuses on individual wants and goals
Problem-based	The individual is seen as the problem	Strengths-based	Acknowledges an ability to perform many tasks
Deficit focus –	Look out for things that the person can't do	Skill acquisition	Provides opportunites to try new things
Professional dominance	Views of individual and family are heard but may be ignored	Collaboration –	Encourages professionals to work together and with individual
Acute treatment	Only try to solve immediate problem	Community integration –	Looks for long-term solutions using different resources
Cure/improvement	A medical approach	Quality of life	A social model
Facility based	Focus on hospitals and care home solutions	Community-based	Focus on support in or close to the home
Dependence	Robs people of their capabilities	Empowerment and choices	Enable individual to become more independent
Episodic	Assumes care is delivered in discrete blocks	Continuous –	Allows flexible approaches and interventions
Reactive	Respond to emergencies or incidents	Preventive	Encourages early intervention to avoid future incidents

The new approach leads to a range of improvements on the old eligibility rules:

- The focus is on outcomes with eligibility thresholds to ensure quality and equity;
- There is more access to services that are essentially preventive in nature – a long term strategy that will prolong independence;
- Consideration is given not only to physical well-being but also to mental and emotional issues that can restrict or impact on quality of life;
- Promotion of information, advice and new (digital) channels to provide ideas on independent and innovative approaches to care and support; and
- An integrated approach to ensure that there is maximum choice available through a combination of public and independent providers.

These are all examples of how a ***Person-Centred Approach*** can change the old ways of doing things to people rather than encouraging them to become partners in their own support.

Personal Profile

The starting point in assessments must be the development of a personal profile. This was once the result of asking dozens of questions to collect enough information on an individual to understand their unmet needs and their relevant support options. Early social media sites, such as Facebook, present everyone with an opportunity to produce an alternative profile that can reflect their interests, their connections, their job, their likes and dislikes and their education and employment history. This can be an inspiration for profiles that aren't about disability and old age, but about the skills, goals, friendships, family bonds, and quality of life factors that are such a big part of person-centred approaches.

Table 15: Information Gathered on "All About Me" Form

Focus	Information	Examples
Name	Full name and preference for general use	Mrs Davies; Liz; Auntie E
Close people	Person who knows how you like things	My daughter Jenny
Your home	Address and type of home	Lyndon Park supported housing
Family	Do you have parents, brothers, , children?	3 children – Bob, Gwen & Tom
Friends	Names of special friends or carers	Alice and Peter
Pets	Animals that are looked after	Sheba the cat
Activity likes	Hobbies, entertainments and past-times	Crosswords, and classical music
Activities dislikes	Places and tasks that aren't enjoyed	Church; day centre visits
Clothing	Preferred dress at home and going out	Jumper and jeans all the time
Food preferences	Types of food or meals that are enjoyed	Pizza; Chinese takeaways
Food disliked	Foods that are disliked or that mustn't be eaten because of allergies or health reasons	Sea-food; peanuts
Drinks enjoyed	Favourite hot or cold drinks	Coffee; strawberry milk shake
Drinks disliked	Least favoured drinks	Tea; beer
Your medication	Daily tablets, medicines, creams and drops and how they are administered	Metformin; sleeping tablets; eye drops; dosette box
Washing/bathing	Preferences in how and how often they bathe	A bath every Friday night
Sleep & rest	Routines for bed at night and for taking a rest	Bed time 11pm; get up at 8am
Communication	Preferred methods of asking people for things	Telephone; shouting; writing
Sensory aids	Use of hearing aids or glasses	Reading glasses
Favourite possessions	When jewellery or other items are carried, and where are they kept	Wedding ring worn during the day; in jewellery box at night
Independence	What can you do without help	Washing and ironing clothes
Life threads	Place of birth, schools attended, first job, first love, special places, memorable holidays. (plus, anything else that describes someone's life, and which is important to them)	Born Halifax; attended Pennine School; worked in Smiths factory; married Pete; 2 children; had holidays in Blackpool.

Forms and form filling are time-consuming and unpopular, and, if the information isn't then shared with everyone who needs it, lead to repetition – perhaps the last thing that someone needs when they want help, especially if they are in a crisis. Fortunately, it has been recognised that there is a role for an alternative approach which allows individuals to say more about themselves, what matters to them and what might be done to improve their lot.

In Scotland, this has led to the "All About Me" approach which was developed for people who have a learning disability, but which is clearly relevant to a wider population, including people who have cognitive impairments. It has led to collaboration between the Scottish Government and the Alzheimer's Society to produce "Getting to Know Me" forms designed specifically for people with dementia who are being admitted to hospital. It gives nursing staff an effective resource to help them understand the preferences, likes and dislikes of people concerning hobbies, food and things that are important to them.

Table 15 shows some of the guidance in completing the forms. It may be simple – but it's still a form, and if it's not used carefully, it could become another tick-box approach to profiling. On the other hand, if this can be used as a springboard to start a conversation, and to find suitable activities or to guide a person away from a situation when they are scared and need a distraction, it becomes an ideal resource for person centred care delivery. Its success depends both on the type of information provided, and on it being readily available for staff to use – records are easily lost or damaged. Electronic records on a memory stick or on a server are easier to update and access remotely.

At the same time, the Greater Manchester Health, Innovation and Education Cluster funded a partnership between the University of Manchester and two NHS Trusts to develop the "Getting to Know Me" resources and training materials which focus on giving staff hints on overcome some of the problems faced by patients when they are in a strange environment. It includes a simple card that can quickly filled in and then viewed at the bedside when someone is in hospital, giving staff ideas on how to talk with a patient, and how to help him or her to pass the time and be made to feel at ease. As well as providing information on names, it offers 6 more areas for discussion:

- Events/places that are important
- People/pets/items that are important
- How time is spent (enjoyably)
- Things that matter in daily routine
- Help when worried or upset, and
- Likes and dislikes.

The information collected in this way provides an opportunity to get to know someone better, and quickly, but falls short of finding out what people actually want from life and may need to be completed again every time they move between wards or between hospitals and other care facilities. An "About Me" passport of key facts has recently been launched by NHS England to help younger people (and children) with mental health conditions such as autism to pass on details of their history and preferences in the form of a simple document that they hold. This builds on work by The Royal College of Paediatrics and Child Health and by various charities and relates initially to epilepsy.

The information obtained using the 'All About Me' profile can be extended to include the individual's home environment and its appropriateness. A survey of the home, with a focus on features shown in Figure 50, might also be used to identify the availability and use of technological devices and utilities (such as a mobile phone and broadband, and DVD players, digital photo-frames, music players, TVs and computing equipment), and the attitudes of other family members to options for support. The relevance of the home relates both to the presence of hazards such as poor lighting, worn carpets, clutter and changes in level, and to the state of the utilities. The latter includes the electrical wiring, and the plumbing, but also the age and repair of appliances including the boiler, space heaters, and kitchen devices such as the cooker and microwave oven. These are all potential causes of future accidents but also carry tell-tale signs of past incidents such as fires or misuse that can be indicative of poor short-term memory or a lack of reasoning, both of which can be indicators of possible cognitive issues or limitations in dexterity or control. These are important factors that can be used to support a professional view of needs but should only be used in a final overview assessment when more is known about the individual's goals and motivations.

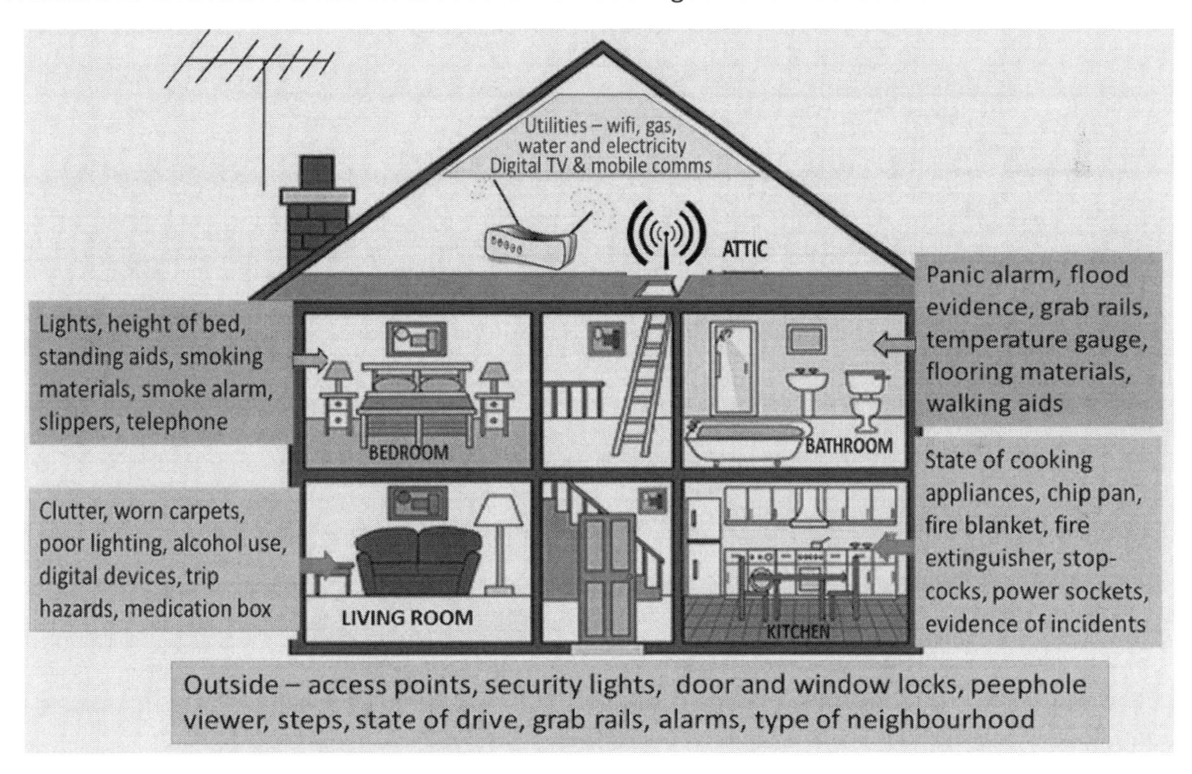

Figure 50: Considerations during a survey of the home environment

Finding Out about Aims, Goals and Ambitions

Many working people spend so much time working, performing daily routines, looking after relatives and surviving from day-to-day that they forget about their future, the bigger picture and what they want from life. The result is that they are at risk of achieving nothing significant in a day, in a week, or even in a year (though sometimes people can't see how much they have achieved). They may die feeling that their lives have been worthless, and that they have underachieved, or that they have specific regrets. Under such circumstances, they may not feel that their any services add significantly to their quality of life. It follows that service provision is not appreciated, and therefore resources can be wasted.

We need to recognise the fact that all people can have hopes and aspirations, even if they will never be fulfilled. We can and should dream about improving our lives and of doing things that we might have missed, or which would be significant. Many of these may not be relevant to health and social care, but the majority can improve both Quality of Life and one's own impression of well-being. Some may be single event ambitions – items that could be on a ***Bucket List***. Some common items are shown in Figure 451 – all were suggested by members of an audience of people aged over 75. This may serve to demonstrate that health, social care and housing professionals may not appreciate fully what some older people may want, and why they are unhappy with the help and support that may be offered to them.

Figure 51: Some Bucket List items suggested by people aged over 75

Other ambitions can relate to doing things in different ways, or of aiming to avoid outcomes that they might dread but have not previously mentioned. A bucket-list offers an excellent means of detailing ambitions that are yet to be realised. Older people tend to be less ambitious in the ideas that they articulate but can be highly imaginative and honest when asked the right questions. This can lead to a list of important opportunities that could make a significant difference to their quality of life if realised. Table 15 lists examples of such ambitions under various headings, which need to be considered when offering help or support, or when trying to help people have what they want in life (as well as what we think they need). They combine the need for support other things that they might enjoy.

The Maslow Hierarchy of Needs was originally proposed in 1943 to explain human development and motivation but has been extended into a popular framework for psychologists to use to understand the needs of individuals. It can readily be used to identifying people's unmet needs, and some of the interventions that might improve their Quality of Life. It is often drawn as 5 tiers as shown in Figure 52. The lowest level of this pyramid represents the basic physical needs of air, water, and food that are essentially the metabolic requirements for all animals. Similarly, clothing and shelter from the elements should form part of human rights, as would a requirement for people to feel safe in their own homes and in the wider community (level 4). This level may be fulfilled through alarm telecare services. Most local authority telecare services focus their resources at this level.

Table 15: Some Ambitions of People that We Need to Consider

Category of Wants	Examples
Housing	Different neighbourhood; Different house-mates
Routines	Going to bed later; Skipping meals
Going out	New places to go; People to accompany them; More often
Leisure activities	Dancing; Cinema; Sports; Travel
Company	More men; More visitors;
Music	Jazz; Choral music; Instrumentals; Chopin
Spiritual needs	Church attendance; Confession
Food and drink	Chinese; Puddings; Sherry; Beer; Gin
Care and support	Help with dressing; help with eye drops; more appliances
Pain relief	Stronger medication; electronic devices
Gadgets and gizmos	Tablet computer; smartphone; drone
Learning	A new language; a craft skill, playing a new musical instrument

Higher levels of need are more associated with well-being, occupation, mental health and happiness. It may be apparent that government (both central and local) focuses primarily on the lower tiers even though attention to the needs shown as tiers 3 to 5 in Figure 43 can greatly reduce future costs by improving Quality of Life and well-being. Individuals may use personal budgets, or their own money, to buy service elements from the higher levels

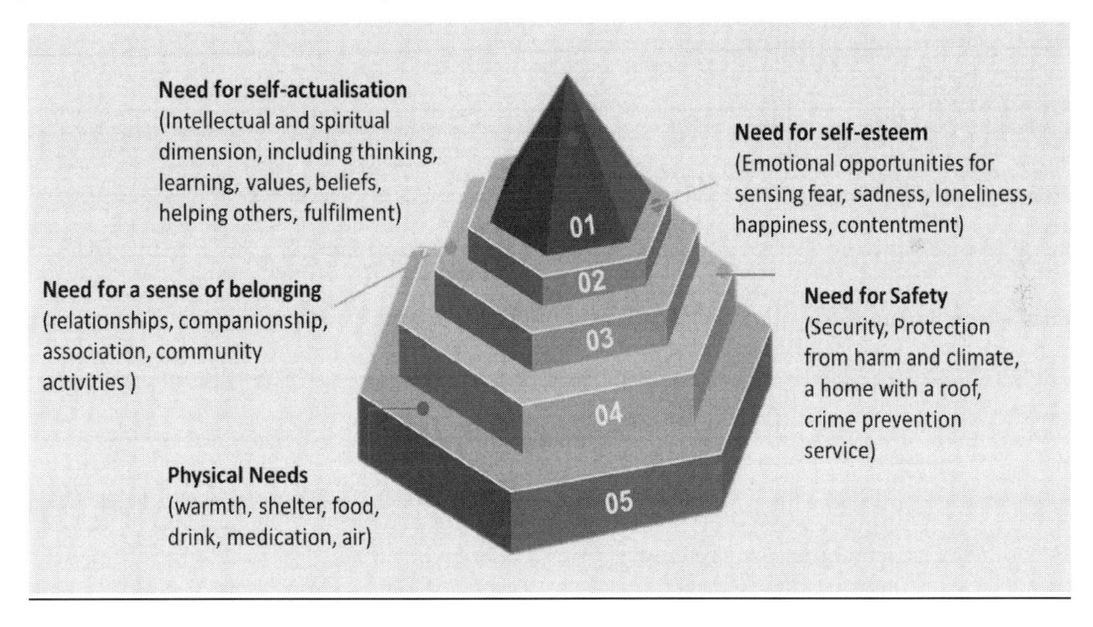

Figure 52: Maslow's Hierarchy of Needs Pyramid

Because every individual is different, a successful intervention is likely to involve the introduction of bespoke services that include housing, elements of social care, health care and various therapeutic interventions. This would have previously involved a large team of people, and might well have been fragmented, inefficient and costly. There would have been gaps that could not be plugged at a local level because of a lack of appropriate services. Technology can address some of these needs, often replacing elements of a conventional and costly care package using standard, or sometimes novel devices and applications.

The types of intervention using technology will expand as new technologies appear, or as existing devices are repurposed for market applications. Most of these applications will make use of digital developments, and will therefore appear only slowly as they mature, and as more of the obstacles to their introduction disappear and access improves for sections of society that are currently digitally excluded. They will be described further in Part 2 of this book. But, for completeness, some examples of technology interventions in line with the five levels of the Maslow Hierarch of Need are shown in Table 16. They will also be described in the next chapter as components of the 7 waves of technology enabled support at home.

Table 16: Needs that May be Addressed using Technology Enabled Support @ Home

Level	Need	Technology Examples
5	**Physical (and physiological)**	Smart thermostat for heating system; Blood pressure or glucose level monitor; Medication reminder or dispensing system; Fridge use and content monitor; Remote controls for appliances and doors
4	**Safety (and security)**	Intruder detection and alarm; Fire, flood and gas detection, alarm and intervention system; Automatic fall detectors, alarm pendants/sensors & alerting system; Video door-bell and teleconcierge system;
3	**Sense of belonging (and social inclusion)**	Advice and information portals including single points of contact; Virtual events and video-based services; Social networks and community notice boards; Voice and other communication interfaces;
2	**Self-esteem (and independence)**	Internet shopping, electronic banking; Activity and meal planners, schedulers and reminders; Queries, jokes and humour generators; Public transport timetables and on-demand taxi services;
1	**Self-actualisation (and fulfilment)**	Augmented and virtual reality techniques/applications; Video reminiscence tools and services; Researching and learning new skills/languages Organising and attending virtual meetings and seminars

Things that Get in the Way – the Obstacles to Achieving Life Goals

Sometimes it is difficult to distinguish between what we want and what we need. We might then have unrealistic expectations on what can be provided, especially if we believe that local authorities and the NHS have unlimited resources. In the future, we are likely to have to fund rather more of our own support needs; we will need to be more realistic when decided our priorities and be more understanding of market forces that might impose restrictions on the frequency and availability of resources. However, some existing low-cost services, as well as low level interventions based on low technology can be used to overcome these obstacles.

Many life goal ambitions may be limited by practical issues, especially those that can make a sudden impact following, for example, an accident or illness. At such times, they, and/or their close family members, can be forced to make knee-jerk decisions in order to achieve an immediate outcome, such as a safe discharge from hospital. These decisions can be made without considering the long-term consequences. People who had wanted to remain in their own homes, might fail to achieve their ambitions due to a failure to put in place actions that could make their homes more suitable for them; this might relate to size, location, changes in removing steps, or heating and lighting requirements. Table 17 describes some of the factors that can become obstacles to them being able to achieve some of the ambitions shown in Table 15. There are many more obstacles than those described in this table. Not all of these can be managed or overcome using technology, but understanding the obstacles is the first step in considering solutions.

It should be noted that these issues are a mix of unavoidable factors, many relating to health and well-being, and a small number of essentially lifestyle or life-choice options. Whilst medical staff can provide rehabilitation and support for illnesses, and occupational therapists can suggest modifications to the home, overcoming a fear of technology is a 21st century requirement which society is only very slowly attempting to overcome. However, it is an opportunity for family members to provide a means of support that is possible when they live some distance away, while technology companies themselves can improve uptake, utilisation and outcomes by improving the design of new assistive technologies. This is especially true of those that provide connectivity and a digital approach to helping to satisfy some of the ambitions shown in Table 15. This will be a theme continued in Part 2 of this book. Nevertheless, an assessment process needs to address these obstacles and prescribe solutions that help to overcome them or, at least, reduce their impact. None of us should expect everything that we want to be provided by the authorities s– that would be unaffordable, and a description of a "nanny state". It isn't only a case of rationing limited resources, and the lack of money – it's also a whole range of practical factors that prevent us satisfying everyone's needs.

Table 17: Six Groups of Obstacles that Challenge the Goals and Ambitions of People

Accidents or Illness	Long term conditions	Disability	Home Environment	Family support	Fear of Technology
Fracture recovery	Cancer	Physical or mobility	Stairs or steps	None close by to help	Not IT literate
Surgery recovery	Neurological diseases	Cognitive or learning	Damp or dirty	Won't ask them to help	No computer equipment
Recuperating after illness	Arthritis	Sight or hearing	Lack of heat or light	Don't trust them	No broadband
General frailty	Depression	Speech or communication	Poor facilities	None living	Can't see how it works

There might also be issues, concerns and risks, often expressed by family members or by professionals rather than by the individual, who might actually neglect or discount them. The challenge is to identify them, and then address the most significant of them including the risks to well-being and safety. Risks can be measured and classified as high, moderate or low depending both on the likelihood of adverse events or accidents occurring and on the severity of the outcomes in such cases. It is more difficult to define issues and concerns in an objective manner, but when they can be articulated clearly, with examples of when and how they might occur, they can be treated in the same way as risks. This opens the door to them being addressed in ways that tackle issues and reduce concerns in practical ways. Services need then to develop in ways that focus on helping people to achieve their goals by either removing the obstacles or by reducing their scale.

As people become older, they are more likely to have long term conditions that they will have to overcome for the remainder of their lives. Some conditions, or at least the symptoms associated with the conditions, can be managed through medication or through other therapeutic interventions. Many can limit the ability of the individual to perform certain activities; for example, diseases of the lungs or airways can lead to a shortage of breath which prevents the individual from being able to undertake physical activities and may prevent them from being able to walk further than 100 metres, or to go upstairs. Oxygen can provide relief but is inherently dangerous and curtails the individual's ability to be away from home for extended periods of time; its use may depend on whether or not the individual continues to smoke.

It follows that many conditions lead directly to issues that need to be resolved. Table 18 shows examples of issues that are relevant to common conditions. These are not intended to be exhaustive but, rather, can serve to illustrate the need to define the issues before looking for solution options. In many cases, there may be co-morbidities and multiple issues that need to be resolved. It follows that a range of technology applications may be needed in most cases. The challenges are to identify those interventions that are likely to be used, as well as those that are going to be cost-effective.

Table 18: Some Health and Disability Conditions and Possible Issues

Stroke	Low vision	Diabetes	Dementia	Learning Disability	Frailty
Speech deficit	Risk of falls	Risk of falls	Communication problems	Bogus callers	Risk of falls
Limited limb movement	Bogus callers	Poor nutrition	Temporal orientation	Fire safety	Poor nutrition
Reduced mobility	Cooking accidents	Lack of stamina	Becoming lost	Social isolation	Reduced mobility
Incontinence	Social isolation	Foot care	Poor sleep patterns	Seizures	Home safety
Memory deficits	Hygiene issues	Lack of exercise	Fire safety	Hygiene	Low muscle strength

Whilst some conditions might lead directly to identifying particular issues that might be relevant, no two people are the same, and are unlikely therefore to experience the same issues because the impact of one or more conditions will be different according to the severity, the stage of disease, as well as their environment, their support and coping mechanisms and any contextual circumstances. It follows that a more personal approach to identifying issues might involve looking at the obstacles that prevent a person from fulfilling their goals and ambitions. Table 19 identifies a range of issues that might prevent an individual from being able to follow some goals that could be relevant to someone who lives on their own in the community, and who wants to improve his or her Quality of Life. Some of these issues coincide with the groups shown in Table 18, which reinforces the importance of understanding what needs to be addressed in order to support an individual. It is clear that neither assessment nor prescription are trivial issues!

Table 19: Groups of Issues that May be Relevant to People Living Alone

Home Safety	Health & Well-being	Security	Loneliness	Anxiety
Unstable on feet	Lack of energy	Unlocked doors	Poor transport	Fear of being alone
Poor vision	Breathing problems	Bogus callers	No family living close by	Poor sleep patterns
Low dexterity	Pain	Rough area	House-bound	Phobias
Poor short-term memory	Smoking	Secluded property	Few friends	Tendency to panic
Old appliances	Poor diet	No way to call for help	Sensory issues	Perceived acute ill-health

Each of the groups of issues shown in Table 19 leads to a set of more specific concerns. For example, being unstable on one's feet, leads to the identification of several possible issues (that will be considered in more depth in Chapter 10) including:

- Falls while outside of the home
- Falls getting out of or getting into bed at night
- Slips in the bathroom
- Trips or stumbles over steps, rugs or carpets
- Collapse getting up from a chair

Each of these can cause serious injury or fractures, but can also lead to a loss of confidence which undermines a determination to live independently; an inability to get back to one's feet without help can lead to many hours being spent on the floor which itself can be demoralising with evidence suggesting the need for a longer period of recovery, often in a hospital.

It would be wrong to suggest that all the issues that can be identified for an individual can be solved whether by medication, by home modifications, by interventions of carers and friends, by new therapies, or by technology. However, there is evidence to suggest that most issues can be addressed, at least in part, by developing an innovative package of care and support that includes a mixture of all of the above. In particular, the potential role of technologies is becoming clearer and, as devices become more miniaturised, more powerful, more connected and more aesthetically pleasing, their role is being extended to provide early interventions, reducing the likelihood of adverse incidents, and providing mechanisms to make life easier for all stake-holders.

The assessment should therefore lead on to a full care plan, starting with the simplest intervention options, before including technologies. Only then should more expensive conventional services be considered along with more sophisticated technologies. The assessor needs to provide a balanced prescription that fits in with the service user's preferences. Commissioners are increasingly requiring assessors to offer technology first but also with a need to achieve their outcome requirements.

Matching Technology Applications to the Needs of the Service User

The assessment process can be seen to be in five parts:

1. Understanding the individual, their health, their abilities and their environment
2. Recognising their life-goals and ambitions
3. Identifying the issues that limit these ambitions; and
4. Proposing interventions that will achieve the required outcomes for the individual (and for the commissioners when the service is to be funded by a local authority or some other public body such as the NHS).; and
5. Choosing the device or technology application that is most suitable, aesthetically pleasing and/or cost effective.

Table 20 describes some more practical and readily achievable goals that may be relevant to people who are currently not independent and whose Quality of Life may be enhanced by technologies. It is not intended as a complete list but to illustrate the fact that there may be an appropriate intervention available in most cases, if only a service provide can understand these goals and provide services that extend from offering simple, standalone items through to interactive service elements that require significant imagination and use of all available technologies. The assessor needs therefore to know the person and know ALL the options.

It may be evident that the task facing an assessor involves being able to consider a wide range of different people from those with relatively few, common needs and issues, through to more complicated cases where there are a multitude of deficiencies and gaps in their ability to be independent. The assessors need also to understand each and every available technology application, their requirements, their limitations and their particular strengths and weaknesses. As technological applications develop rapidly, and new digital opportunities emerge, the training agenda rapidly grows, and there is a risk that few professionals will possess the required range of knowledge and skills. The need to be able to identify functional needs in the way that an Occupational Therapist might do, along with the practical issues that an installer might possess.

Table 20: Examples of Matching Goals to Technology Opportunities

Goal or Ambition	Examples of Relevance	Technology Opportunities
Stay put	Avoid need for residential care Avoid move into level access home	Telecare services Stair-lift or exoskeleton
Get out more	Visiting local amenities and friends Attending social events	Mobility scooter Go online to discover events
Be more independent	Perform domestic tasks without help Take charge of making decisions	Buy a robotic cleaner Online banking/shopping
Become physically fitter	Take more exercise Lose weight	Use a treadmill Wear a fitness tracker
Take up a new interest	Learn a new language Play strategy games	Web services/conversation Virtual games forums
See more of my family	Share birthdays and holidays Follow grand-children	Video calls (e.g. Skype) Social media (Facebook)
Improve Quality of Life	Learn relaxation techniques Enjoy films and music	Virtual yoga or meditation Reminiscence
Feel less anxious	Feel safe at home Feel more secure	Telecare sensors Video door-bell
Improve well-being	Improve stamina Overcome pain	Kinect video games TENS machines
Become a good cook	Take a course in cookery Invite friends for dinner	Electronic cook-book Virtual dinner party
Enjoy better sleep	Sleep for longer Wake up refreshed	Monitoring of sleep quality Provide relaxing music

Furthermore, the ability of an assessor to access the whole range of options will diminish over time, especially if the service has been outsourced to a provided who operates a limited inventory, and which doesn't have the processes in place to test and improve the options on a continuous basis. Figure 53 is a four-level approach to assessor qualifications, skills and knowledge appropriate to different groups of people. It enables higher numbers of people to be supported by technology without the risk that people with more complex issues will be wrongly assessed and offered inappropriate technology applications.

Table 20 describes some more practical and readily achievable goals that may be relevant to people who are currently not independent and whose Quality of Life may be enhanced by technologies. It is not intended as a complete list but to illustrate the fact that there may be an appropriate intervention available in most cases, if only a service provide can understand these goals and provide services that extend from offering simple, standalone items through to interactive service elements that require significant imagination and use of all available technologies. The assessor needs therefore to know the person and know ALL the options.

It may be evident that the task facing an assessor involves being able to consider a wide range of different people from those with relatively few, common needs and issues, through to more complicated cases where there are a multitude of deficiencies and gaps in their ability to be independent. The assessors need also to understand each and every available technology application, their requirements, their limitations and their particular strengths and weaknesses. As technological applications develop rapidly, and new digital opportunities emerge, the training agenda rapidly grows, and there is a risk that few professionals will possess the required range of knowledge and skills. The need to be able to identify functional needs in the way that an Occupational Therapist might do, along with the practical issues that an installer might possess.

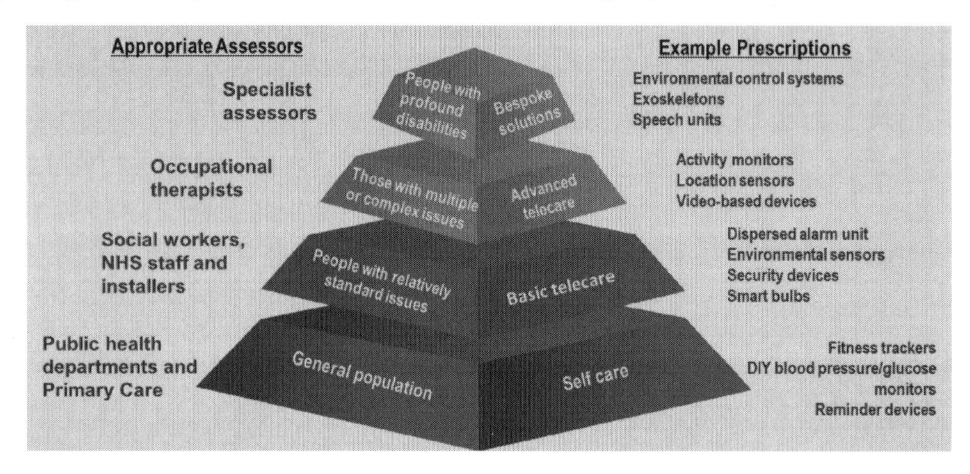

Figure 53: Segmenting Service Users into Different Groups for Assessment Purposes

There are now many alternative products available to provide the remote sensing and communication functions that may be appropriate to remotely support an individual in the community. This means that assessors must choose which type of technology intervention and then also select the most appropriate product to meet this need. Unfortunately, a lack of interoperability between the various in-home devices (and specifically between dispersed alarm units and peripheral sensors) means that options are limited as soon as one device is in place. Further linked devices would need to be compatible with the alarm unit (the system hub) irrespective of the quality of the offering and of their aesthetics. Consumers had only a small role to play, and the usability and popularity of products was not considered by commissioners nor by providers and installers.

In recent years, new products have emerged which are smaller, more efficient, and which can operate outside the narrow range of circumstances that were a restriction to the first generation of device. Furthermore, they can be identified with a wider range of application areas so that some might suit people who live alone while others are designed for applications where the user lives with other members of the family, or support workers either some or all of the time. The result is that there are over 20 different devices for determining the occupancy of a bed, for example. The most popular devices work successfully for standard divan beds but, because they determine changes in pressure, fail to operate correctly if the bed has a reclining or profiling function, or if it is used with an alternating pressure mattress to reduce the risk of developing pressure sores. There are also 3 to

4-fold variations in purchase price and monthly service charges ranging from 0 to £40. The great range provides users with choice, but also opportunity to select the wrong product unless they have access to the whole market and are able select the device that meets their individual needs.

Assessors, as well as private purchasers including the families of end users, need help from independent and trusted sources i.e. decision support tools. Some manufacturers provide basic support documentation and online resources, but these only help select which of their products are most appropriate for a particular circumstance; they don't allow comparisons that would enable assessors to make informed choices based on a range of parameters. It may be apparent that device (and technology-based application) must therefore be identified also by a range of codes or tags that reflect their features. Table 20 shows a number of technical features, and also some examples of usability features that may be relevant in general when selecting a solution. To these we should add other practical factors such as price and manufacturer. Some of these factors may be used as filters to ensure that the discovery journey doesn't become too lengthy, and also to allow users to quickly see how restricted their choice might be if they demand a particular feature.

Table 20: Examples of Product Features of Technology Support Solutions

Technical Feature Examples	Usability Examples
Standalone item	Easy to install
Links wirelessly to a fixed telephone line	Needs a local computer for operation
Links wirelessly to other products	Body worn
Links to other products using wires	Needs to be recharged frequently
Programmable through monitoring centre	Water-proof
Links to a web-site	Recyclable
Mains operated	Low-cost
Battery operated	Needs little maintenance
Uses a mobile connection and SIM card	Works with any smartphone or tablet device
Generates or uses video signals	Suitable for people with visual impairment

Health and social care professionals who are already expert in assisted living technologies and solutions, (including telecare and linked devices), may not need to find products through a full process of discovery, especially when they are reviewing provision or if they are dealing with someone who has been in receipt of a service for many years. They are more likely to understand what's needed and will need to search for a group of similar products in order to compare their features in a direct manner. This means that the entry point for their searches is likely to be through a more direct search of products. They will be helped if products are collected into different groups. With potentially thousands of different products, these groups will need to be split into sub-groups though, with appropriate filtering using the features described above, a sub-group containing 50 different products could quickly be reduced to a handful for comparison.

Table 21: Examples of Product Groups for Consideration in Decision Support Tools

Telecare & Peripherals	Products that Can be Used Away from or Outside the Home	Home Support Devices
Alarm/alert sensors	Location devices	Reminder devices
Telephone receivers	Devices with a SIM	Kitchen products
Telephone peripherals	Security lighting/cameras	Bathroom products
Video-based devices	Access control	Bedroom products
Security devices	Apps	Sensory aids
Control devices	Weather/external environment alerts	Hygiene aids
Medical sensors	Navigation aids	Robotic devices

Table 21 shows various groups which have been collected under the headings of Telecare & Peripherals, Products that are suitable for use away from the home, and devices that are not linked into telecare systems, but which support assisted living by making domestic tasks easier or by reducing the likelihood of an accident or other adverse incident occurring. It may be apparent that some products could be suitably listed under more than one of these headings – but this would not be an issue when looking for a particular product to meet a potential user's needs and under a defined set of circumstances.

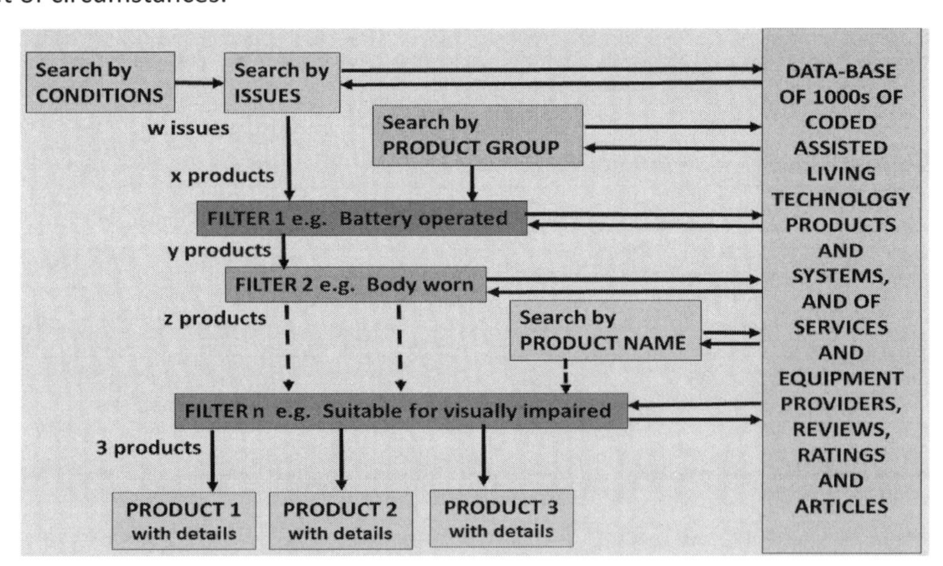

Figure 53: Schematic Diagram of Product Discovery Journey

As more people fund their own care and support or receive personal budgets to buy their own services and products, a decision support tool may need to be appropriate to both professionals and to members of the public. It may therefore be important to provide sufficient choice in the "patient journey" to enable the individual to select different starting points. Figure 53 shows schematically how such a search may be performed using as many, or as few, filters as appropriate. Each filter allows the number of products that are suitable to be reduced until only 2 or 3 remain. At this point, the details of each product can be shown so that they can be compared for suitability and cost.

The details provided for each product need to include any dependencies or additional requirements, so that the potential user is aware of how it may be part of a system, and how issues of interoperability can be resolved (or not) as the case may be. This also provides an opportunity to offer opinions of other users or of professionals. The former can be consumer accounts of success or failure which cannot be verified but can be aggregated to overcome individual bias. The latter can include an evidence-based approach which could involve independent reviews from named individuals, and a method of scoring which combines key features so that products can be compared more objectively. This is essentially a measure of usability which can be developed to be of considerable value to all stakeholders, and which can also help manufacturers to improve designs. It also provides an opportunity to offer case studies, vignettes or articles to explain the merits of specific technologies.

Reviews, Usability and Scoring Systems

It is possible to review any gadget, device, system, or website according to a number of different criteria, only some of which may be relevant to any individual user or potential user. In the case of telecare or other smart Assistive Technology devices, high level systems may be used by disabled people for whom a failure could be catastrophic. On the other hand, an individual item of telecare, such as a sensor or actuator, may be installed as part of a large package designed to cover an array of needs, many of which may not be needed on a day to day basis. It follows that professional review needs to be clear on the criteria used to make a judgement, especially if it involves a combination of scores from other categories. Table 22 shows a set of scoring categories and some of the rationales that may be adopted for use in reviews. These are derived from the system used originally for Group Comparison Reports within T-Cubed's EPG (Electronic Prescription Guide) as well as in the successor Vivo Guide which was developed both as a website and as an app for handheld devices for inclusion within the Liverpool *dallas* project.

Table 22: Examples of Categories That Could be Used for Standardising Expert Reviews

Scoring Category	Scoring Rationale
Features:	This reflects on the key functions supported by the device in relation to the issues that the manufacturer claims the product is designed to address
Design:	This considers how well the unit has been designed in terms of appearance, size, weight and its appeal from an aesthetic point of view. It also considers any technical innovations or design features that result in the product looking or offering more than its competitors.
Performance:	This considers how well the units perform in 'real-life' functional tests concerning sensitivity, efficiency, range or other factors that allow objecting testing.
Value for money:	This considers the features and performance of the product in relation to its costs (including on-going service or rental charges) as a measure of what is provided for the stated cost.
Usability:	In this context, usability refers to the ease with which it is possible for the device to be setup and installed either by professional personnel or by carers and other family members, and in the overall effectiveness and satisfaction of the user.

In some respects, the *usability* category embraces fundamental principles of assistive technology design and could be the subject of separate reviews. It refers to the quality of a user's experience when interacting or learning to interact with products, systems or applications, including devices, software or websites. It has become a term used extensively and increasingly in information technology as organisations begin to understand the importance of websites to enable content to be browsed and search on-line, and to support a reduction in the nee to print brochures and catalogues that are so often binned after the first read. However, the usability principles that are being developed can be used more generally with respect to any assistive or enabling technology products.

Usability is not a single, one-dimensional property of a product, system, or user-interface but, rather, is a combination of a number of factors including:

1. Learnability: How easy, and how quickly, can users accomplish basic tasks the first time they encounter the design or product?
2. Utility: refers to the product's functionality and whether it does what users need
3. Intuition: implies a nearly effortless understanding of operation and options
4. Efficiency: how quickly can an experienced user who understand the functions perform or accomplish defined or advertised tasks?
5. Memorability: when users return to the product after a period of not using it, how easily can they remember enough to re-establish proficiency?
6. Errors: how many errors do users make, how frequent or severe are they, and how easily can they recover from them?
7. Satisfaction: Does the user find the device or system pleasing to use?

Thus, the 7 points described above may provide the basis for general expert reviews, while the categories shown in Table 22 might enable a professional user score to be generated. This could, and should, be kept apart from end user reviews which may be useful as an endorsement from the perspective of a manufacturer or a distributor, but which may lack the objectivity needed for an evidence-based account of a product's quality.

Moving from Product to Purchase or Service Procurement

In order to move forward from recommendation to a prescription, an assessment must be associated with a location where the service user lives (or spends most of his or her time). This is to ensure that all the discovered products are available in the relevant area, country or territory. Fortunately, rapid changes in the way that web-based retailing has become a global phenomenon means that old restrictions on access have largely disappeared, though the delivery time and the cost of importing may remain a barrier in some cases. This implies that individuals may be able to select a product and then buy it immediately if the product details discovered include procurement information. It should be possible to "click through" to a web store from a product page.

In the same way, many assisted living products will not be suitable for self-installation and, indeed, may be useful only when procured as part of a full package such as an alarm telecare arrangement using all the elements described in Chapter 3. In such cases, the focus should be on finding a local service provider that can both include the identified product within a service proposition and can provide evidence of its previous use to achieve desired outcomes. If the assessment was being performed by a professional from an existing service provider, then their service may become the default option. If the product is not currently available from this service provider then other options may be offered, or the original service provider can add the product to their inventory. In each case, the result is an improved matching of device to the individual, and of greater choices.

The impact of this approach is likely to be that forward-looking service providers will expand the range of products that they have available, in order to grow their business and achieve better outcomes. In general, a potential service user will need to search for half a dozen or more products in order that they might address all their issues, but they will be able to directly relate each element to an issue, and will, furthermore be able to exercise considerably more choice in developing their own package of support.

In order to achieve an assessment using the above principles, each provider should:

- Create a database of products that have been coded in such a way that it can be searched online using a range of simple filtering tools;
- Develop a web portal that can be used by service providers, payers and the public to discover the most appropriate products for their needs; and
- Engage with both service and equipment providers so that all their offerings are included in the product database.

Some on-line resources, or apps, are likely to be produced by some equipment providers to mimic the Vivo tool developed for Liverpool CCG through their dallas project. The risk is that they will only show products that are sold by their sponsors, thus restricting choice. Ideally, an independent agency should be responsible for overseeing the product list and for ensuring that descriptions and limitations are accurate.

The Future of Assessment

Technology advances at such a pace that the half-life of a new technology support application applications may be as little as 2 years. Products may then be overtaken by new devices or systems that offer more advanced features, improved usability and greater reliability at a lower price. It makes it almost impossible for staff with assessment responsibility to keep up with all developments. Thus, whilst the goals, needs, risks and obstacles facing individuals may be similar, the choices and the optimum solutions may change. Assessors cannot function effectively and efficiently if they aren't aware of all the opportunities. Furthermore, they need to be confident that the new solutions do as they claim and that they are suitable for all user groups. The claims of manufacturers and distributors may not always be consistent with the reality of introducing products to vulnerable people who might face a number of challenges including reduced dexterity, vision impairment and hearing loss.

There may be a need for future online resource tools but also for more innovative approaches to learning based on collective judgement and gaming principles. The Vivo Pro game was developed to provide health and social care professionals with a method of learning about new products in a more interactive manner. It leads players through a simulated assessment using a range of vignettes that have been designed to present issues and circumstances that can be interpreted in various ways, all of which could be addressed, at least in part, by the use of technology enabled support applications. Each player (or team) is presented with a Person card, together with a number of Issue cards, and some Product cards (see Figure 54 for examples). Each is required to identify the issues, and then the products that are most appropriate to address them for the person that is described. By offering several products that could all appear to satisfy the needs of the individual, teams must make choices and be prepared to justify them to the rest of the group. In practice, it has been found that different assessors view problems differently and, hence, choose different solutions, all of which may be deemed to be 'correct' depending on local priorities, including cost.

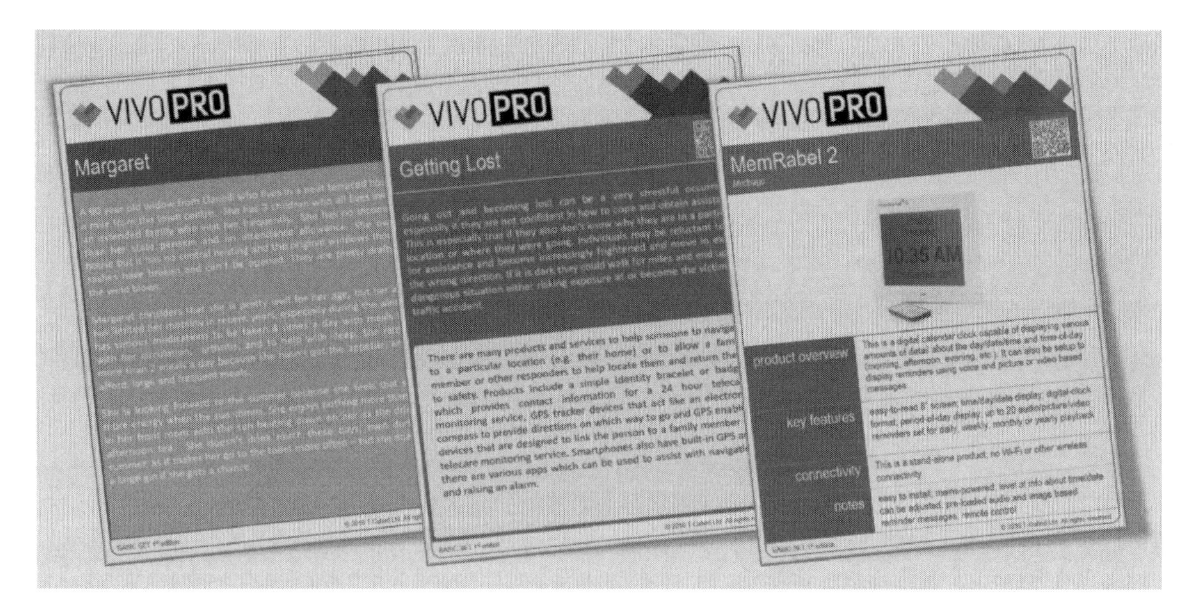

Figure 54: Different Cards Employed in the Vivo Pro Game

It has been found that this the use of specific colour schemes for Dispersed Alarm Units and sensors can underline the importance of ensuring compatibility of linked products, which can help avoid installers subsequently being asked to take out wrongly paired items of equipment. Similarly, the game can be used specifically to introduce new products in order to help assessors to understand the benefits that they can offer and, consequently, change their pattern of product selection. This process may become increasingly important as more products from the seven waves of applications are mainstreamed (see Chapter 7). In each case, a QR codes on the game cards enables players to access the Vivo database on their smartphones or to find out more information when necessary.

In conclusion, it should again be stated that good assessment requires both an in-depth knowledge of the individual and his or her lifestyle, as well as an understanding of what products and services could make a big difference to an individual's. But in practice, there will be significant post code lotteries about the equipment and services that are available in a particular area. The device inventory will be restricted by costs and by particular limitations imposed by commissioners and service providers. Local knowledge is therefore essential so that expectations can be met, but also so that shared concerns can be identified, which might offer the possibility of shared solutions.

Personal budgets allow consumers to choose products and services, and perhaps shared assistants; sign-posting will become increasingly important, but not relevant to NHS prescribing staff because they must provide products that are free and available, at least for short periods of time.

Concluding Remarks

Good assessment has always been a requirement for providing any form of care, from medical treatment of an acute illness through to long term support for an individual who has a disability. The requirements are a full knowledge of the individual, their lifestyle, personal goals and existing support network as well as the experience and skills to match their needs to a range of interventions that include assistive technologies, probably in a wider care package.

The challenges faced by assessors include a shortage of time to get to know the individual; snapshots may not be representative of the long-term picture, while members of the family may not be aware of the real concerns of the individual and might also offer opinions that might not be in the best interest of the individual. This might be especially true when an individual has significant issues associated with frailty or major neurological disorders where the impact is exaggerated in order to influence the outcomes to benefit family members rather than the individual themselves.

However, perhaps the greatest difficulty facing assessors is one of keeping up-to-date with technologies that improve so quickly, making applications redundant in a matter of months. They need to have a constant refresh of available solutions but need also to know which ones are available in their locality. Few local authority telecare services have an expanded inventory of technology solutions available. Most may have only one option to address a particular issue, and that may not be the optimum one for the individual being assessed. Worse than that, some telecare services limit their operations to a particular narrow range of products, perhaps those supplied by one manufacturer. The assessor must then recommend sub-optimum solutions that will, inevitably, fail to result in the desired outcomes. Commissioners can address this deficiency by moving to an outcomes-based model, but this works only if they too understand the relationship between benefits and a good technology fit. Improved and regular training may be the answer – but unless it is provided by independent trainers who have an extensive knowledge of the technologies, progress may be limited.

Chapter 7

The Journey to Technology Supported Living

Introduction

When the term Technology Enabled Care (TEC) was introduced in the UK in 2012, it was considered by many to be an important step in overcoming problems associated with historically poor definitions of terms such as telecare, telehealth, and telemedicine. The misuse of these terms for over a decade, mainly by industry, but without being corrected by academics, had plagued the care and support industry for decades, and possibly held back developments and applications. However, TEC is itself a generic term that must include the whole spectrum of technologies that are relevant to care and support. These can be described under the four headings shown in Figure 55.

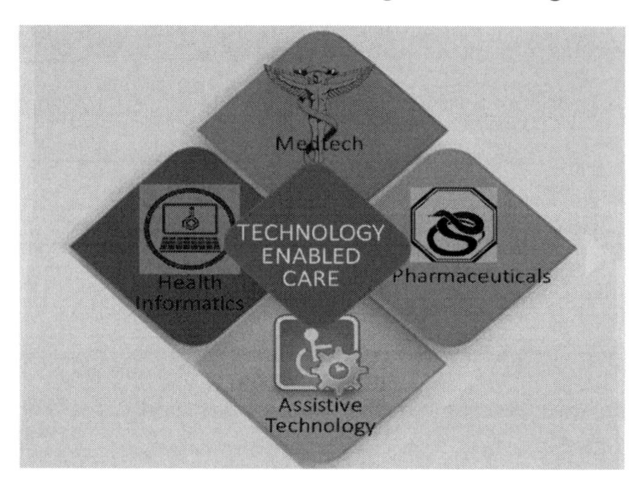

Figure 55: The Four Pillars of Technology Enabled Care (TEC)

MedTech is itself a generic term for a wide range of systems and devices that include MRI scanners, surgical robots, radiotherapy devices and other expensive items that are essential for acute hospitals and treatment centres. They currently have little place in primary care, let alone in people's homes. Pharmaceutical technologies play an important role in treating illness, and research into new drug treatments are so expensive and sophisticated that they are usually the subject of quite separate budgets both for research and for approaches that help to manage conditions of ill-health and disability in the community.

On the other hand, Health Informatics lies at the heart of shared information and decision making, whether through providing emergency clinicians and paramedics with the data that they need to provide efficient crisis response, or through electronic prescriptions and scheduling. The data held in these systems will also become the backbone for advanced data analytics and the use of Artificial Intelligence (AI) that will drive future agendas, and which will be considered further in Part 2 of this book. Personal health records will increasingly be at the core of treatment choices and self-management of our health and well-being. The use of this information will enable people to make decisions on how best to cope with the challenges of old age including long term conditions and the almost inevitable loss of physical, sensory and cognitive function that accompany it; they may choose either personal care and/or, increasingly, the assistive technologies and tools that can help them to do things for themselves more easily, relying less on carers. This is likely to be the driver for independence that provides the motivation to move more quickly into the next era of care.

Thus, it may be appropriate to combine Health Informatics and Assistive Technologies into a new term. This could be either ***Technology Enabled Community Care (TECC)*** or, ***Technology Enabled Support (TES)*** which may then be abbreviated to ***Technology Support (TS).*** It will include connected and standalone technology applications that are based on one or more devices or items of equipment, usually in a system. TS applications are themselves enabled by informatics infrastructures at a national level. Local TS provision should focus on personal items of TS that can be worn on the person (or carried by them at all time) or which can be installed in the home, either on walls, ceilings or floors, or within appliances or items of furniture. Many of these devices are available now and have been used as the basis of telecare services for the past 10 to 15 years. Others will appear as a result of new designs and more efficient processing chips, especially miniature sensors, remote data storage and battery improvements. But, more significantly, their applications within a health and support system have been made possible due to improvements in the telecommunications infrastructure especially mobile data services.

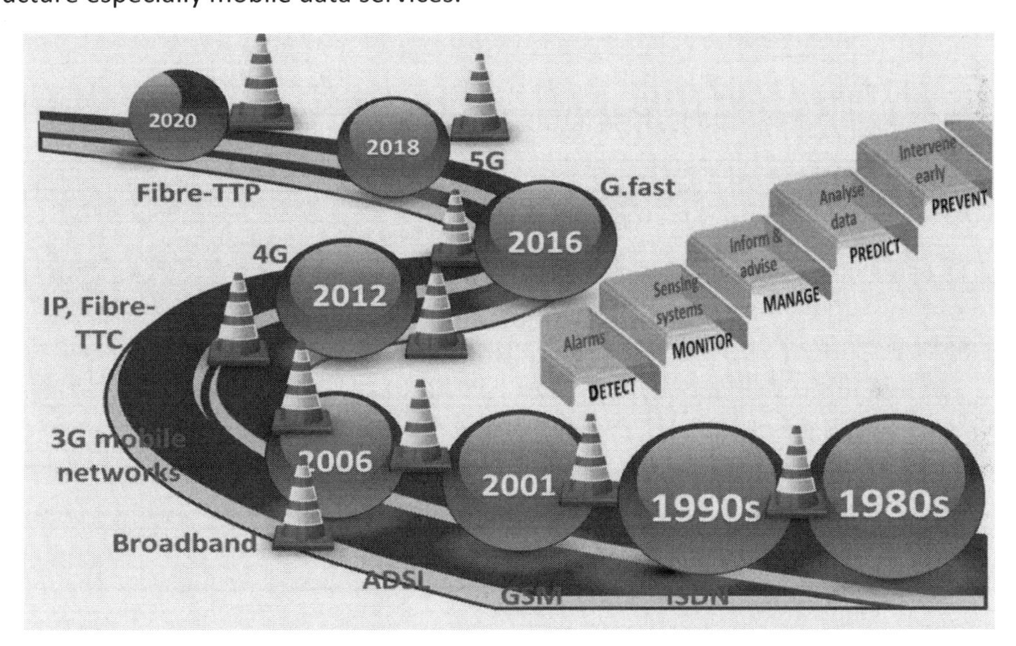

Figure 56: Telecommunication Developments that Have Enabled TS Services

Figure 56 shows an extended time-line for both mobile and fixed line telecommunication speed and access with significant developments marked with a cone. These are providing the options for TS applications that will extend provision from alarm notification and detection, through to monitoring, prediction and, ultimately, prevention (as previously identified as the direction of travel in Chapter 4). There remains, however, a significant lag between the availability of the telecommunications infrastructure and the maturing of TS applications. Developers of applications need to be aware of the commercial realities, including acceptance by older people, and commitment by funding agencies, if they are not to experience commercial failures. Indeed, the quality and reliability of many of the original social alarm and telecare systems has been so great that there will continue to be great resistance to change.

The Rise of Mobile Communications

As described in Chapter 2, ownership of a landline and telephone were something of a luxury in the 1950s and 1960s when sheltered housing was being developed. Indeed, as shown in Figure 57, only 35% of households in the UK owned a landline telephone in 1970. But the numbers increased rapidly over the 25 years; over 90% of homes had such a phone by 1995, peaking at 95% at the turn of the Century. Since then, the percentage has dropped away slightly and has now reached a near constant level of 88%.

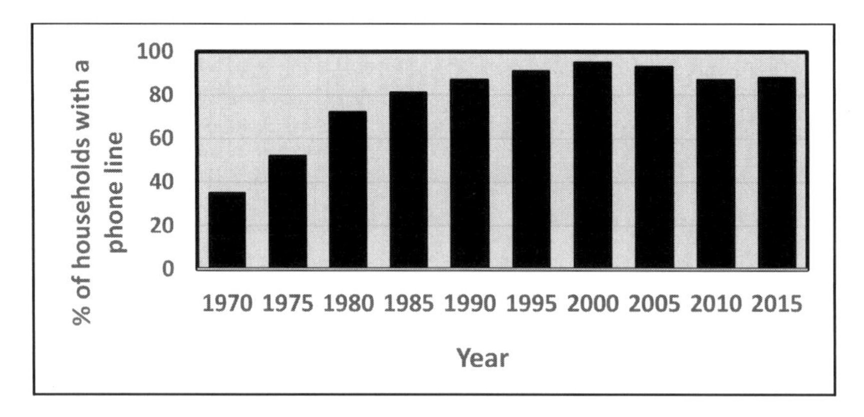

Figure 57: Trend in ownership of a landline by UK households

The reasons that the percentage increased from 1970 are mainly associated with greater comparative wealth, and an emerging significance of communication, especially as the population continued to disperse in search of higher education or improved employment prospects. The subsequent decline has been fuelled by the growth in mobile phone ownership, especially amongst younger people who are less likely to have a property of their own, but who need to remain in touch with their friends. Some older people have chosen to give up their landlines in favour of a mobile device paid for using a Pay as You Go tariff. This is mainly to avoid having to pay for a line rental and the need to pay for 2 different telephone arrangements if they choose also to own a mobile phone. The levelling off of landline ownership may now be the result of people using the landline as a mechanism for having Broadband access to the Internet, enabling them to have a Wi-Fi hub.

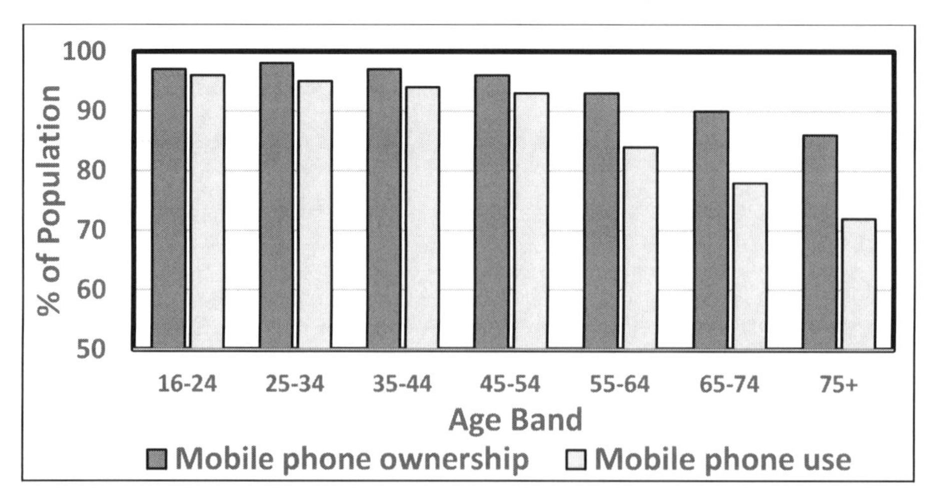

Figure 58: Age Dependence of Mobile phone ownership and use in the UK

95% of UK households now have a mobile phone. However, as mobiles are associated with individuals rather than with the property, there are many households that have many mobile devices, but also many individuals who don't have one. Figure 58 shows the age dependence of mobile phone ownership. Both ownership and use amongst people aged between 16 and 54 is over 90%. However, ownership falls off beyond this age; although over 85% of people aged over 75 have a mobile phone of their own, their use is restricted and is little over 70%. The reasons are complex and may be related to the fact that modern mobile devices (i.e. smartphones) are more sophisticated and can be used for a wide spectrum of applications. Indeed, it has been suggested that their use for data now exceeds their use for voice communication. However, older people also find the devices more difficult to learn to use, and struggle to justify the effort, especially if they are challenged by poor dexterity and vision. Nevertheless, many do appreciate the benefits of a communication tool that is small enough to be carried with them everywhere and at all times, and increasing numbers use their devices to improve access to other generations, especially through the sharing of photographs, which has been possible at scale as new generations of mobile communications appeared.

Table 23: Characteristics of Mobile Telecommunications Generations

Property	1G	2G	2.5/2.75G	3G	4G	5G
Dates of Introduction	1980s	1990s	Late 1990s, early 2000s	Early to mid-2000s	2010s	2020s
Typical speeds (kbps)	2.4	64	64 - 384	1000 - 3000	100,000 – 20,000,000	1,000,000 – 1,000,000,000
Service	Basic voice	Improved coverage & capacity; SMS also	Data-enhanced WAP, GPRS & EDGE	Voice, data & video - UMTS	Designed for data and mobile ip	Plans are developing for fastest provision
Key features	Analogue	Digital	Enhanced service provision	Allows internet use	Offers true mobile broadband	Not yet available

The features, speeds and service provision offered by different mobile phone generations are described in Table 23. 6G is proposed to integrate 5G with satellite networks for global coverage in the future. The dates shown for introduction of 1G to 5G are for the UK and Ireland, but should be treated with some caution, because they generally refer to the situation in big towns and cities. Due to the limited range of signals at the frequencies used for mobile telephony, there are areas where reception has been poor or non-existent, irrespective of which commercial network is chosen. The presence of 'not spots' continues to be a threat to service expansion and application in many areas. This topic, including consequences and possible solutions, are discussed further in Chapter 11.

Nevertheless, the relatively rapid development of services, handsets and deployment of mobile technology has been the major enabler for a new range of applications amongst the public. In particular, the introduction of digital communications in the 1990s delivered initially at slow speeds to support voice and music, led the way for faster speeds that allow mobile use of the Internet and video telephony. These have become the basis for genuine person-centred support propositions.

The Seven Waves of Technology Support

It may be apparent that new applications need to be proved before some people will accept their role in supporting older and vulnerable people. There is no process for gathering such evidence, at least not at the level required for new drugs for example, so manufacturers of equipment and suppliers of service have faced unprecedented challenges to introduce innovation, despite the benefits being blindingly obvious in many cases.

Figure 59 shows the Seven Waves of Technology Support, against a time-line with the first wave having arrived before 2000. The 7th wave, which according to folklore is going to be the biggest, will be arriving within the next 5 years. The model was introduced initially for West Sussex's Technology Enabled Lives project but has subsequently been deployed across the UK and internationally. Associated with each wave are different types of application, each enabled by improved telecommunications (see above) and by advances in sensors and in system developments. These allow the collection and merging of several data feeds, the storage and analysis of bigger data sets, often in real-time, and the displaying of dash-boards, alerts and trending information. Part 2 of this book will discuss future opportunities, but service changes based on their emergence will be described below.

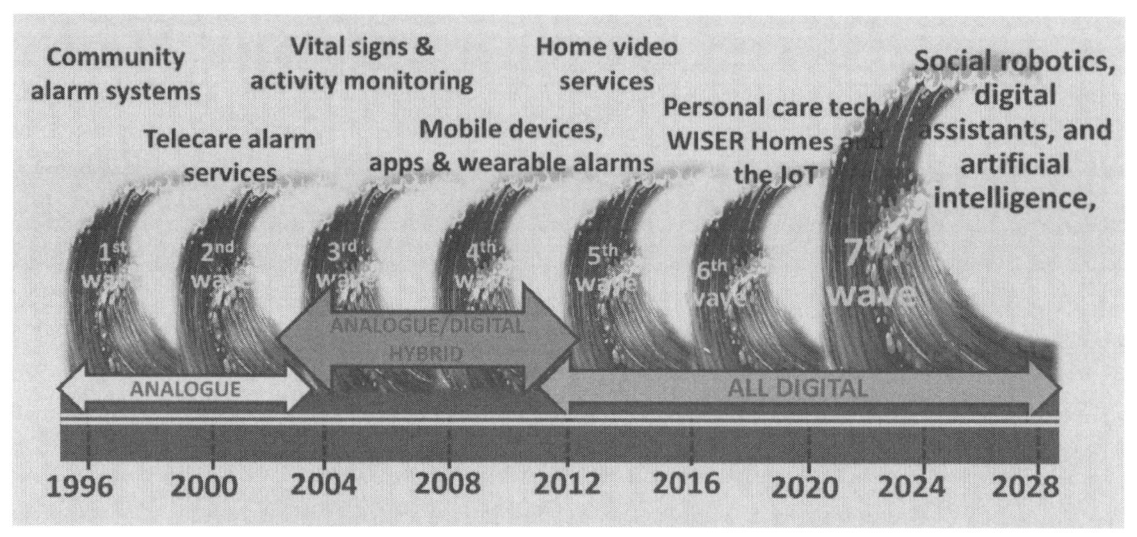

Figure 59: The 7 Waves of Technology Support and Community Care

It may be significant to note that over three-quarters of current telecare service users have no more equipment than a button and a box i.e. a community alarm system. Some will have simple items of assistive technology such grab rails, automatic lighting and walking aids (see Chapter 5) but are unlikely to have benefited from an integrated approach to managing their risks and unmet needs. Many will live in sheltered housing schemes; their TS system might therefore be associated with where they live rather than with their own preferences and issues. These problems can be overcome by using improved assessment processes (see Chapter 6) and person-centred approach to provision.

Figure 59 also shows the 7 waves have split into 3 sectors – analogue, hybrid (analogue to digital), and all digital. It may be apparent by comparison with Table 23 that there is a considerable lag between the introduction of faster, all digital mobile communication networks and hardware, and the wholesale deployment of the applications shown in Figure 59. Nowhere is this more apparent than in the 3rd and 4th waves of technology support where the mobile communication requirements, all digital in nature, have been in place for nearly 20 years for much of the country. Yet, the replacement of the old analogue methods of communication with digital ones based on mobile phone technology had been so slow that hybrid arrangements needed to be put in place to provide appropriate services. These were initially inefficient and comparatively costly but have improved significantly in recent years as the benefits of enhanced mobile provision and use of the Internet have matured. The initial services activity and vital signs monitoring applications have already been described in Chapter 4. Further improvements will follow rapidly as the digital agenda is adopted.

Applications for 5th wave devices, based on video communication services, will be described further in Chapters 10 and 11. Similarly, the 6th wave and 7th wave applications, which are wholly reliant on digital communications and devices, will be major topics for Part 2 of this book. However, opportunities for using mobile alarms and devices based on mobile technology are mature and form the basis of many important applications to support people with cognitive impairment and those, including carers, who have become lonely as a result of social isolation. These applications are described in Chapters 9 to 11. The remainder of this chapter will therefore be devoted to the 4th wave of technology applications, including the development of apps and some wearable devices for use in social care.

4th Wave of Applications

The use of a basic (2G) mobile arrangement for simple alarm signalling is relatively trivial. There are a number of low-cost mobile phones that have fast-dial facilities, some through a big red button at the rear, which can be programmed directly to a monitoring centre to replicate the role of a Dispersed Alarm Unit but in a mobile format. Figure 60 shows a number of mobile handsets that are proving to be attractive to older people and to others with limited vision or dexterity issues. Guide prices are also shown for comparison with dispersed alarm units that are generally in the range £100 to £200 depending on features. All dispersed alarm units available in the UK have a Class 1 radio receiver, are sold with a wireless radio alarm trigger (see Chapters 2 and 3) and comply with EU standards for social alarm systems but are restricted to use within the home.

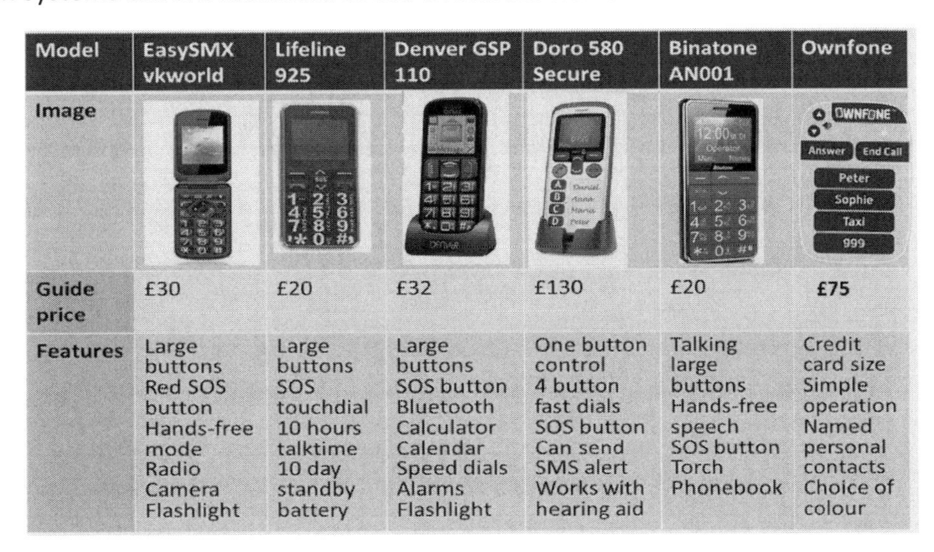

Model	EasySMX vkworld	Lifeline 925	Denver GSP 110	Doro 580 Secure	Binatone AN001	Ownfone
Image						
Guide price	£30	£20	£32	£130	£20	£75
Features	Large buttons Red SOS button Hands-free mode Radio Camera Flashlight	Large buttons SOS touchdial 10 hours talktime 10 day standby battery	Large buttons SOS button Bluetooth Calculator Calendar Speed dials Alarms Flashlight	One button control 4 button fast dials SOS button Can send SMS alert Works with hearing aid	Talking large buttons Hands-free speech SOS button Torch Phonebook	Credit card size Simple operation Named personal contacts Choice of colour

Figure 60: Six different mobile phone units that could be used for alarm purposes

Of particular relevance are the two more expensive units shown in Figure 60. The Doro 580 has only 4 buttons and benefits from universal design features which makes is attractive to a wider audience, while the Ownfone credit card sized phone can be worn around the neck and is available with a range of different button options. Both can be preprogramed with family numbers (and photographs) but have also been used by reablement services to be provided to people who have recently been discharged from hospital. They seem to overcome the stigma associated with the alarm pendants of social alarm and telecare units, whilst having the advantage of working inside and outside the home.

Thus, in principle, a simple m-care service may be introduced either alongside telecare or, in some cases, in place of it. This low-cost option has been resisted by most monitoring centres and service providers for a decade or more partly for reasons based on reliability and availability issues, but also on more negative commercial issues associated with transformation of business, and the role of existing (analogue) technologies. Emergency help and support lines need be trusted as essential services for use by vulnerable, older and disabled people. The telecare industry has developed robust requirements that service providers have to meet for accreditation. These ensure both that the equipment used within the home, and that at the monitoring centre, meet the requirements of EN 50134, a European norm that has developed and been extended over the past 20 years (see chapter 2 for more details). Although many would claim that it is becoming redundant due to the surge in modern (mainly digital communications and hardware), it provides a baseline that many commissioners use as a starting requirement in contracts.

As more people are expected to choose and fund their own social care in the future, they will need to consider more options. This will also be true in sheltered housing schemes (see Chapter 2) where the cost of technology within rents has to become more transparent. More significantly, those who wish to continue with a social alarm scheme will have telecare equipment similar to that used in dispersed housing. Both a shift to dispersed alarms and a move to mobile will be an important one for Registered Social Landlords and for tenants. The weaknesses that remain for the use of telecare service are:

1. Individuals – they will need to pay their own bills to ensure that services can continue, and also be to interact appropriately with the system both to test its performance and to operate the alarm at times of crisis; and
2. The telecommunication network – which may be subject to accidental outages caused by electricity line problems, and damage to overhead or underground cables, as well as issues with cabinet equipment in exchanges or at distribution points.

For m-care services, two additional issues arise:

3. Availability of a network in the scheme and when tenants are outside; and
4. Finding the location of a service user when they are raising an alarm but are not in their own homes.

Existing sheltered schemes that use hard-wired alarm systems are at particular risk of system faults within the scheme as such problems would probably affect all tenants. Consequently, there is a need for a rapid repair option as part of a maintenance scheme. On the other hand, when a hardwired scheme is decommissioned, although there remains a chance of a common mode failure which would affect every tenant's telephone line, the most likely failures will be with individual numbers. It can sometimes take 24 hours for such a failure to be identified. Repairs can be prioritised by BT (or similar) if the individual has been flagged as vulnerable, but delays can sometimes run into days.

While mobile phones can also be subject to service problems, these are usually at an area (or cell) level rather than with individual numbers - unless a handset is faulty or has not been charged. However, some networks and SIM card providers are able to check on lines regularly and frequently using a "heart-beat" arrangement which will provide an alert if a particular mobile number is out of action. This arrangement has been used for several years in the security industry. The quality of mobile networks varies from place to place, so it is essential to understand the strength of signal available at the sheltered scheme for each of the major networks. In some cases, the strength will be fine for all networks, and this will mean that any network can be selected. However, it is more usual for only one or two of the networks to provide adequate indoor coverage at any location, especially if it is outside of the big centres of population.

Figure 61: Dispersed alarm units (DAUs) with integrated mobile connectivity

The larger providers of dispersed alarm systems have moved to provide alternative to the units that required landlines. Figure 61 shows the DAU models from Tunstall, Tynetec and DORO respectively, all of which are compatible with the range of wireless sensors produced by their companies. They have been (typically) more than £100 more expensive than the landline versions, but are more attractive, both in price and in appearance, than the option of using a landline version with a separate GSM modem which was the common solution from 2006 to 2012. These will eventually give way to internet protocol units as system move to an all-digital infrastructure. The impact will be discussed in Part 2 of this book.

Roaming SIM cards can play a role generally in mobile telecare systems, especially when coverage is needed as tenants move about the community and into and out of areas of high signal strength for a particular network. Depending on the extent of network "steering" employed, this means that the SIM can switch quickly between different networks to provide a good signal. The switching time and the amount of "searching" performed before a switch will determine the practicality of this function. It follows that roaming is not needed when a device is used at a fixed location and is unlikely to add benefit unless its purpose is to provide an alternative network in the event of a network failure. It is currently unclear how often such failures occur and, when they do, how often it affects all networks rather than one. It should be noted that if a SIM offers a "heart beat" function, then telecare service providers will quickly be made aware of a failure and will be able to take appropriate action to protect the tenants or service users involved.

Complex Telecare Using Mobile Technology

Many functions of telecare that are achieved using peripheral sensors are based on the home and a dispersed alarm unit but could be replicated using an intelligent hub such as the SenCit which contains a SIM card and a built-in movement detector and electronic thermometer as shown on the left in Figure 62. It allows alerts to be generated and sent to multiple recipients using SMS. The use of additional components such as a door contact sensor allows the system to be used to detect incidents of nocturnal wandering for example. When a means of raising an alarm is needed without a speech option, Figure 62 shows other options.

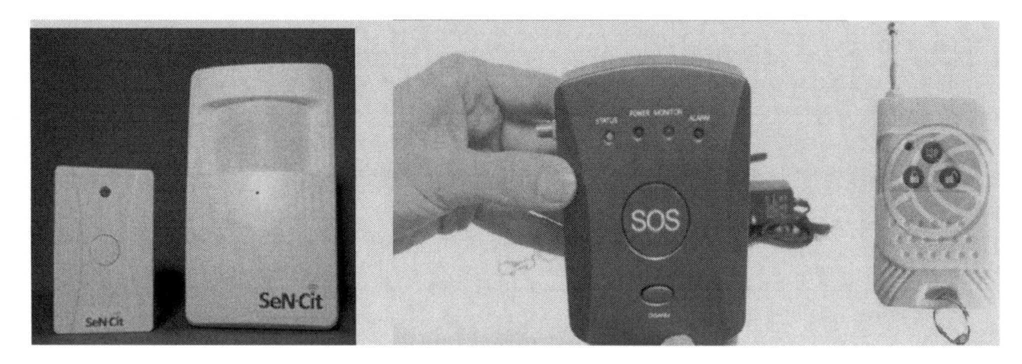

Figure 62: Mobile alarm hubs systems without speech options

Other conventional telecare peripheral sensors and alarm options can operate using their own integrated mobile SIM card arrangements so that a hub may not needed. These devices may be particularly useful when there is a need for only a single sensor solution to manage the identified risks. Thus, the water detector and temperature alert devices shown in Figure 63 may provide the owner or manager of a property with an SMS alert using GSM when a flood is detected or when either a high temperature (fire risk) or a low temperature (freezing pipes risk) are detected. There are also mobile-based detectors for bed occupancy, door status and medication concordance, which all extend the opportunities for risk management because they allow for remote programming or reprogramming of time parameters. Opportunities will be discussed in Chapters 8 and 9.

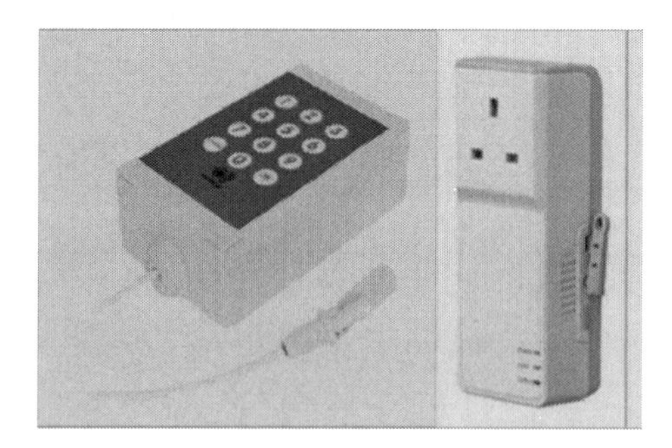

Figure 63: Flood detector and temperature alert devices with internal SIM cards

There are many smoke detectors available that can provide alarms using GSM to send SMS to a monitoring centre, the fire service, relatives and/or neighbours without the need for a hub device. This is a much lower cost solution if the focus is on fire prevention and safety, though there can be no guarantee of the delivery speed of SMS (especially at busy periods). It does, however, offer an additional means of sending alerts from communal areas in schemes, including vestibules, corridors and reception areas. Examples are shown in Figure 64. Some of the smoke alarm options, such as the

KP Mini Wireless system, can also provide panic and security options using the same SIM card and a 433 MHz wireless communications link which some may regard as less secure or appropriate than the arrangements used in conventional telecare systems, though it is unlikely that this can be proved.

Figure 64: Examples of Smoke Detectors with SMS Alerting Capability

This approach may be seen as insufficiently robust for individual tenants, though some tenants may choose the mobile approach over hard-wired or telecare methods of protection because they need no landline, more a large dispersed alarm unit. Other environmental sensors are available that could be deployed in this way to detect dangerous gases including carbon monoxide, ether, methane or propane.

Smartphones and Apps

In 1996, Nokia introduced the Nokia 9000 Communicator which was widely considered as one of the first smartphones in the market. This was long before BlackBerry would include a physical keyboard on a device which would be adopted by businesses as a secure method of providing communications and other smart functions to employees who were often out of the office and who needed a personal digital assistant that didn't need a stylus. Ericsson was on a similar timeline with the launch of the R380 launched in late 1999 which was the very first device to be marketed as a genuine 'smartphone'. It was also the first mobile device to use Symbian OS, an operating system that would dominate the business market for nearly a decade.

The landscape changed completely with the first launch of the iPhone by Apple in 2007 (see left of Figure 65), the first 'smart' device aimed at the man or woman in the street. It had a touchscreen surface that could be navigated with a finger rather than using a stylus like every other smartphone that came before. Steve Jobs knew exactly what it could do for mobile computing and for bringing information and communications to the whole population and committed enormous marketing resources into 'reinventing the phone''. Meanwhile, Andy Rubin was developing his own version of a mobile operating system called Android using the open source Linux system. Following Google's purchase of the Android operating system, the first Android phone was the HTC Dream, released in 2008. It is also shown in Figure 65 (in the centre) with a future version without the separate keyboard, shown on the right.

The iPhone was an immediate success, Google's Android took a couple of years to prove to many manufacturers that an open operating system would have long term advantages. Despite attempts by Microsoft to introduce Windows to mobile phones, Android became the dominant operating systems for smartphones by 2012, though new Chinese entrants have become increasingly popular in recent years. Few basic mobile phones are now manufactured, as people become more addicted to their smartphones. OFCOM, the regulator, suggested in 2018 that the average Briton checks their phone every 12 minutes during the day. This is no longer about making and receiving phone calls and text messages but is about online activities. The average time spent using the phone has increased to nearly two and a half hours, suggesting potentially addictive behaviour that may need to be addressed to avoid future public health issues.

Figure 65: The first iPhone and Android Smartphone devices

As shown in Table 24, modern smartphones are packed full of features that enable them to be used for versatile and useful applications of relevance to a wide range of contexts and industries. In particular, they boast powerful processors and embedded sensors available that can provide the necessary platform on which a number of "apps" can be launched. Apps are effectively software elements that support the user in specific ways. They can be downloaded (usually from the Apple or Google Play stores) when the user needs help in managing their lifestyles or conditions.

Table 24: Properties and Uses of Modern Smartphones

Features	Sensors, actuators, connectivity and applications
Powerful processors	Sensitive microphone - voice and sound recognition
Large touch-screen colour display	High quality audio – for use with internal/external speaker
Large memory and storage capacity	3 axial accelerometer – to detect movement or falls
Telephone and text messaging	GPS location – for directing user to destination
Versatile apps and screens	Magnetometer – for magnetic field detection
In-app notifications and messaging	Cameras with optical zoom – for imaging and video calls
Familiar interface options	Vibration unit – for creating alerts and rhythms
Lightweight and always available	Bluetooth – for connecting and syncing with peripherals
Long and improving battery life	Wifi – for wireless Internet access using Broadband
Contactless and biometric ID	NFC and proximity sensor – for contactless payments
Hands-free operation	Gesture and pressure control through screen

There are literally hundreds of thousands of apps available but only a fraction of these are relevant to potential vulnerable service users. Indeed, they need to be separated into those developed for health and social care professionals, and those that are intended for use in telecare or self-care services. Their use isn't trivial because they are:

(a) specific to the operating systems of the smart phone to be used i.e. iOS, Android or others; and

(b) designed to be managed by the user (or their carers) so that they are only launched at the appropriate time and don't interfere with other apps that are also running.

It follows that, in the short term, many of the older people who currently benefit from telecare will not want to use smartphones and will therefore not use apps successfully. In the future, this situation is likely to change as the apps mature and as users become accustomed to using and manage their smart phones (or tablet devices and smart televisions). It is apparent that there is already a need for greater vigilance and for a level of curation and testing so that both users and prescribers have the confidence to use or recommend options.

Several organisations, such as Orcha, have offered professional accreditations for apps; these typically involve testing features such as usability, security, limitations for use and value for money that are considered in reviews of hardware devices (such as described in Chapter 6), and when endorsement or accreditation are possible. However, this approach may not always be useful for apps because the majority of those available are free of charge; this means that neither the developers nor the distributors are in a financial position to fund such work. Therefore, alternative and more innovative testing approaches are needed.

Figure 66: myhealth apps partners and research collaborators

Two organisations lead in this work in the UK. One is myhealth apps (http://myhealthapps.net/) which claims to bring together the world's favourite healthcare apps (using healthcare in its most general form) – *tried and tested by consumers – and supported by a group of organisations and partners shown in Figure 66, The second is the NHS Apps Library which has been in a beta release -* https://apps.beta.nhs.uk/ *during most of 2018.*

The myhealthapps.net approach, managed by PatientView, is based on recommendations for apps by appropriate healthcare communities internationally, and involves:

- empowered consumers
- patients
- carers
- patient groups
- charities and other not-for-profit organisations,

This is intended to bridge the gap between the public (represented by the groups above) and the app developers. This enables the latter to improve the content, relevance and usability of apps so that quality, impact and trust in apps can be improved. Patient/user groups can submit apps for inclusion directly via the website. PatientView confirms reviews from over 100 professional bodies and other information before uploading details to the site, thus ensuring a spectrum of user opinion and covering a range of languages.

Apps may be classified in several ways, depending on how they are used. From a self-care perspective, and recognising different perceptions of risk, myhealthapps.net separate apps into 3 classes broad classes, though many of the apps will feature in more than one class:

- Disability — these are apps that enable people to cope with daily living and provide support to people with any type of disability, including physical, mental and sensory impairment.
- Health, wellbeing and care in the community - these apps include most lifestyle apps and aim to support individuals to manage their health and care without the need for medical assistance; and which do not result in clinical decision-making by the user or require input from a health professional.
- Medical apps - these involve clinical decision-making, diagnosis or treatment.

Some of the health apps are further classified into 1 or more different sub-groups under which recommended apps are displayed in the form shown in Figure 67. Many of the most popular apps appear under many of these sub-groups. Each display shows the availability for different operating systems, the cost, and the relevant languages, along with one or more comments by various reviewers. 24 sub-categories are shown for Senses, Mobility and Learning.

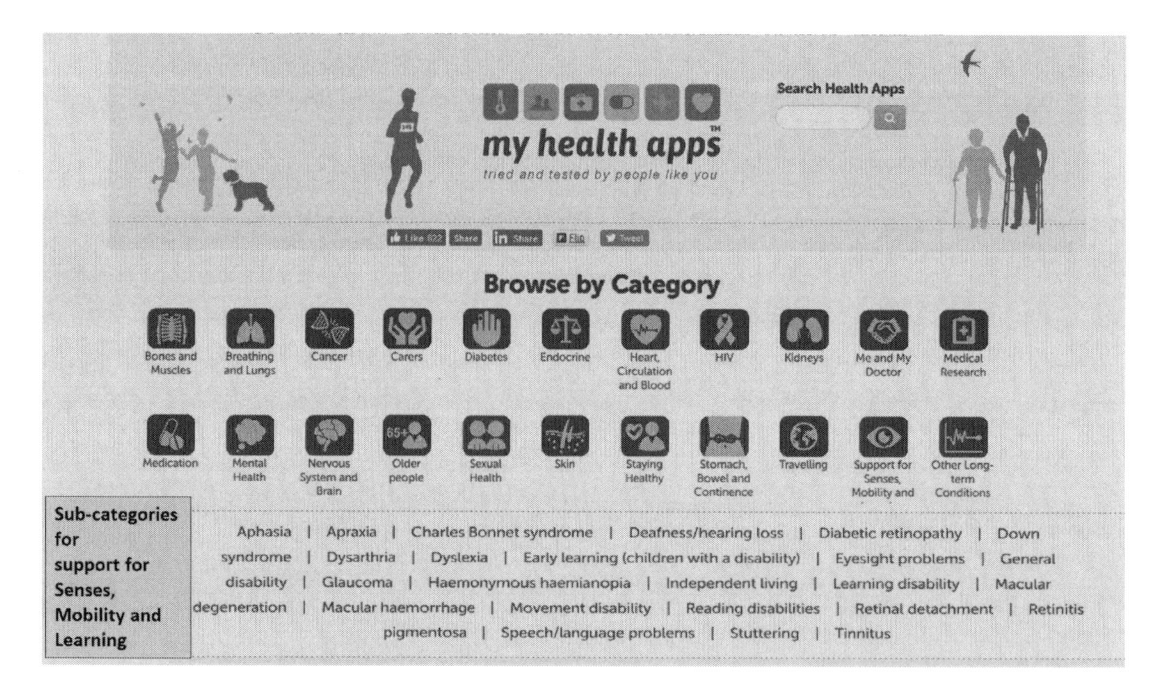

Figure 67: App Classifications for my health apps

Wearable Devices for Use in Social Care and Overcoming of Disabilities

Portable electronic devices, and especially those that are worn on or carried about the body, overcome many of the concerns associated with conventional telecare, but especially the restriction of operating only within the home environment. This might prevent them from going out as much as they might like, resulting in:

- a loss of exercise opportunities and future mobility concerns about not using muscles in the ways needed for their health,

- a reduction in the benefits to their well-being afforded by fresh air, open spaces, and the freedom associated with being outside; and
- a risk of social isolation and loneliness if they become house-bound (as will be discussed further in Chapter 11).

Wearable devices are being introduced within the 4th wave of technology support as part of the mobile agenda, especially as they can be controlled by smartphones when running dedicated apps that can collect and analyse the data produced . However, their use to support independence has been restricted in many areas due to a lack of clarity of their benefits, and some concerns about their reliability and application for specific user groups. Some of the major benefits and concerns involving the use of wearable device for social care and personal well-being benefits are described in Table 25. The significance of linking a device to a smartphone should not be underestimated.

Table 25: Some Benefits and Concerns in Introducing Wearable Social Care Technologies

Benefits	Concerns
Continuously available	Many have significant power needs
Programmable to satisfy individual needs	User interfaces may not be simple and intuitive
Multiple functions within the same device	Need discrete personalisation to avoid stigma
Can be worn in the bath or in the shower	Designs may not be robust and waterproof
Allow direct feedback to individual	Information may be accidentally shared
Can manage some risks associated with cognitive or intellectual disability	May be used as an electronic tag and for tracking and surveillance
Avoids the user becoming lost	Incomplete mobile phone coverage
Can synch with smartphones for remote monitoring and alerts	Many older and vulnerable people are unable to use and manage smartphones
May detect falls and convulsive seizures	Lots of false alarms
Unique identification	Exploitation by criminals
Reminders for tasks and medication	Users need training in their features
Attractive designs can increase utility	Devices may be stolen

Because medical devices are specifically excluded from this section, the cost of providing such devices may be relatively low, especially if they are produced at scale and incorporate standard sensing and communication techniques. Some applications are discussed below focusing on devices that may be worn on the wrist, the location that appears to be most attractive to most users because of the similarity to a watch.

<u>Tracking and Location using GPS</u>

There are dozens of devices on the UK market that are suitable for managing the risk of a vulnerable individual becoming lost. The majority are small enough to be attached to a key-ring or kept in a pocket or a hand-bag but may be left behind if the individual has short-term memory problems or simply chooses to wear different clothing. This has resulted in new designs that integrate the electronics and the GPS technology into small devices that can be worn with different sets of clothing. The uses of GPS locators will be considered further in Chapter 9 in the context of people who have cognitive impairments due to head injuries, learning disabilities or cognitive disorders.

Several studies have concluded that GPS watches are more likely to be accepted by men than any other form factor, while women are happy to wear devices on their wrists provided that there are both choices of colour and style and that the devices have smaller profiles. Many wrist-worn GPS devices shown cost under £100 and are designed mainly for children and for their tracking by concerned parents. They have a range of functions which are, in many cases, too complex for people with limited cognition and dexterity to operate successfully. In general, GPS devices of this type **DO NOT** communicate with the user's smartphone. Indeed, they are usually provided for people who do not have (or cannot use) a smartphone reliably.

Sleep Analysis and Improvement

Good quality sleep plays a key role in promoting and maintaining good health and well-being at all stages of life. It's needed to protect physical health, mental health, emotional health, as well as the quality of life and factors relating to safety. Sleep is about more than rest and freedom from disturbance; it also relates to the length of sleep and when periods of deep sleep and REM (Rapid Eye Movement) might occur. Sleep deficiency affects the level of tiredness, concentration and the ability to think, remember, react, learn, make sensible decisions, and interact with others. It can increase the risk of depression and self-harm or other dangerous behaviour. The examples shown in Figure 68 all have powerful sleep analysis capabilities. Improvements to battery life and the associated apps are being made on a regular basis.

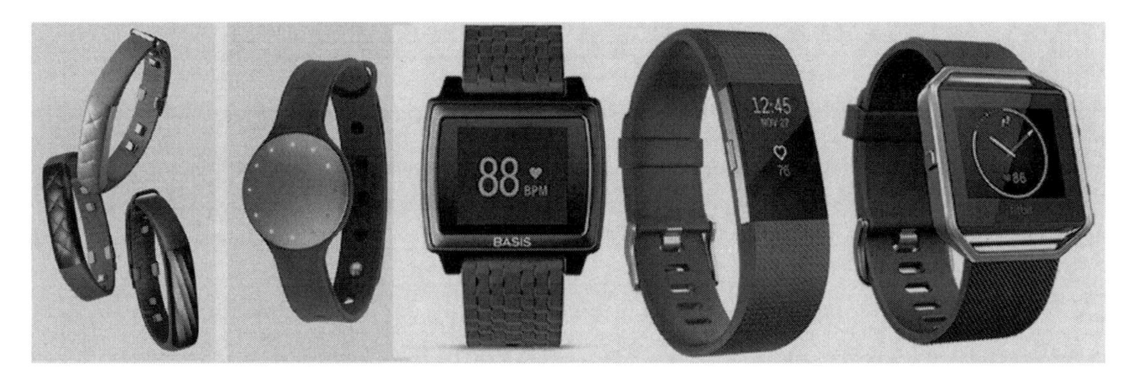

Figure 68: Jawbone Up3, Misfit Shine2, Basis Peak, Fitbit Charge2 & Fitbit Blaze Sleep Trackers

A lack of sleep restricts the body's ability to repair itself and has a long term impact on several conditions including heart disease, kidney disease, high blood pressure, stroke, obesity and Type 2 diabetes which all tend to be age-related. The lack of sleep affects how the body reacts to insulin, leading to higher than normal blood glucose levels, and the onset of diabetes. It follows that tracking sleep can provide an insight into changing patterns of activity and help to identify remedies such as a need for medication review. In extreme cases, it might indicate a physiological problem such as obstructive sleep apnoea (OSA) which is a fairly common medical disorder where the upper airway repeatedly collapses during sleep.

Worn sleep analysing devices can also be used more routinely to identify periods of rest during the day, and confirm when an individual goes to bed, and gets up during the night or in the morning. This is possible because they have sensitive accelerometers that are capable of detecting micromovements at the wrist and relating them to other physiological parameters such as heart rate and breathing rate, that vary according to the stage of sleep. The analysis opens up access to key information of relevance to assessment, especially for people who may have mild cognitive impairment and who may be unable to provide accurate information on activities and time spent in bed. This approach is fundamentally more satisfactory for people who live with a partner or with their family as it is individual and enables the actions of individuals to be uniquely identified.

Tracking Activity Levels

Devices that can monitor sleep and disturbance levels can also monitor more general activity levels. Most have been optimised to correlate the frequency and amplitude of the signals received to the type of activity, the speed of motion, and the number of calories consumed. The absolute number of steps measured with a wrist-based band may be significantly different to that measured with a conventional pedometer (worn on the waist) or with devices worn on the legs or attached to socks or shoes. This is due to errors in distinguishing between walking (or running) activities with those do to other tasks, such as brushing one's teeth, or knitting, and whether the device is worn on the dominant or non-dominant arm. Nevertheless, the trend information is ideal for daily comparisons and for meeting personal goals in exercise, especially when personal and discrete reminders or prompts are offered when no activity or movement has been detected for a period of time. Their role as part of a weight reduction programme, is well recognised though the evidence for weight reduction (without supportive changes in diet) is perhaps surprisingly weak.

The potential of these devices for monitoring activity to demonstrate both total activity and hour by hour action profiles has yet to be used in assessment nor to support programmes of reablement or rehabilitation following accident, surgery or illness. However, as these devices develop into smart watches, they could be used in real-time to detect anomalies of a medical nature (see Part 2 of this book. The major developers of these systems also integrate weighing scales (and body composition analysers) into their reporting mechanisms. The devices shown in Figure 68 are packed full of additional sensors which could enable analysis to support personal exercise goals in an unobtrusive manner. Figure 69 shows the author's data dashboards which include an automatic input of data from a linked weighing scale.

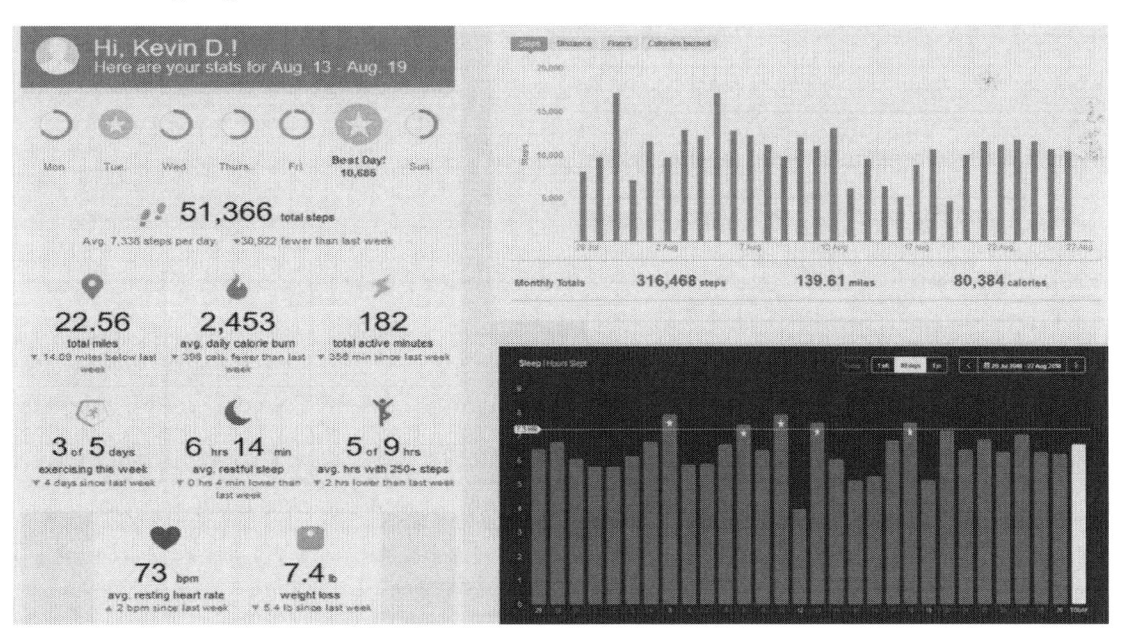

Figure 69: Typical Weekly and Monthly Dashboard Activity and Sleep Information

Detecting Epileptic Seizures

About 5% of all people will experience an epileptic seizure at some time in their lives. Not all will have epilepsy but over 30,000 people in the UK will receive a diagnosis every year. The two most common types of seizure are absence seizures, where the mind goes blank for a few seconds, putting them at risk of an accident if they are operating machinery, and convulsive seizures which are the most common. But there are many other types of seizure, and these can affect people irrespective of age both during the day and during sleep at night.

Fortunately, once diagnosed and treated with appropriate medication, nearly three-quarters of sufferers can manage or control the symptoms and their conditions. However, they, and their families and loved ones, may remain anxious about the risk of future incidents occurring, especially when alone and away from their homes. Because convulsive seizures usually involve shaking of the whole body, they can be monitored by wristband devices of the type described above and used for sleep and activity tracking. However, to provide reassurance, they need also to reliably detect all incidents, and then provide an alert so that help can be summoned. This detection needs to be achieved without a high level of false alarms (often caused by oversensitivity or a lack of local device intelligence) which could lead to more distress and a waste of emergency services resources.

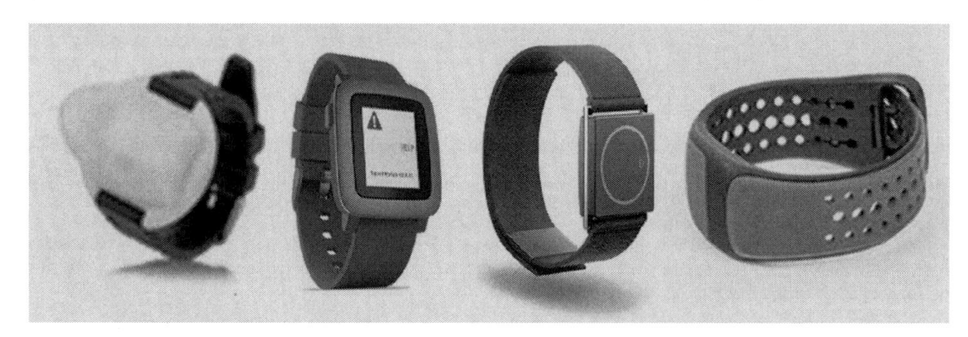

Figure 70: Four Examples of Smart Wrist-worn Devices Used to Detect Epileptic Seizures

A number of new wrist-worn designs have emerged recently that are showing great potential for the reliable detection of convulsive seizures. Figure 70 shows four of the most popular devices available currently in the UK, though it is anticipated that some top of the range generic smartwatches may soon claim similar features. The Epi-Care Free, on the left of Figure 70, was featured in the TV soap Emmerdale during 2016 and is being used in some NHS services currently following extended clinical trials. The Epilepsy Solutions SmartWatch, second left in Figure 70, is an epilepsy alarm built into a wrist-watch. It automatically raises an alarm if its built-in accelerometer detects repetitive shaking movements that are characteristic of an epileptic seizure. In each case, the device must be linked to a suitable smartphone (using a Bluetooth connection) for communication with a suitable responder.

The Embrace watch (third from left) monitors Electrodermal activity (EDA), sometimes known as Galvanic Skin Response (GSR), which is related to emotions such as fear, anxiety, and excitement. It can therefore detect many seizures (as well as tracking activities and sleep levels). It also works with an accompanying app which allows a smartphone to send alerts as appropriate. Finally, the PulseGuard technology (fourth from left in Figure 70) monitors an individual's heartbeat, looking for any sudden increase that might indicate a seizure is about to occur. This device can detect absence seizures in some people. These devices have the potential to transform the lives of many people with epilepsy as well as those of their carers, giving them the confidence to live more independently. For completeness, the Epi-care and the Brio, both from Epilepsy Alarm UK, are also available as wearable devices that link to the user's own smartphone. However, none of these named devices can be considered low-cost purchases; prices start at about £250 and some options are £1400. The lower cost devices also need to be supported by monthly monitoring charges, and for some people (including those who live alone), by a telecare support service that can offer maintenance and physical response in the event of an emergency.

Identification of People Who Have Special Needs or Responsibilities

It is recognised that the well-being of most people can be improved by spending time in the open air, appreciating the benefits of nature and of meeting people. This does, of course, increase the risk of them suffering an accident or exacerbation of illness while they are away from their home and, potentially, alone. The risks of an adverse outcome may be further increased if they are unable to communicate with those people, including the emergency services and hospital staff, who may be on hand to help them. To support appropriate responses (including medical treatment) it may be necessary for them to be quickly identified, along with clarity on their condition and needs. This was the origin of *In Case of Emergency* (ICE) and *S.O.S. Talisman* tags, which have since given way to a range of bracelets such as those shown in Figure 71.

Figure 71: A Range of Emergency Identification Bracelets

The simplest approach is to provide all relevant information on the tag, thus enabling any responder to recognise any issues and to act appropriately. This does, unfortunately, emphasise vulnerabilities and offers only limited opportunities for providing sufficient information. The responder must therefore use the listed contact telephone numbers in order to ensure that actions are timely and correct. So increasingly, responders are invited to use web resources in order to provide relevant information. This can be limiting, so a number of more innovative approaches are available that all have the benefit of being low cost and capable of reducing access to information so that confidentiality can be maintained.

A range of silicone wrist-bands is shown in Figure 72. The Welbeing OneCall is on the extreme left as the worn component in a life-saving identification service for individuals and their carers. Each member of the scheme receives a silicone wristband printed with a unique reference number and the phone number for Welbeing's 24-hour 365 days emergency telephone service. The number allows emergency and medical professionals to access vital medical information details when required, for example, if the individual is lost or disoriented and is found by a member of the public or if they are involved in an accident. Personal information is not given to the caller but to the person's pre-arranged contact when they are notified and informed of the person's whereabouts. If no responders are available, then the Emergency Services will be called to attend.

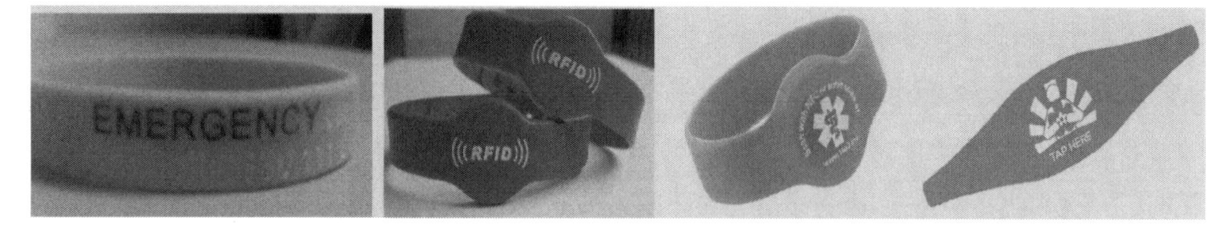

Figure 72: Examples of Wrist-band Devices for Identification

The electronic version of the printed wrist-band contains an NFC (Near Field Communication) device which combines a passive digital communication method (Radio Frequency Identification) with advanced human engineering enabling it to be easy to wear and used for identification in all sorts of environments from hospitals through to theme parks and point of sale. They can be read only by readers in close proximity to the device, ensuring that information isn't accidentally intercepted by fraudsters. The security offered is sufficient to enable such devices to be used in payment systems such as the Barclay BPay bands. The device that is third from the left in Figure 72 is the Tap2Tag which is being used by the emergency services in the South West of England to identify people who require ambulance or paramedic services, to identify relevant personal details of people wearing the tags using an app that they have on their smartphones. These simple tags are used extensively in Ireland by people who suffer from epilepsy. The final device in Figure 72 is the Guardian Angel or Dementia Buddy device being used in Wigan. It supports dementia-friendly town and city initiatives. These devices are issued free of charge to residents who have dementia, and the accompanying app enabling a smartphone to directly phone the person whose number has been included on the NFC chip. Other councils, in partnership with third sector organisations, are considering the use of similar devices that could enable trained people, and especially those employed by local authorities, the NHS and the police, to identify vulnerable individuals so that they can be taken to a place of safety if they are found to be unsafe, lost or distressed. Such initiatives are discussed further in Chapter 9.

Concluding Remarks

The rapid changes in telecommunications, sensors and local intelligence are continuing to make available new and improved ways of enabling vulnerable people (and their homes) to signal to external carers or to relevant authority that they, or their properties, are in distress and in need of intervention. The beneficiaries are not limited to older or disabled people; everyone's home is at risk of environmental issues or of burglary. Wirelessly linked devices might also be relevant to lone-workers, to children and to victims or previous domestic violence. All groups can benefit from the knowledge that help is close and can be summoned rapidly by a simple action.

Fortunately, the information that can be collected by a wearable or by a mobile linked device may not need to be processed for alarm of other emergency purposes. It can be used within a personal self-care and management plan, giving the user both advice and warnings about their behaviour or the status of their vital signs. Over extended periods of time, the systems can use nudge theory to slow change behaviours that can help to avoid some long-term conditions such as cancers and diabetes. This approach may be one of the few affordable ones to national health systems as costs increase due to an ageing population.

In the short term, people must learn to accept, and then embrace, the use of these technologies; they must not become the playthings of the young. Fortunately, they cannot be seen as a badge of dependency, and might therefore become a symbol of a 'cool' outlook. Their extended use must be a responsibility of all stakeholder groups including families. But they are linked – little point having a Fitbit unless the user has a smartphone to record the data and to display the dashboard. This is an opportunity for younger members of families to engage with their parents and grandparents to ensure that they buy the right ones for their needs, ensure that they are kept charged, and that they synchronise regularly with their apps on smartphones. The devices make ideal Christmas or birthday presents.

It remains an issue for debate whether hard-up local authorities or the NHS should fund these devices for some groups. Without close support for managing devices, it would be money wasted. Perhaps there is a big role for the 3rd sector and for digital volunteers generally. This is part of the digital transformation of care that is the main driver for Part 2 of this book.

Chapter 8

Technologies for managing frailty and its effects

Introduction

Frailty is related to the ageing process, that is, simply getting older. It describes how our bodies gradually lose their in-built reserves, leaving them vulnerable to sudden changes in health triggered by seemingly small incidents such as a minor accident in the home (such as a spilt pan), a fall, a simple infection, a change in the weather or a new medication regime. In medical terms, frailty is a syndrome that defines a group of older people who are at highest risk of losing their independence due to a hospital admission, a disability, an injurious fall, or an illness from which they are unlikely to make a full recovery. These are associated with poor outcomes including the need for long term care, and premature death.

The British Geriatric Society refers to five 'frailty syndromes':

1. Falls – whether due to collapse, slump, legs giving way or unexplained;
2. Immobility – usually a sudden change in capability, such as being able to transfer from the bed or toilet (which might well lead to more falls);
3. Delirium – occasional acute confusion (and not related to dementia);
4. Incontinence – a worsening or new onset of urinal or faecal incontinence
5. Susceptibility to side effects of medication – e.g. sickness, or hypotension.

Table 26: The Canadian Clinical Frailty Scale (2008)

Score	Description	Features
1	Very fit	Active, full of energy and well-motivated. They usually exercise regularly and are the fittest for their age
2	Well	No current disease symptoms but not as fit as those in category 1. Might exercise occasionally
3	Managing well	May have medical issues that are well managed. Do not exercise other than routine walking
4	Vulnerable	Independent, but lifestyle may be limited by their symptoms so that they have slowed up
5	Mildly frail	Need support with some IADLs, especially with going out, preparing meals and performing housework
6	Moderately frail	Need support and practical help with some ADLs as well as with bathing and climbing the stairs
7	Severely frail	Medically stable but dependent on carers for all ADLs including personal care
8	Very severely frail	Completely dependent for all aspects of life. Unable to recover from any form of infection or illness
9	Terminally ill	Approaching end of life within 6 months

Canadian researchers have considered a range of variables to be combined in a frailty scale of the type shown in Table 26. The components include the ability to perform many Activities of Daily Living (ADLs) and Instrumental Activities of Daily Living (IADLs) without help, but also elements of mental well-being, changes in status, and health. The hand-grip strength (and shoulder strength) are also included, along with usual and rapid walking speeds and their body mass index (BMI). These components have been the subject of considerable research interest many years, and therefore give confidence in the validity of their inclusion in an index.

It may be apparent that individuals who have been classified as being in groups 4 to 6 are the ones who are most likely to benefit from technology support. Those classified as 7 or higher will already be receiving high levels of personal care, possibly in a care home. The potential for making care home residents more independent using technology is discussed in Part 2 of this book. It would be useful if people who are 'pre-frail' could be identified, monitored and then supported using technology in a genuinely preventive mode. Telecare systems that are capable of providing these early warnings are under development and will be discussed more fully in Part 2. Tools, such as the Electronic Frailty Index (eFI), might also be to identify patients over the age of 65 who are living with moderate and severe frailty. This is measured through access to the electronic patient record (EPR), and again be considered further in Part 2.

Many older people who are frail are easy to identify – they may have weak muscles and energy levels, as well as having several other conditions and disabilities such as poor eyesight and hearing, arthritis and diabetes, and breathing or memory problems. This means that older people who are frail will probably be housebound and will typically:

- walk slowly and with an unbalanced or unsteady gait,
- become exhausted quickly and after a short period of activity,
- struggle to transfer to their feet from a chair, the bed, or the toilet or to climb stairs
- show poor muscle strength (including hand grip), and
- have lost 5 kilogrammes or more during the past 12 months unintentionally.

If 3 or more of the above 5 factors are present, then the individual is likely to be frail, and can be recognised through a social care assessment as well as by a full and comprehensive medical geriatric assessment. Frail people are at a heightened risk of accident and illness, especially during the winter months when it's likely to be colder. These risks can be addressed by a combination of interventions involving a multi-agency approach. The benefits can be measured both by a reduction in the number of accidents and hospitalisations that are experienced, and by improvements in their quality of life.

Technology can help manage the risks associated with frailty in specific ways:

1. Using continuous monitoring of activities (see Chapter 4) to detect rapid or slow changes in lifestyle that indicate a general slowing down that requires interventions
2. Monitoring specific activities such as eating and drinking, as well as use of the bathroom, and measurement of body weight, to identify nutrition and hydration issues, and poor control of the home environment;
3. Measuring and improving medication adherence so that the potential of pharmaceutical therapies can be optimised; and
4. Managing falls to minimise potential physical and psychological damage.

It is the final one of these risks that is perhaps most directly associated with frailty, and which can have the most immediate and devastating effect on independence. In practice, the likelihood of an individual falling is increased as they become physically weaker, poorly hydrated, less compliant with their medication therapy, or if they become less active. For this reason, this chapter focuses on the management of falls and on the role that technology can play to make people safer, especially in their own homes, by identifying the people who are most at risk, attending to the hazards, and managing the risks by either reducing the likelihood of a fall, and/or reducing the likely impact of a fall.

A Process for Falls Management

Falls are a common problem for older people; over 30% of individuals aged 65 or over are likely to fall at least once a year; half may become recurrent fallers. Although most falls do not result in serious harm, many falls do result in fractures as well as bruising and concussion; the psychological damage is also significant, especially if the fall leads to a long lie waiting for help to arrive. In such cases, confidence is reduced resulting in an unwillingness to go out and to perform everyday tasks and activities. This can initiate a fear of further falls and a spiral of decline that includes reduced mobility, balance, and social inclusion; these lead to premature dependency and admission to institutional care – which some people dread.

Although the focus of this book is on community living, many people attend hospitals at some point of their lives and are also at risk of suffering falls as inpatients or as outpatients. This is a particular issue for the NHS and has led to NICE issuing guidance in the form of the quality statements shown in Table 27. These were originally published in 2015, but statements 1 and 2 respectively were revised in 2017. Statements 7 to 9, inclusive provide the guidance that is central to most community approaches to falls management.

Table 27: NICE Guidance on falls in older people

Statement no.	Guidance
1	Older people are asked about falls when they have routine assessments and reviews with health and social care practitioners, and if they present at hospital.
2	Older people at risk of falling are offered a multifactorial falls risk assessment
3	Older people assessed as being at increased risk of falling have an individualised multifactorial intervention
4	Older people who fall during a hospital stay are checked for signs or symptoms of fracture and potential for spinal injury before they are moved
5	Older people who fall during a hospital stay and have signs or symptoms of fracture or potential for spinal injury are moved using safe manual handling methods
6	Older people who fall during a hospital stay have a medical examination
7	Older people who present for medical attention because of a fall have a multifactorial falls risk assessment
8	Older people living in the community who have a known history of recurrent falls are referred for strength and balance training
9	Older people who are admitted to hospital after having a fall are offered a home hazard assessment and safety interventions

It follows that actions and interventions that can reduce either the likelihood of an individual experiencing a fall, or the potential harm that results from a fall, can play a significant role in supporting the independence (and improved quality of life) of older people and their families. This would help in cutting the overall cost of falls to the health and social care system whilst making older and disabled people more confident in their ability to live independently (or on their own) when challenged by declining health and mobility. Interventions include, but are not limited to, strength and balance training. These may be offered in a falls clinic which should also be used to demonstrate and consider technological support measures.
But the challenge of reducing the effects of falls, and their frequency, is enormous. Falls can be caused by a combination of factors that are associated with:

- the home and outside environments (as in statement 9);
- the health and fitness of the individual;
- their gait and level of frailty; and
- their lifestyle and habits.

But assessments are needed BEFORE a fall occurs, rather than following an incident. Only then will it be possible to maximise the prevention opportunities.

Thus, whilst some fallers can be identified on the basis of the factors described above there can also be accidents that can be caused instantaneously by something unpredictable such as the blowing of a light-bulb, or a piece of paper that blew in on the breeze when the door was opened causing someone to slip. Nevertheless, the risk of falling can be managed in most cases by understanding the risk factors and by modifying as many of those that are possible. This needs to be done without impacting on an individual's freedom to such an extent that they deliberately avoid taking sensible actions to avoid accidents. The 6p process, shown in Figure 74, approaches the problem at different levels which make it applicable to all groups and, in some respects, all ages.

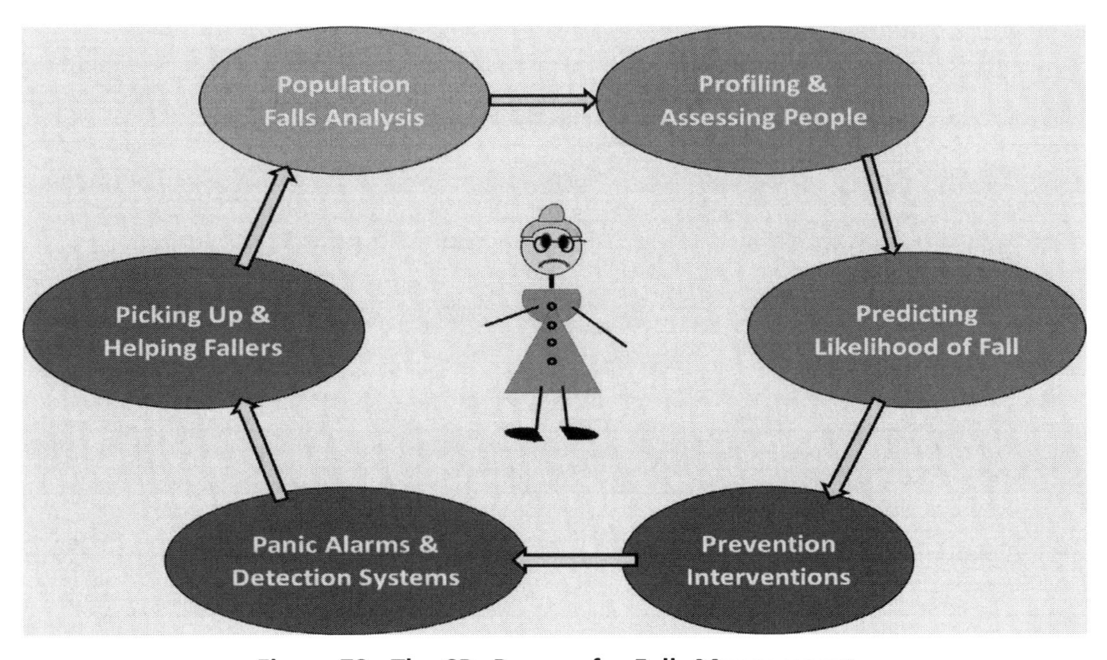

Figure 73: The 6Ps Process for Falls Management

1. <u>Population Falls Analysis</u>

There is a wealth of information available on people who present at Hospital Accident and Emergency Units due accidents both in the home and outside. This includes the ages of people who fall, where they fall, when they fall, what their injuries were, their falls history, the cause of the fall, and how long they had to stay in hospital. It is also possible to study what happens after hospitalisation and whether they received any form of rehabilitation including the option of attending a falls clinic. Although every individual is different in terms of their relative strengths and weaknesses, their home environment and the level of support that they receive, important lessons can be learned at a population level.

This can be used to identify groups of people, such as those aged over 85, who are most likely to fall, but can also become as the basis of Public Health information that can be made available to everyone to help them avoid dangerous situations and circumstances. People can then be made aware of the specific extrinsic and intrinsic risks to their safety. It follows that some small modifications to their lifestyle and their home environments can help to keep them safe. The data from GP records systems can also be used in this way. For example, data from Systm1 in Newcastle was analysed for several generations of patients. It was possible to identify those who are statistically most at risk of, for example, breaking their hip, through an analysis of medical histories including those of family members. Significant savings could then be made by planning interventions specifically for those people identified to be at enhanced risk.

Unfortunately, data on falls has been difficult to collect at hospitals because of the lack of a standard reporting methodology. Some of this may be due to different definitions being used for what constitutes a fall, and the significance of the different types of fall event described in Table 28. All these types involve "an event in which a person unintentionally comes to rest on the ground or another lower level such as a chair, a toilet or a bed" but the mechanism for getting there may be very different between a trip and a crumple, for example.

Table 28: Types of fall event suffered by older people

Name	Description
Slip	A sudden loss of footing on the floor resulting in an unintended slide of the foot; tend to occur quickly and result in one or both feet moving in one direction while the head and trunk move in the opposite direction
Trip	Characterised by a sudden loss of footing that is caused by a significant and abrupt change in elevation. It can also be attributed to an obstacle or obstruction near the surface of the floor
Stumble	A trip that is followed by one or more steps as the stumbling person attempts to recover balance
Tumble	When a body's centre of gravity is suddenly and unexpectedly shifted or displaced to some point outside of the body e.g. when transferring
Crumple	Total and complete collapse of the body, often the result of loss of sensation in the legs or dizziness caused by a cardiovascular incident

2. Profiling and Assessing People

Some people will be more likely to fall than others. This may be because of their physical frailty (and associated physiological factors including the way that they transfer and walk), their sensory capabilities, the domestic tasks that they undertake, and the hazards that they must overcome in their lives. These factors all contribute and can be used to provide a static assessment of falls risk based on the **THELMA** elements described below:

- **T**ests – there are a number of validated tools for assessing the risk of falling based on gait, walking speed, ease of transfer from chair or bed, and level of activity. Some are designed for use in assessment centres while others can be performed in the individual's home and over a period of several days so that the results are not a "snapshot" in time. Wearable devices such as the McRoberts DynaPort MoveMonitor (left in Figure 743) are designed specifically to fulfil this need and to track activity as do the plethora of wearable bracelets, such as Fitbit, Jawbone, Vivofit, and Fuelband (right in Figure 74) and which were described in Chapter 7. The Kinesis Quantitative Falls Risk Assessment uses computer analysis of leg movement and speed during a timed get up and go test to produce a risk score. GaitSmart works in a similar manner, linking to a processing system for analysis.

Figure 74: Wearable Activity Tracking Devices

- **H**ome **E**nvironment – occupational therapists and other health, social care and housing professionals have great experience in using surveys of the home to identify risks. They look for problems associated with clutter, steps and carpet edges as well as poor lighting, trailing wires, other trip hazards and the need for handrails and mobility aids. Reminder messages (such as post-it stickers) can also indicate recurrent problems with memory and a lack of awareness of the risks associated with climbing on chairs to reach inaccessible items. These issues will be addressed in Chapter 9.
- **L**ifestyle and disability – falls can occur at any time, though statistically they are most likely to happen during transfers (including in the bathroom) when physical limitations can cause issues. Many leisure and domestic tasks require the use of all the senses if they are to be performed safely. People who have visual impairment and/or other sensory impairments are more likely to struggle with unexpected issues, including flooding, responding to alerts or door-bells, especially if their disability is relatively recent or if their senses are in rapid decline. Their hydration levels and cooking habits are also good indicators of lifestyle, and changes can be detected non-invasively using activity monitoring using simple movement, door status, temperature and light levels such as those shown in Figure 75. This shows an appropriate level of use made of both the bathroom and of the kitchen. Experience enables assessors to determine whether an individual is drinking and eating enough for the weather conditions.

- **M**edication – many medications taken by older people are designed to lower blood pressure and can therefore lead to postural hypertension if they get up suddenly from a chair. Other medications are associated with dizziness and with an increased risk of falls. Others can make people drowsy. Poor medication adherence can lead to underdosing or overdosing both of which can lead to instability. Assistive devices and monitoring systems (see chapter 9) can simplify processes and improve compliance.
- **A**ctivities – people who are physically active protect themselves from many of the issues relating to falls, provided that they do not overexert themselves or take risks with stretching and climbing. Thus, the risk of falling is associated with low levels of activity but can also be associated with higher levels of activity, especially if tasks are being repeated and there is evidence of purposeless movement which might be indicative of a cognitive impairment. People who have a dementia are both more likely to fall and more likely to suffer a fracture.

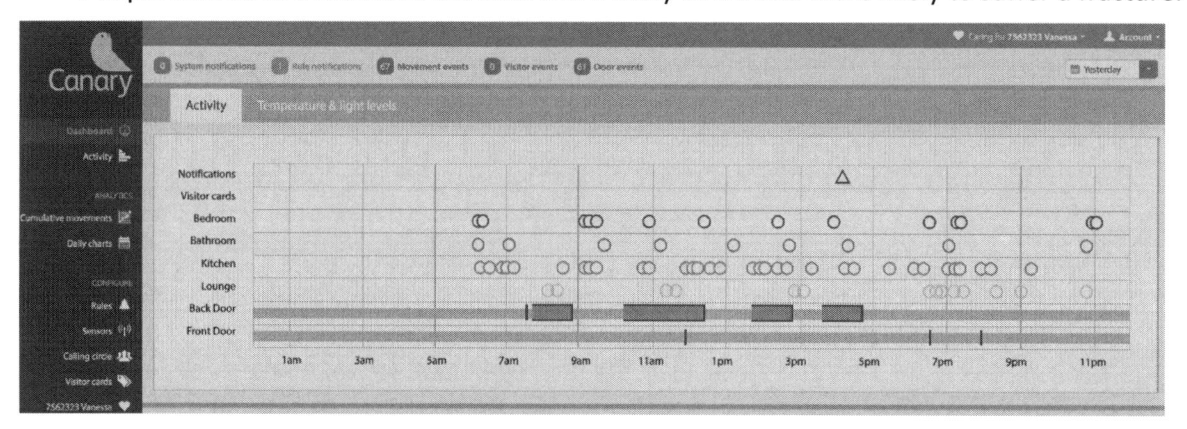

Figure 75: An Assessment of Activities to Reveal Use of Kitchen and Bathroom

The problem is that assessment requires individuals to be prepared to undergo testing, which many people who have not yet experienced an injurious fall may not be willing to do without some form of compulsion. Indeed, activity monitoring requires the home to be fitted with sensors that some will feel to be an intrusion into their lives.

3. <u>Predicting likelihood of a fall</u>

Falls prediction tools are algorithms that can provide an indication of the likelihood of a fall event occurring over a certain time-frame. Existing systems are quite crude in that they can operate only over extended timeframes and lack a dynamic element that would allow prediction to be changed following a sudden change in any of the many variables that need to be included in a sensitive and responsive system. However, systems such as STRATIFY, when applied in hospitals and other care institutions, can predict up to 93% of fallers by considering history, transfer and mobility capabilities, need for toileting and levels of agitation.

Currently, more advanced prediction systems use many of the parameters that are measured through testing to produce a result in which the weighting of these elements might initially be arbitrary, but which can then be adjusted over time as evidence is collected. In the future, real-time measurements or situations can be included, especially if the home and/or the individual is being monitored by a telecare or mCare system. This would both look at trends and identify behavioural patterns that might indicate a suddenly changed risk e.g. unaccompanied use of the stairs, failing to switch on lights or forgetting to use a walking aid or slippers.

There is no current knowledge available on the relative weighting of static and dynamic factors in fall prediction algorithms, but this might also depend on the lifestyle of the individual. A future telecare system should have a mechanism for alerting either the individual or his or her carers when the risk prediction threshold exceeds a defined limit. This could be the use of a smart home system that provides audio alerts and messages, with opportunities to use the Internet of Things to gather additional information such as external weather conditions, and the presence of hazards on the scheduled route or programme for the day.

Setting risk limits for such prediction systems will also require more and extensive testing and research. Initially, the predictions will be based on knowledge of those intrinsic risk factors that cannot quickly be modified, but which can provide a baseline. This may then be improved by considering the number of extrinsic risk factors, and how they can be mitigated over a matter of minutes (or perhaps seconds) by providing information to the individual that might cause them to be more careful or use to remember to use their walking aids or spectacles. Examples of both intrinsic and extrinsic risk factors are provided in Table 29.

Table 29: Risk factors than can be used to predict the likelihood of falling

Intrinsic Risk Factors	Extrinsic Risk Factors
Decreased visual acuity (cataracts, glaucoma etc)	Slippery surfaces (especially in bathroom)
Medical conditions (diabetes, postural hypotension etc)	Poor levels of artificial or natural lighting, especially on stairs
Problems with gait	Low levels of activity
Loss of muscular strength	Rushing
Poor balance	Stairs and steps
Effects of medication	Worn-out carpets
Dehydration	Inadequate footwear
Poor eyesight	Lack of handrails and grab rails
Declining short-term memory	Reluctance to use walking aids

Ultimately, the two most important predictors of falls are:

1. Age – chances of having a fall increase rapidly with age beyond 80; and
2. History of falls – people who have suffered a previous fall are likely to fall again.

Unfortunately, people are reluctant to admit to having had falls in the past. It follows that the quality of existing algorithms for prediction may be compromised through a lack of access to reliable data. However, there is usually sufficient information available to be able to predict falls as highly likely. moderately likely, or unlikely, which may be suitable for a traffic light system.

4. Prevention interventions to reduce levels of risks

Strategies for reducing the risk of falls have been considered from an evidence perspective by the National Institute for Clinical Excellence (NICE) and by Cochrane reviews. Strength and balance training are recommended along with other tests to identify specific problems. Most other physical interventions, such as brisk walking and other low intensity exercise training approaches, lack the evidence to be supported, but Yoga, T'ai Chi and dance activities that promote slow stretching and use of muscles have not been proved to be safe and beneficial. However, it is becoming increasingly clear that all forms of exercise are beneficial, and that performing activities in groups is most popular.

Similarly, common sense approaches to dealing with risks, including regular eye tests and the use of correction lenses, as well as wholesale deployment of assistive technologies are worthwhile interventions provided that they are acceptable and usable by individuals. Thus, whilst a range of aids to mobility can help with walking and stairs, some worn devices may be unpopular, stigmatising and restrictive. Assisted technologies and devices can reduce the risk of falls both by reducing the likelihood of a fall (by dealing with a hazard) and by reducing the harm caused by a fall. Most telecare services have focused on the latter, through rapid detection of falls. They are part of the connected service provision on which telecare services are based; but for maximum impact should be used alongside other assistive devices both in reducing the harmful effects of falls, and in preventing them from occurring as much as possible Figure 76 shows simple approaches to cushioning a fall, giving some protection to the body.

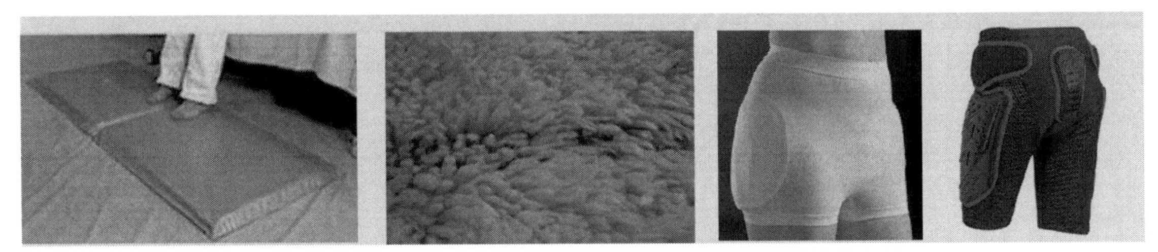

Figure 76: Methods of Cushioning the Impact and Reducing the Harm Caused by a Fall

They include the use of mats to reduce the impact, and hip protectors that have been designed to prevent an impact being focused on weak and vulnerable areas of the hip. Fractures of the hip, such as the neck of femur, are common and often have poor outcomes, especially if there is a delay between the accident and the subsequent repair in a hospital. Unfortunately, hip protectors are difficult for a frail person to put on in the morning without help and may be uncomfortable to wearable. Therefore, they aren't generally used successfully in the community other than in nursing homes. Fall detection technologies are discussed separately, and in some depth, in the next section because of their relative importance in giving people reassurance.

There are many more technology intervention options to reduce the likelihood of a person having a fall. Examples are shown in Table 30; their role is in overcoming the issues of poor mobility, declining vision, short term memory loss, and lack of nutrition and/or hydration. Poor eyesight is a major risk factor, especially at night; this can be managed in some cases by using higher intensity lighting i.e. 100-Watt bulbs rather than 60 Watts, that many households have used in order to save on electricity. Low energy bulbs were issued free of charge to many older householders over the past ten years, but their operation has not been popular because of the few seconds that they take to warm up. Modern LED lights overcome this problem, but few people understand how to compare light levels, and are guilty of selecting lower outputs than are appropriate. Perhaps a more significant problem is that older people don't always switch on when it's dark, either to save on electricity, or to avoid waking up others in the household. There is also a more practical problem in many UK homes in that it is often impossible to switch on a hall light from the landing due to the wiring configuration. It means that coming downstairs during the night is a particularly risky situation and can lead to nasty and injurious falls down the stairs.

Table 30: Examples of technologies that can reduce the likelihood of a fall

Intervention type	Examples
Mobility aid	Walking sticks, frames and trolleys, grab rails and bannisters
Stair aids	Stair and through floor lifts, Stairsteady, Quadstep
Bathroom aids	Raised toilet seat, toilet frame, perching stool
Visibility improvements	Contrasting edges to steps, better and automatic lighting
Medication support	Alarms, apps, blister packs, dispensing aids
Rush prevention	Door intercom, automatic door opener, cordless phones
Flooring	Plain fitted carpets, removal of rugs (with curled up edges)
Reminders	To put on slippers, specs; to use aids; to eat and drink
Furniture	Motorised chair (such as Lazy Boy), profiling bed
Exercise aids	Vibration platform, exercise bike, pedometer
Balance/posture wear	Weighted garments; slump alerts

Movement detectors have been used for many years to activate electrical equipment, alarms and lights; but they are usually components in complex linked systems that need to be planned and installed by electricians of engineering professionals. They are also relatively expensive. Recently, smart bulbs of the type shown in Figure 76 have become available in different fittings (including the bayonets that are popular in the UK) and brightness levels. They simply replace existing bulbs to operate automatically but only in the dark and on detecting motion. The light switch can be sellotaped in the 'on' position so that it can't easily be switched off. Prices that started at about £20 in 2016 have dropped to about £10 recently, which is not significantly more expensive than 'dumb' LED lightbulbs.

Figure 76: A range of smart light bulbs with integrated movement detectors

5. <u>Panic alarms and detection systems</u>

The benefit of having a falls detection system is that it provides people with the confidence to perform routine domestic tasks that they might otherwise have neglected for fear of falling and being left on the floor for many hours or days. The good old social alarm (known in other countries as Personal Emergency Response Systems or TeleAssistance) has played an important role in connecting vulnerable people to a 24 hour centre or to their relatives for over a quarter of a century' at the press of a button on a worn pendant switch or the pull of a cord hanging from the ceiling for a quarter of a century, they have used the simplest of user interfaces, and the ubiquitous landline to give people and their relatives confidence in living independently. Their evolution and applications have already been discussed in chapters 2 and 3 of this book.

Their value and effectiveness have been reduced by:

- The failure of people to wear them at all times – many older people are in denial about the risks, and often choose to leave their alarm pendants somewhere inaccessible claiming that wearing such a device stigmatises them. Pendants that are integrated into jewellery provide a more attractive and appropriate interface (left of Figure 77) while "glow in the dark" and different colour options make them more acceptable to younger people, including those with a learning disability
- An unwillingness for people to raise the alarm manually at times of the day (or more often the night) when they feel that they may be a burden to call-handling services, the ambulance service or to their own family.
- An inability to raise the alarm manually due to be the severity of their injuries or the confusion resulting from the accident.

Figure 77: The CAIR Onyx Pendant Alarm and Versions of the Tynetec Touch

The first intelligent automatic worn fall detector was invented over 15 years ago; it was designed to be worn around the waist or in a pocket (left of Figure 78). They have since been improved through more advanced sensing and processing technologies with different devices available for wearing on the wrist (easy to wear but with significant detection issues) and around the neck (visible but well-defined). However, they are unlikely to become universally acceptable and reliable because of the range of events that people describe as fall events (see Table 28). The technology within a mobile phone is likely to offer an effective alternative to dedicated fall detectors through a dedicated app, though there is little agreement on how the quality of the algorithms employed can be validated and compared. The components of a fall need to be considered if the wearability, utility, and success of worn devices is to be improved. These design issues will be discussed later in this chapter

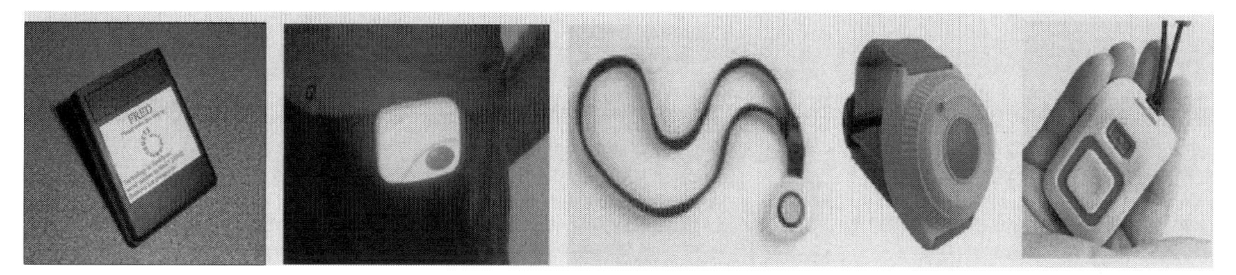

Figure 78: Early Versions of Worn Fall Detectors

Fortunately, fall events may be detected by using non-worn technologies that are either embedded in the property, or within items of furniture. The former approaches are increasingly being deployed in more institutional settings where central processing and the use of infra-red, video or acoustic monitoring and Wi-Fi can be used to identify and raise alerts in many rooms in multiple apartments, thus reducing overall costs. These techniques are considered in Chapter 18 in Part 2 of this book. The latter are often deployed in the bathroom, because a service user is likely to remove their worn devices in there when they are undressing, especially if they are going for a bath or a shower. Popular solutions involve easy-press versions of pendants several of which are attached to appropriate surfaces on the walls or at the side of the bath or shower. Figure 71 shows both the easy-press trigger and an alternative approach based on a panic alarm strip that can be installed along the entire length of one or more walls of the bathroom or shower cubicle. It can be activated by pressing any part of the strip using finger, toes, elbows or nose.

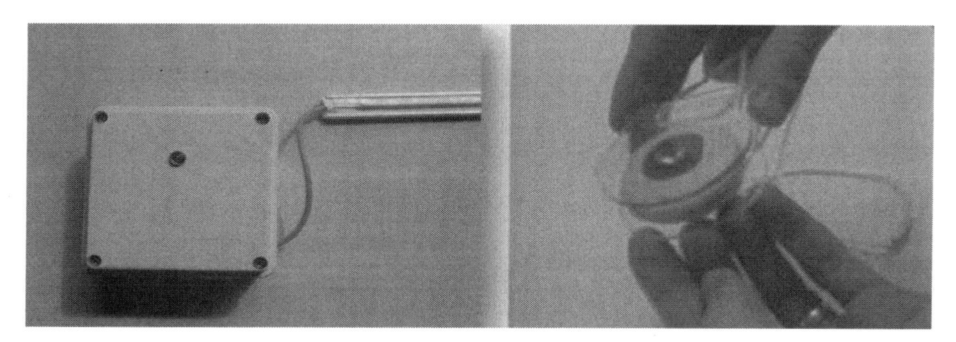

Figure 79: A panic alarm strip and an Easy Press alarm trigger for use in a bathroom

A lack of movement in a particular room either for an extended period of time or during certain times of the day might also be used for fall detection using exception principles. Such an approach may be possible using activity monitoring systems (such as Canary, TextCare, Just Checking. ADLIfe, iCare and

Lively) but inevitably result in a delay of potentially several hours in raising an alert. For example, no movement in the bathroom by, say, 10am could be used to indicate a problem, but the accident or illness responsible may have taken place or started many hours before this. The systems can respond more quickly if the outputs of 2 or more sensors are combined into *virtual sensor* combinations. More sophisticated systems using Artificial Intelligence will be considered in Chapter 14 in Part 2.

Falls can occur at any time of day or night, but those that occur between about 10pm (i.e. at night) and 7am (in the morning) are perhaps the most frightening as far as older people who live alone are concerned. This is because they are more likely to be visiting the bathroom, where many falls occur, and because they are at risk of being left on the floor for many hours before they can be found. This is, of course, an important reason for carrying an alarm pendant or having access to a device such as a pull cord that can be pulled to raise the alarm. It has already been explained why many older people are reluctant to use these technologies. It follows that one of the most successful methods of automatically raising an alarm during the night is by monitoring the bed occupancy status using a special sensor. These devices are so important to fall management that they will be discussed separately at the end of this chapter.

6. Picking fallers up quickly and safely

The benefits of detecting falls rapidly and reliable are maximised if the individual knows that their condition has been detected and if help is on its way. In the vast majority of cases, there is no physical injury – and therefore no need for the person to be taken to hospital. However, it is common for the faller to be unable to get up from the floor without help, and for them to be shaken and in need of physical support and comfort before they can continue to live independently. Integrated services therefore need a "recovery" or "response" element as described in Chapter 3. This should include at least 3 different elements:

- A means of accessing the property quickly and without having to call the police to break down the door;
- A method of checking for injuries to determine whether paramedics need to be called; and
- Technologies to enable staff to help people to be lifted safety off the ground an helped back into bed or into a chair without causing them unnecessary discomfort and without compromising the health and safety of response staff.

Integrated telecare services that include the recovery element, are available in some geographic areas. They are usually run by local authorities or by housing associations, but often in association with third sector providers and other community services; they can save their NHS partners many thousands of pounds per annum in reduced payments to ambulance trusts and hospitals. Some of the larger managed service providers are able to harness the resources of local organisations to offer fully integrated services. It follows that by avoiding unnecessary ambulance call outs and A&E presentations, services can become more efficient with fewer delays and shorter treatment times.

Mechanical key-safes, such as those shown in Chapter 2 are frequently used despite their limitations. In the digital world, these are likely to give way to more intelligent devices that use Bluetooth or Near Field Communication to 'talk' to smartphones carried by the responders. This will enable an audit trail to be established for greater security and for verifying time of access and departure. Similarly, automatic door opening systems, or key storage facilities similar to those used for collection of parcels from 24-hour supermarkets, may be used to offer choice to consumers.

When inside the property, telecare services that only detect emergency situations, and without a confirmation of the problem, can overload responder services with potential false alarms. There may be an increasing need for responders to deploy remote video teleconferencing (secure versions of Skype, Google Hangouts, Face Time, Fring, ooVoo etc) to connect them to paramedics or to staff at A&E for support in making decisions. It follows that a Rapid Response and Recovery (3Rs) service should be available across all regions with financial support through individual charges as well as block funding from other organisations that will benefit by keeping people in their own homes without the hospital presentation.

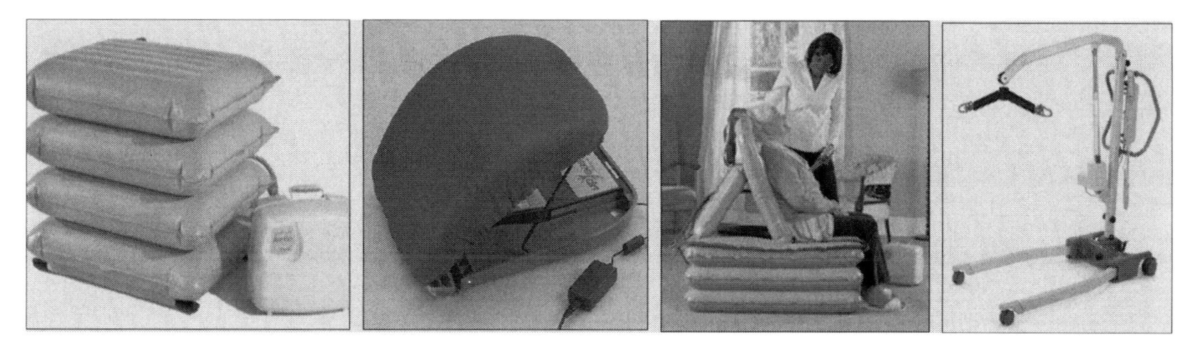

Figure 80: Examples of Portable Lifting Equipment

Finally, when it is confirmed that a service user is not in need of medical attention, and at their request, the 3R service should carry and use portable lifting equipment such as inflatable and powered cushions and powered raisers and hoists such as those shown in Figure 80. These are designed to be used by one person, following appropriate training.

Worn Fall Detectors

To understand how fall detectors are designed, it is first necessary to consider the phases of a fall. Although they can be caused by slips, trips or mechanisms described in Table 28, they are not instantaneous events; they involve the 4 shown in Figure 81, described in Table 31. This model includes a final (fourth) stage of recovery which can be achieved eventually with or without help. The timings, T_0 through to T_3, may be arbitrary. However, it may be fair to suggest that the quicker the individual goes from upright to fallen (i.e. T_0 to T_1) then the more likely it is that they have suffered a severe fall. This phase may last less than half a second. The 'post-fall' phase (T_1 to T_2) can be any length of time, but the longer it takes before recovery, then the greater the chance that help may be needed due to injury or immobility.

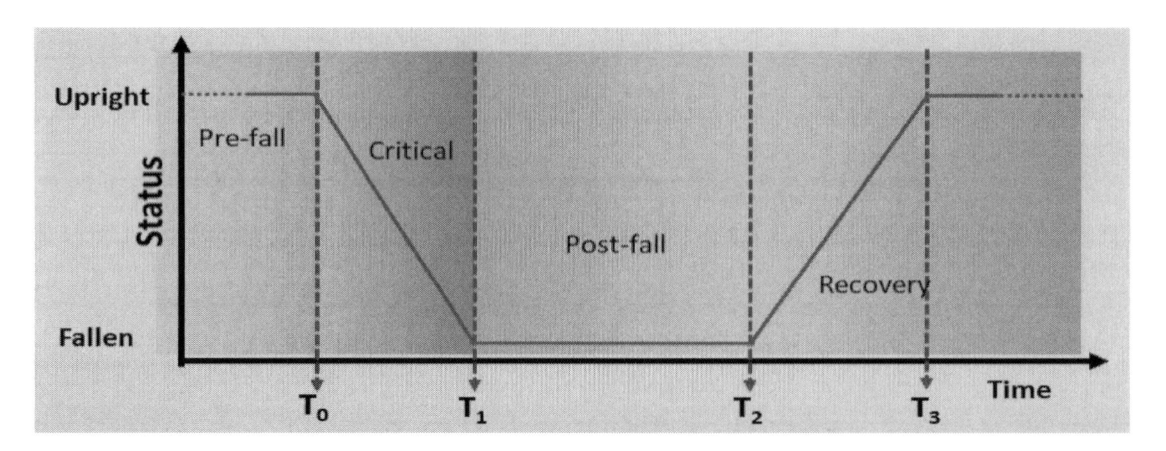

Figure 81: The Four Phases of Falls Model

It may be apparent from Figure 73, that a fall is characterised during the critical phase when status changes from upright to being horizontal on the floor. This may be distinguished from simply taking a lie down on a bed by both the speed of descent (i.e. uncontrolled and under gravity) and by the resulting impact. It follows that a worn fall detector can detect either:

- a sudden change in height
- the change in acceleration
- the impact on reaching the ground;
- the horizontal state (orientation) of the body following the impact; or
- the lack of movement following the impact.

Table 31: Stages in the 4 Phase Model of a Fall Event

Phases	Description
Pre-fall	The individual is going about their everyday activities including potentially dangerous movements such as transfers from a chair or a bed
Critical	This is the relatively sudden movement of the body towards the ground
Post-fall	The individual will be at a lower level, perhaps on the ground or on the stairs, and may be inactive for a short period as they realise what has happened
Recovery	There is an attempt to rectify the situation by getting up

Of course, if only one of these parameters is chosen to trigger an alarm, it will yield significant levels of false activations. The first smart worn fall detectors on the market – the FRED device shown on the left of Figure 78 – used impact and orientation as the trigger, both measured with individual devices that were low power and reliable in operation. The fact that it was still being used commercially some 15 years after its introduction demonstrates that the original design principles were valid. Yet, it failed to overcome wearability issues for both men and women. For men, it would trigger false alarms when they dropped their trousers to go to the toilet or when getting undressed at night, whilst the holster designed to hold the device in position at the waist of women was one of the least attractive elements of telecare ever seen!

The next generation of fall detectors used a triaxial accelerometer to detect changes in acceleration during the critical phase, the impact, the orientation and the lack of subsequent movement. However, the power requirements were excessive, and finding the optimum thresholds for an individual was virtually guesswork, because factors such as impact were also dependent on the surface on which the fall occurred.

Figure 82: Wrist-worn fall detector designs

Table 32: Comparison of body locations for worn fall detectors

Location	Advantages	Disadvantages
Waist	Close to the centre of gravity for detecting genuine falls; may be integrated into a belt	Difficult for ladies to wear if they don't have a belt; may be visible and stigmatising
Wrist	A natural location for a watch and other items of jewellery such as a bracelet	Poor location for detecting movement during and after fall; size and design may be unpopular
Around neck	Used in social alarms for decades	Stigmatising and easy to remove and forget
Ankle	Good location for hiding device and for monitoring mobility	Resembles the electronic tags used to monitor the movement of prisoners
Thigh	Can be used to detect attempts to get out of a chair or a bed; easy to conceal	May be uncomfortable to wear; can slide down to the knee
Foot	Sensors can be embedded in socks or hidden in shoes	People change their socks often and may own several different pairs of shoes and slippers
Ear	May be embedded into hearing aid or the frame of spectacles; good for fall detection	Many people need to switch between reading glasses and distance glasses
Head	May be integrated into a cap or a hat	Unlikely to be worn indoors by men

Wrist-worn detectors were then introduced. They were more acceptable for wearing (especially by men) but the sensors could not determine orientation unambiguously because of the many degrees of freedom available at the wrist. The latest devices recognise the importance of using a standard position, whether on the torso or on the wrist, but also include barometric sensors to detect changes in elevation that might be expected during a fall. From a technical and sensitivity perspective, it remains the case that the optimum position for a fall detector is at the waist, but for commercial success, usability issues must be resolved whilst retaining some of the advantages of other locations shown in Table 32. Examples of wrist-worn devices are shown in Figure 82.

There are other arrangements that can be offered as alternatives for dealing with both gender and wearability issues depending on individual preferences. Figure 83 shows an integrated belt-worn device alongside a detector attached to spectacles frames. In the former case, the device switches off when the belt is undone, thereby avoiding any false alarms. In the latter case, the device is at a position of high sensitivity for impact, acceleration, height and impact detection. If the spectacles become dislodged and fall off during the fall, the device would be able to measure the same factors needed for detection.

Figure 83: Two alternative worn fall detector approaches

A smartphone could, in principle, be made into a useful fall detector by utilising an app and its in-built accelerometer. However, some people carry their smartphones in a back-pocket and others in their hands, while many women carry theirs in their handbag. Thus, neither its location nor its orientation may be known with any certainty which immediately creates issues relevant to both setting alarm thresholds and providing opportunities to quickly cancel the alarm in the event of a false alert. In the same way, frail older people are amongst the least likely people to carry a mobile phone or a smartphone, thus the appeal in the market place may be limited.

Nevertheless, several fall detection apps have been produced and tested, while many commercial mobile fall detector devices have been produced using the form factor of a mobile phone. Some examples of devices currently available are shown in Figure 84, and some of these are already displacing fall detectors within telecare services when individuals are given a choice. It is likely that future mobile systems will combine sensors located within a smart watch, or other wearable devices, with the communication and processing capabilities of a smartphone (and an app) to provide more complete solutions within a mobile support ecosystem. Indeed, some of these devices already make use of the smartphone for relaying alert information to a responder. It should be noted that the accompanying literature for these devices states explicitly that they will not detect 100% of falls. Indeed, many mobile devices that can detect falls are not advertised primarily as fall detection systems but as multi-function support devices that offer a fast-dial S.O.S. button and GPS tracking capabilities. The useful time between battery recharging is an important usability parameter that is often overlooked, as is the ease (and speed) of charging. Such issues will be discussed more fully in Chapter 9 when worn GPS location devices are considered.

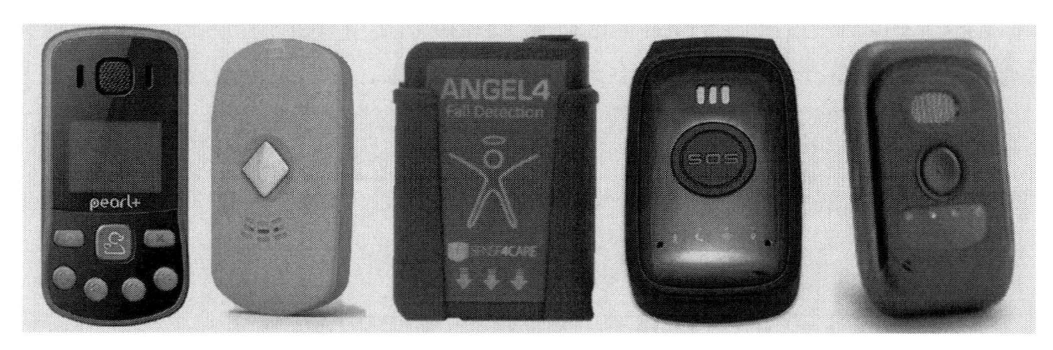

Figure 84: Examples of Mobile Fall Detection Devices

Bed Occupancy Sensors

All worn fall detectors suffer from potential reliability issues due to the variability in the wearing location as well as having to operate in various home (and external) environments where floor coverings are different and where individuals may walk at different speeds with or without the support of assistive devices. Fortunately, the availability of fall detection techniques based on bed occupancy and simple logic/times have provided an excellent alternative to worn fall detectors for use at night. Figure 85 shows how these technologies can be separated into the following groups:

A. *Under the bed/mattress or patient* pressure devices – these use the change in overall weight introduced by the person lying on or in the bed to create a pressure change on a sensing element somewhere underneath the person or the bed;
B. *Off the bed* devices – they use a variety of techniques including (beam breaker) infra-red systems, passive infra-red (PIR) detection, primitive pressure mats and sophisticated systems based on digital imaging;
C. *In the bed* non-pressure devices – these include devices attached to the user or elements that sense their presence within the bed.

All these groups of devices are optimised to detect the change of status (i.e. occupied or unoccupied) when someone gets into or leaves the bed. This enables timing functions to be initiated, and alarms or alerts to be generated if a rule has been broken e.g. if the user has left the bed during the night and not returned within a reasonable period of time. In people's own homes, the benefits, or otherwise, of these approaches, and of different types of element, are dominated by reliability. This includes not only of the designs but also of their resilience in overcoming practical issues.

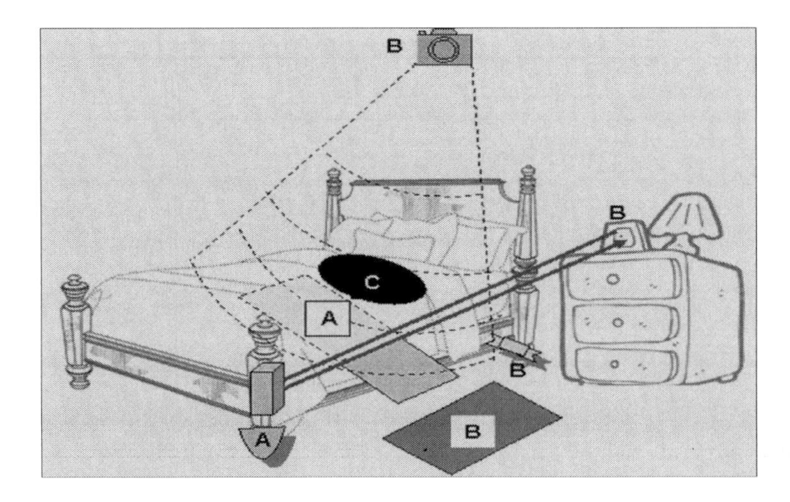

Figure 85: Examples of the Three Types of Bed Exit/Occupancy Technology in Use

These issues are, for example, people moving around the bed during the night, and exiting from different sides (or at the base) of the bed at various times. There is also the possibility of devices being displaced or interfered with by carers when they change the bedding, or by the user, especially if they have a cognitive impairment. Devices that fit under the mattress of the bed tend to be the most successful and cost-effective though a full assessment should be performed in order to match the lifestyle, habits and bed of the user to the selected assistive technology. This may be particularly important for people with a cognitive impairment (see chapter 9). The useful lifetime of devices, set-up and programming arrangements, and the initial cost might also be relevant.

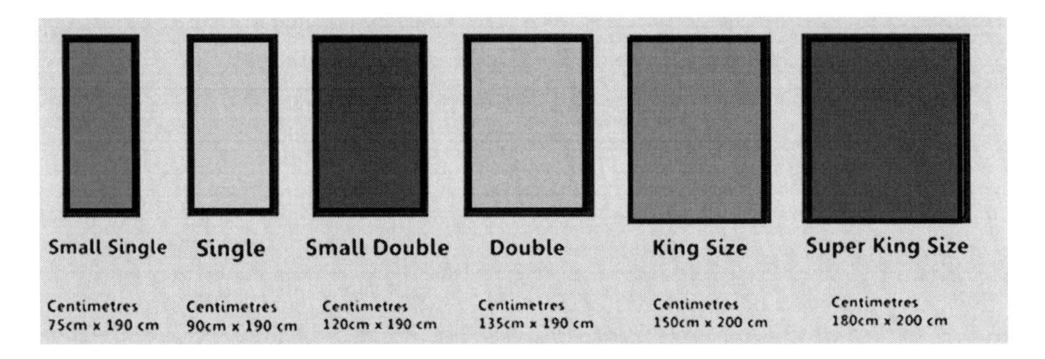

Figure 86: Bed and Mattress Sizes Used in the UK

A number of different options for Under the bed devices are shown in Table 33. Their relative simplicity is apparent though they have a number of disadvantages as described in the table. Nevertheless, they are the basis of sensor options provided by the majority of the major Telecare supply organisations; the sensors are connected to electronic amplifiers, processing and interface modules that also have transmitter to link them wirelessly with a dispersed alarm unit. These arrangements can usually control lights and offer delays in activating an alarm when appropriate. Most have limitations which prevent them being used effectively for all beds and for people of extreme weights. For some, the sensor is accurate in a repeatable manner only for patients weighing from 40Kg to 150Kg due to their elastic properties. However, many of the lower cost devices work appropriately in most cases, but are not suitable for all situations, and all types of bed. In particular, although the majority of older people live alone, many choose to continue to sleep in a double or larger bed for comfort or personal reasons.

Table 33: Examples and Features of Sensors that Fit Under the mattress/bed or patient

Images	*Descriptions*	*Group Advantages*	*And Disadvantages*
	The narrow **sensor pad** by **Tunstall** and is the current market leader	The method employed by these devices is relatively simple as is replacement	Devices generally have both min and max weight that can be monitored
	This **pad** is available in sizes to fit a single bed or chair. Mainly sold through **Chubb**	Installation can be performed without the need for special tools	Most devices need to be calibrated with the weight of the patient
	Thin **sensor pad** – available in the UK and Europe through Legrand	They can link to existing interface modules for programmable alarms	Devices need to be removed and replaced to change the bedding
	Tape-switch sensor from **Alert-It** to monitor large bed	Interfaces can be used to switch electrical devices e.g. lights & speech units	May be limited by the type of bed and may need extra boards
	Frequency Precision's **air mattress** to cover a large bed area	Can be used to monitor 1 side of a double bed	Most lower cost bed pads need to be replaced after 6 to 12 months
	SensorCare element sandwiched between 2 solid metal sheets	Two devices can be employed for single occupancy of double beds	Some pad solutions fail when used with alternating pressure mattresses
	Alert-It's bed leg, foot, wheel or castor **weight sensor**	Battery warnings can go through to the Telecare monitoring centre	Some pad solutions lead to false alarms with profiling beds

Figure 86 shows that there are many different sizes to consider, from Small Single through to Super King size. Although the sensing elements may fit in any of these beds, their ability to integrate weight across the entire mattress may be compromised if the sensing element is smaller than the size of the mattress. This can lead to false information on status being presented lead to false alarm situations and unnecessary responder call-outs.

Similarly, there are different styles of bed to consider. Figure 87 shows three popular designs. The bed on the left using an arrangement of slats to support the mattress, making it unsuitable for using a narrow sensing element without having to attach it to a wider piece of wood (or MDF) which will lead to a reduction in comfort levels. The central picture is of an electronic profiling bed that can aid comfort and help people to sit up or get into or out of bed more easily. They are popular with frail people and with their carers. However, their design means that many flat sensing elements will fail depending on their position in the bed with respect to the profiling function, and where the weight of the body lies.

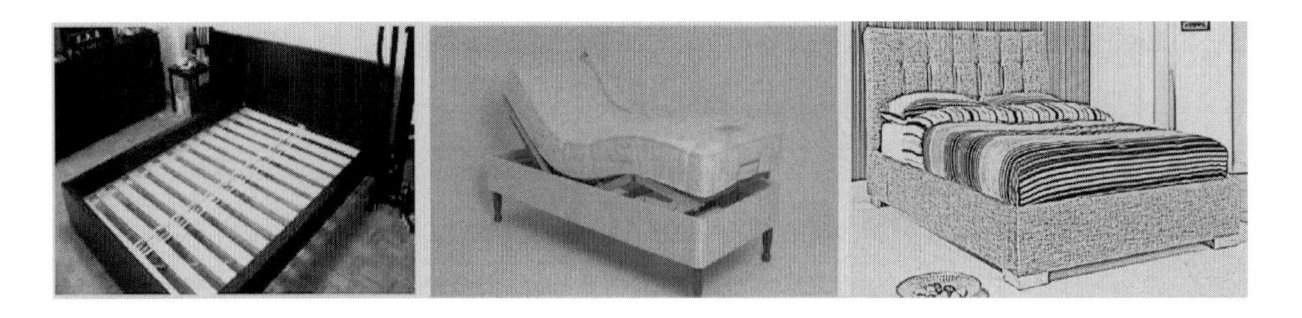

Figure 87: Three Styles of Bed that Can Cause Problems for Some Sensing Solutions

The picture on the right of Figure 87 is of a modern bed design featuring very wide legs/feet. It is an example of a low-profile design that can reduce the impact if someone falls out of bed (though bed raisers are more likely to be used by people with mobility problems following a hip fracture for example). These designs make bed occupancy sensors that sit under a leg difficult to implement.

Table 34: Examples and Features of Sensors that Are Off the Bed

Images	*Descriptions*	*Group Advantages*	*And Disadvantages*
	The **Doro Nemo** is a PIR based sensor used on the floor or attached to a bed leg.	Detection method uses movement across a visual field, enabling tailored to bedroom	Devices need to be mounted at the most appropriate location
	The **Alert-it P162B** is a free-standing, motion detector used for monitoring when an individual leaves bed.	The technology is generally mature and reliable, requiring no calibration	They can be moved accidentally by carers when cleaning or changing the bed and not relocated
	This "**curtain PIR**" looks for movement in a narrow field running parallel to the bed.	Unlikely that patient or service user can interfere with device	False alarms can be generated by falling quilts or bed clothes
	Uses active reflector-based **Break Beam system** to detect individual moving next to bed.	Used to monitor any size and type of bed and for double occupancy without weight restriction	Most devices need additional wiring or transmitters for linking into pagers etc.
	WeSpot uses digital image capturing and processing to detect bed exit attempts	Insensitive to profiling or to the use of alternating pressure mattresses	Battery management is required - there is no direct connection to a monitoring centre

Alternative approaches are needed especially those that are 'off the bed'. These are more independent of the type of bed, the weight of the user, and the use of electronic technologies. These technologies may be used to profile the bed, to create an alternating pressure to reduce the likelihood of pressure sore formation, and to allow the use of lowering mechanisms that enable carers to more easily manage personal care needs without the help of a second carer.

It should be mentioned that the latest developments in bed occupancy monitoring combine a PIR movement detector with an under-mattress sensor. The Frequency Precision device shown in Figure 88 is designed for positioning the sensor across the edge of the bed so that it can trigger when the legs swing out of the bed, and before they touch the floor. The second alert is generated when the individual's weight is lifted from the bed. This approach enables the device to be used with any type of bed, and when alternating pressure equipment is being deployed. It is flexible and can therefore operate successfully with profiling beds.

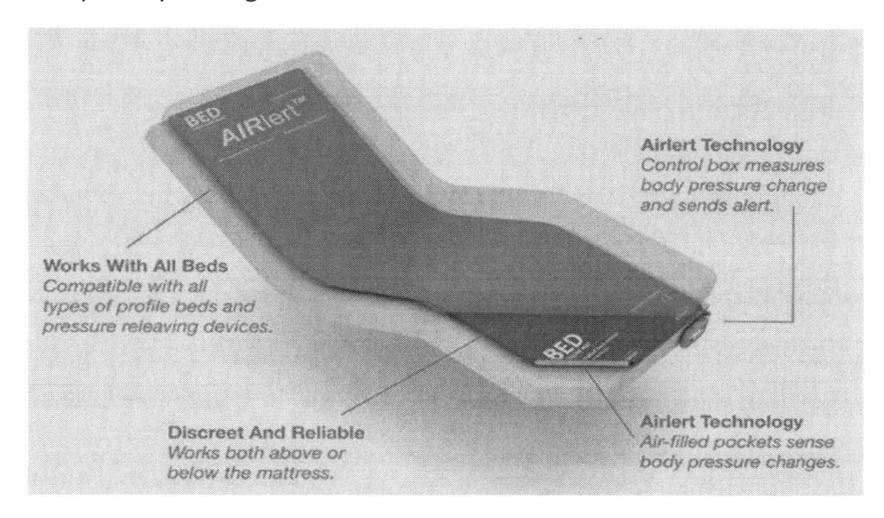

Figure 88: The Frequency Precision Airlert Bed Sensor

A new generation of in-bed sensors has recently appeared which do not require attachment to the person nor to the bed, but which can be placed under the patient in an entirely unobtrusive and non-invasive manner. They do not rely on pressure for monitoring purposes. They are manufactured using thin sheets of 'electro-active' materials whose electronic or dielectric properties are changed when the bed is occupied. Two different types of device are shown in Figure 81. Both can be fabricated into thin sheets for placement under the individual or the bed-sheet. Both are resistant to moisture and could be used to produce a novel form of bedding which has integrated sensing properties.

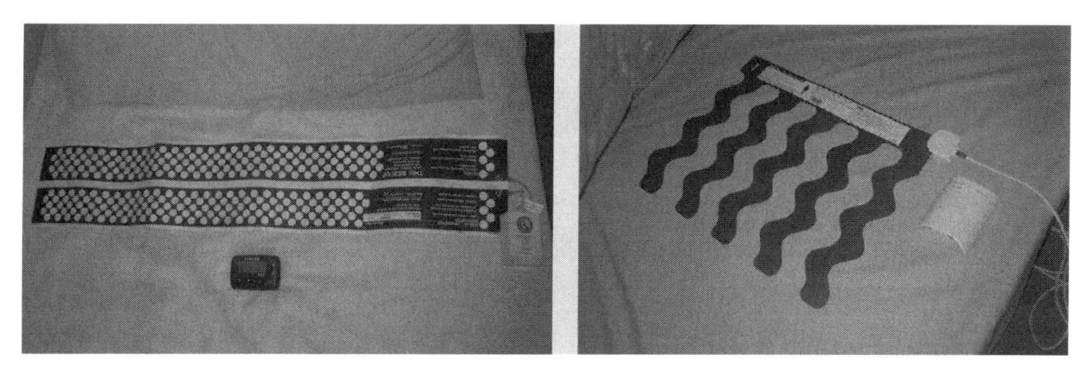

Figure 81: Novel bed occupancy sensing devices – the Bed-Ex and the Emfit Sensors

The Bed-Ex material has five features that are relevant for applications in health and social care, both in the community and in hospital or nursing home settings:

1. It is low cost and can therefore be a disposable item; although an individual sheet may have a useful lifetime of several weeks or months, it could be replaced frequently if contamination occurs or if a new patient occupies the bed;
2. It is plastic coated and impervious to moisture, easy to clean and tear resistant; it has no creases or crevices where bacteria can be deposited, or which can become an irritant to the skin of people who are prone to the development of pressure sores;
3. It can be used with patients of very low or very high body mass because it senses capacitance which is not linearly related to weight;
4. It is compatible with all types of bed construction, including electronic profiling units, and with all types of mattress, including those that provide alternating pressure to manage pressure areas; and
5. It can be cut to size and shape using an ordinary pair of scissors; this allows the material to be used as a sensor on a bed, a chair, a commode or a toilet seat.

Although most people living in their own homes, and who need to use the toilet during the night, would make their way to the bathroom, this remains a hazardous experience for many people with poor eyesight or whose balance might be impaired as a result of taking sleeping tablets. A commode can be provided next to the bed and is found to offer an easy transfer. Unfortunately, people who might spend several minutes on the toilet or commode struggle to get up again and may need the help of their carer. A sensor on the seat may be used to provide a warning and, thus, reduce the risk of a fall.

The Emfit sensor system on the other hand, is based on an electro-active polymer which produces an electrical signal in response to low frequency movements. When the electrical signal is fed into a suitable high impedance amplifier, it provides a very sensitive detection of vibration, movement, and low-frequency sounds (such as coughs). Indeed, it can detect breathing movements, heart sounds and actions such as turning over. The sensor element is installed underneath either a foam mattress or a mattress-topper and is connected to a bespoke, programmable signal processing and alarm unit. Consequently, bed occupancy is associated with a certain level of background "noise" – a combination of movement, breathing, heart sounds etc. If this normal level of background signal disappears, the control unit can generate an alarm immediately or after one of several defined delay periods.

It follows that the same approach may be applied to provide a record of voluntary movements and of turning by staff, offering a means of managing pressure areas and avoiding the formation of ulcers. Similarly, it may be used to detect bouts of coughing and wheezing and to assess sleep quality, which is another factor that can influence Quality of Life of patients who spend the majority of their time in bed or in the house. This approach is unlikely to be unique to these particular bed occupancy sensors. Similar opportunities may be present for all those techniques and devices that allow continuous monitoring of weight or pressure including those that use an air mattress (Frequency Precision) or a strain gauge within a sensor that fits under a caster or bed leg. Indeed, it is likely that bed occupancy sensors will soon also be able to monitor sleep phases and the quality of sleep.

To conclude this chapter, Table 35 provides an outcome-based falls toolkit which is a summary of the options available to help manage the risk of falls, and as a guide for dealing with the issues that are likely to be present when assessing a frail person. Many of these actions have been introduced in this chapter, but other solutions will be described in the next chapter, because people who have cognitive impairments are very susceptible to taking risks, such as climbing on chairs, and exhibiting a lack of insight that can lead to falls.

Table 35: Summary of Options to Manage Risk of Falls in Frail People

Action	Effect
Detect falls quickly through telecare	Increases confidence in avoiding a long lie
Exercise programme (including Tai Chi)	Improves balance, strength and gait
Improve posture and gait	Reduce likelihood of falls
Training to get up from the ground	Reduces likelihood of a long lie
Perform home survey	Can reduces number of environmental hazards
Improved lighting levels	Compensates for loss of visual acuity
Introduce automatic lighting	Avoids trips in a darkened room at night
Regular eye-checks	Corrects visual problems through lens prescription
Spoken reminders	Reduces likelihood of not using glasses or slippers
Treating postural hypotension	Avoids problems with syncope and blackouts
Providing dietary supplements	Reduces likelihood of bone fractures
Managing medication	Reduces overdosing, underdosing or missed dose
Monitoring of domestic activities	Improves lifestyle and mobility
Use of Powerplate vibration machine	Improves balance and strength in legs
Reminders for nutrition and hydration	Maintains body strength and organ function
Walking aids such as a cane or rollator	Overcomes effects of poor balance
Enhance the edges of steps and stairs	Reduces likelihood of fatal trips and falls on stairs
Personalised advice and information	Helps avoid risky behaviour

Concluding Remarks

Falls are undoubtedly one of the more significant risks to the health and well-being of older and disabled people, as they can often be the first step on a spiral of decline that leads to a loss of independence. The telecare industry has, quite naturally emphasised the importance of early detection, though there has yet to be evidence that automatic worn alarms are more effective than active alarm triggers. However, some of the most dangerous situations caused by falls at night can effectively be detected using bed occupancy alerts.

It may be evident that a more integrated approach to falls management is needed, with more emphasis on fall prevention through more use of assistive technologies, many of which are low technology and unlinked. Telecare services (and other community falls initiatives) should offer an extended range of interventions that include better assessments, and response services that ensure that nobody has to endure a long lie after a fall. That will give people more confidence to carry on positively with their lives.

Chapter 9

Supporting people with major neurological disorders

Introduction

The previous chapter focused on the support needed for people who have become frail and whose health, well-being and independence are affected negatively by this syndrome. Old age is a major cause, and genetics might also play a role, but the fact that frailty doesn't affect everyone demonstrates the importance of preventive measures such as diet and exercise. Research suggests that changes in lifestyle can continue to have a positive effect well into old age. Rather less is known about the causes of cognitive impairment. Yet, the ability to continue to live well and safely in the community when challenged by symptoms such as memory loss, difficulties with thinking, problem-solving or language poses major problems for health, social care and housing professionals both at a policy level and at the coal face.

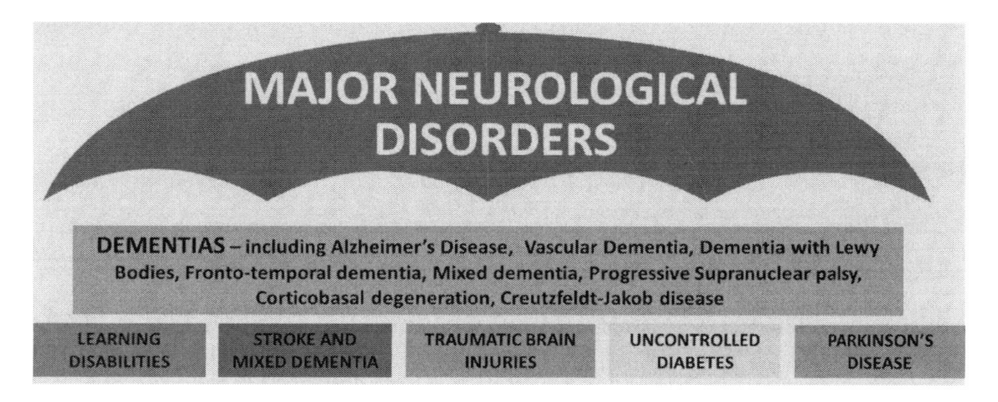

Figure 90: Major Neurological Disorders that Can Result in Cognitive Impairments

Collected under the umbrella of Major Neurological Disorders shown in Figure 90, are a number of causes of cognitive impairment. They are dominated by dementia, a chronic or persistent disorder of the mental processes caused by brain disease (or injury) and marked by memory disorders, personality changes, and impaired reasoning. The term Major Neurological Disorders is increasingly being used to describe the nearly 100 different diseases and conditions ranging from very rare through to very common. They differ from some of the conditions and diseases shown on the lower tier in Figure 90 in that they are progressive; there is no cure and no treatment that slows or stops progression. But there are drug treatments that may temporarily improve some symptoms. Other conditions, such as stroke and brain injuries, can be improved by therapy (see Part 2 of this book) whilst management of diabetes and Parkinson's disease can be improved through appropriate medication and therapy.

Learning disabilities, and conditions such as Down's syndrome, are present from birth; their impact on levels of achievable independence and safety can be reduced through education and support over extended periods of time and can continue through school and college through to gainful employment. Technologies to provide support are considered separately at the end of this chapter. and will improve rapidly as their quality and design become more inclusive. The focus of this chapter is on technologies and strategies that use technology to provide support for people with dementia, especially those who suffer from Alzheimer's Disease (AD).

Diagnosis and Types of Dementia

Perhaps because dementia cannot be cured, individuals, and their families, have been reluctant to face the challenge of seeking a diagnosis often until it is too late to make changes to lifestyle that can help them to achieve some degree of quality to their lives. In 2010/11, in England, less than half (42 per cent) of those estimated to have dementia had been formally diagnosed. This percentage had increased to nearly 60% by the beginning of 2015 and is likely to be close on 2 in 3 by 2020, thanks to measures designed to improve diagnosis.

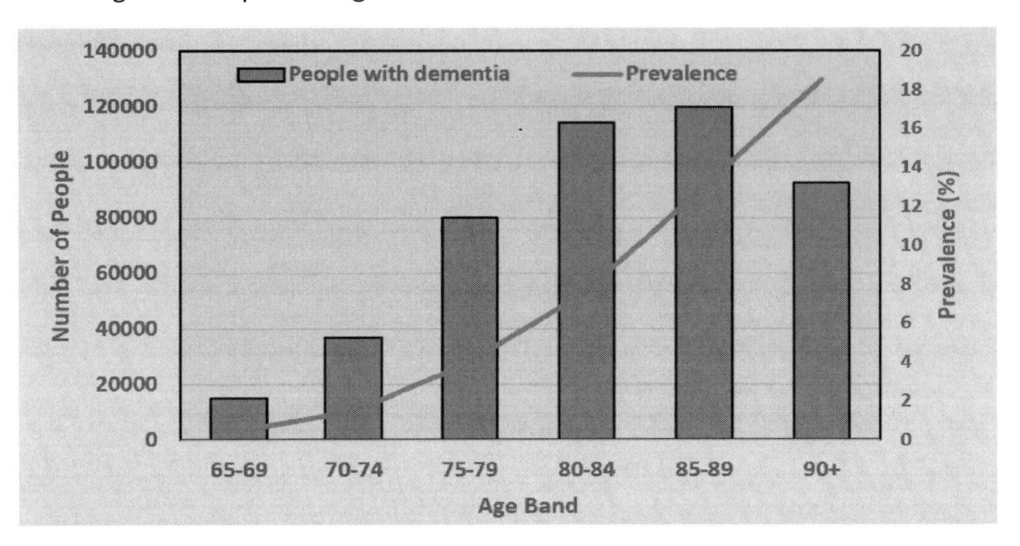

Figure 91: Age Dependence of Dementia Prevalence and Number of Confirmed Cases

This means that the number of people in the UK who are likely to be suffering from dementia is moving rapidly towards a million, but that there may be almost as many who have not been identified as those that are known to the NHS and to local authorities. Most of these will be aged 75 or over, though there are many who are affected in middle age. Figure 91 shows the number of confirmed cases of dementia for each age group; it is unclear in which age groups the missing one third of likely sufferers will sit, but it is likely that they live alone or with a partner who does not wish to share their concerns with their GP or with formal support agencies. Nevertheless, it is now increasingly possible to consider moving forward in ways that address the need for support, though it may take a number of years to remove any stigma relating to the condition. Technology can become an important weapon in this fight, and could offer significant support in some areas, at least until new pharmaceutical interventions succeed in delaying or reversing the effects of the disease.

The major dementia diagnoses are compared in Table 36. It should be noted that the estimates relating to percentages are available only as a range. This is both because of disagreements between experts involved in their research, and because many people, especially the oldest people to be affected may be suffering from a combination of disease forms i.e. mixed dementia. It remains the case that over a half (and perhaps up to two-thirds) of all dementias can be considered to be Alzheimer's Disease, sometimes combined with one of the other forms of dementia which has probably been studied more than any of the other diseases shown in Table 36. In practice, until there are genuine treatment options, it may not be necessary to differentiate between the diseases but to focus on the symptoms and issues, and on their management

Table 36: The Major Forms of Dementia

Type of Dementia	Prevalence	Description	Clinical features
Alzheimer's Disease	50 to 70%	Problems with short term memory, language and reasoning	Amnesia, aphasia, apraxia, agnosia, steady decline
Vascular Dementia	15 to 20%	Impaired judgement, difficulties with motor skills and balance	Memory, episodic but long-term decline
Dementia with Lewy bodies	4 to 5%	Symptoms include disordered sleep and hallucinations	Memory, attention, psychosis
Front-temporal dementia	2 to 3%	Previously called Pick's disease; personality changes and language problems	Visual-spatial and memory skills remain intact
Mixed dementias	10 to 15%%	Various symptoms	Rapid decline

Because dementias are progressive, the symptoms and issues change over time, both in nature and in the degree of severity. It follows that it is necessary to monitor the decline in cognitive function over time using a standard scale, especially as no simple biomarker has been found that allows the status of the disease state to be measured. One of the most popular measurement techniques is the Mini Mental State Examination. This uses a series of questions and tests which, if answered or performed correctly, yield a score up to a maximum of 30 points. Most people will score 27 or more and this can be used as a baseline. Table 37 shows some of the questions and their purpose. MMSE tests have been produced in different languages and therefore offer medical personnel a tool that can be used alongside other tests to give an indication of cognitive decline.

Table 37: Examples of MMSE Test Questions to Assess Mental Ability

Mental Ability	Example Questions	Max Score
Orientation	What is today's date?	**5**
Registration	Repeat the names of 3 objects that are shown and named.	**3**
Attention/Calculation	Spell a 5-letter word backwards	**5**
Recall	Name the 3 objects that were shown above 2 minutes ago	**3**
Language	Read and obey a written command	**1**
	Name 2 common objects that are shown	**2**
	Copy a design using pencil and paper	**1**

In practice, it has been established that an individual's ability to cope with loss of cognitive function is dependent on several factors including their level of education, and other practical skills that they will have learned and improved during their lives. These should be used to assess both the level of risk and the compromises that need to be struck. This is to allow them maximum freedom within a framework of responsibility and ethical acceptance that society expects. This approach appreciates the need for a person-centred approach to assessment (see Chapter 7) that leads to sensible compromises being made.

Figure 92 shows how technology and conventional forms of support must generally occupy the 'sweet spot' of intervention that lies between the extremes of total freedom and total control. Unfortunately, there are no tools that can be used with confidence to determine where individuals might lie on the continuum of acceptance of what they might see as interference. Thus, the views of family members must be sought, where appropriate, and their support used to persuade the individual that a particular solution may be suitable. This is one of the balancing acts that need to be performed by assessors when seeking the optimum interventions.

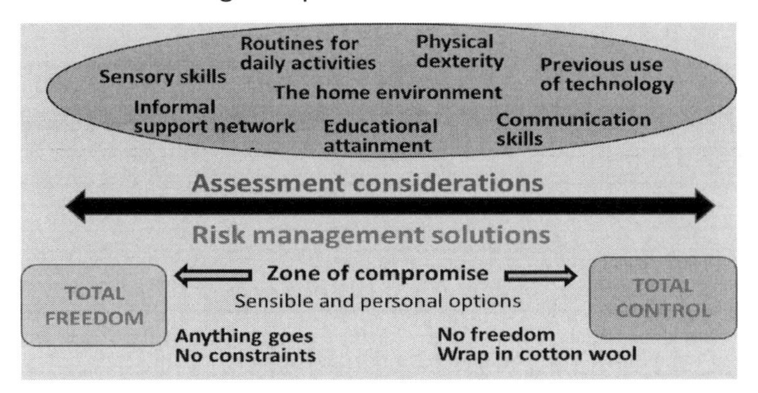

Figure 92: Assessment and Compromises in Supporting People with Dementia

It is often family and friends who first notice the symptoms of dementia, and before a formal diagnosis has been made, and before a visit to a Memory Clinic. Some behaviours noticed in this way are described in Table 38 along with their impact on the individual and their family members. They are rarely collected as an objective account of decline in cognitive function but can become the subject of family discussions. Unfortunately, they are often exaggerated either in their impact or in their frequency, so there may be a considerable delay in the symptoms being discussed with the individual or with the GP or with social services. Without evidence, it is difficult for the authorities to take things further without using, for example, activity monitoring systems (such as Just Checking, Canary or TextCare) to identify or confirm the behaviour that the family feels is dangerous.

Table 38: Examples of Issues Posed by People Who Have Mild or Moderate Dementia

Behaviours Noticed by Family & Friends	Impact on Individual and their Carers
Acting in a different and sometimes strange manner towards familiar people	May fail to recognise or put names to faces of family, friends and neighbours
Becoming slower in performing everyday activities and domestic tasks	Struggling to do simple things and of the sequence of actions that are needed
Asking the same question over and over again despite receiving the answer several times	Effect of loss of short-term memory
Loss of temporal orientation, typically confusing day with night, and experiencing poor sleep	Likely to get up during the night and cook food, phone relatives and perform chores
Being unable to find their way back home after going out purposefully, sometimes at night	Losing a sense of direction to become lost, giving impression of wandering aimlessly
Failing to deal with bills or hiding them away	Potential for utilities to be withdrawn
Leaving items such as spectacles, a purse, keys in strange places and being unable to find them	Telling people that items have been lost or stolen when they haven't

In practice, not all the behaviours and issues shown in Table 38 are dangerous. Indeed, distinguishing between different levels of concern can be extremely difficult, as it depends both on context and on whether the individual lives alone, receives help with domestic activities and tasks, and lives in an area where there are good neighbours and few security concerns. Table 39 lists 20 issues that are common in people with dementia, together with an indication of whether they are simply frustrating, difficult to deal with, or actually dangerous either to the individual or to people that they interact with. The decisions are entirely subjective but gives some idea of priorities for an assessor to deal with in terms of finding solutions. These solutions might involve technology but equally might involve workarounds, the introduction of professional home carers, or changes in the way that tasks are performed. Unfortunately, no two people with dementia are the same; they all show different issues and at different levels of severity. A full assessment is therefore essential.

Table 39: Issues to be Addressed to Manage Behaviours of Some People with Dementia

Issue	Description	Level of Concern
Forgetfulness	Loss of short term memory	Frustrating
Repetitive questioning	Asking for the same information over and over again	Frustrating
Wont change clothing	Wears same clothing for days at a time	Frustrating
Apathy	An indifference to events and surroundings	Frustrating
Hoarding/hiding things	Unwillingness to discard unneeded objects in the home	Frustrating
Short-tempered	Easily angered with little or no reason	Frustrating
Agitation	Physical restlessness, manifested by pacing as an example	Frustrating
Temporal disorientation	Confusion about time and time-keeping	Frustrating
Delusions	Believing in thoughts despite being contradicted by reality	Difficult
Refuses help	Will not accept that they are unwell and need support	Difficult
Lack of inhibitions	Inappropriate public acts e.g. touching people, urinating	Difficult
Fear of water	Wont wash body, hair or clothing	Difficult
Verbally abuses people	Exhibits unusual ranting and raving attacks	Difficult
Lack of security	Fails to close or lock doors and windows	Difficult
Dresses inappropriately	Doesn't consider the weather or occasion for clothes	Dangerous
Wandering	Walking away from home without evident purpose	Dangerous
Medication misuse	Failure to take prescribed medications appropriately	Dangerous
Wont eat	Fails to receive adequate nutrition	Dangerous
Wont use assistive devices	Fails to use equipment to make tasks simpler and safer	Dangerous
Pica	Eating non-food substances including hair, paper and soil	Dangerous

This classification of issues suggests that the risks to independent living are greatest if the individual is likely to show dangerous behaviours. The dangers are primarily of accidents, either to themselves or to others; this includes physical harm due to exposure to inclement weather, poisoning (through medication mismanagement or eating non-food items), or road traffic accidents caused by a lack of insight. However, overcoming the issues identified by relatives and friends does not necessarily help maintain the Quality of Life of an individual with dementia. Genuine solutions must also ensure that they are able to continue to perform both the Activities of Daily Living (bathing, dressing, transferring etc) and the Instrumental Activities of Daily Living (paying bills, buying food, making telephone calls, managing medication etc) with help from domiciliary carers, family members or NHS staff. Adjusting the level of support as needs change can be an incredibly difficult task that requires dynamic assessment, and a recognition that the range of tasks that can be performed independently will inevitable reduce over time, and as the Mental Ability declines.

Figure 93 is a complex diagram that shows the possible decline dynamics of both Quality of Life and in MMSE score over time for someone with dementia. It shows 2 possible trajectories, one for someone who receives the old-fashioned type of care that involves wrapping the individual in cotton wool so that they can do little for themselves. The second curve relates to the decline in Quality of Life when the individual receives person-centred care that include technology support and interventions from carers that help them to be healthy through good diet, mental stimulation, physical exercise, and enhanced communication and engagement with approaches that include reminiscence. On the right-hand side of this diagram are listed ADLs and IADLs in an order (not evidence-based) of declining capability with progressive loss of function (measured by MMSE score).

Two horizontal lines are included in Figure 93. One shows a level of cognitive ability that enables an individual to cope with living independently in their own home without carers. A lower horizontal line shows the point where they might be no longer be able to live successfully at home without 24-hour care. This level of care may be difficult to achieve without moving into a specialist facility, though this does not need to be a care home. The gap between the 2 decline trajectories can be as great as 3 to 5 years and provides an indication of the potential of a person-centred care approach in maintaining quality of life and independence. There is clearly reduced expenditure using technology in this way, but only if appropriate applications can be found.

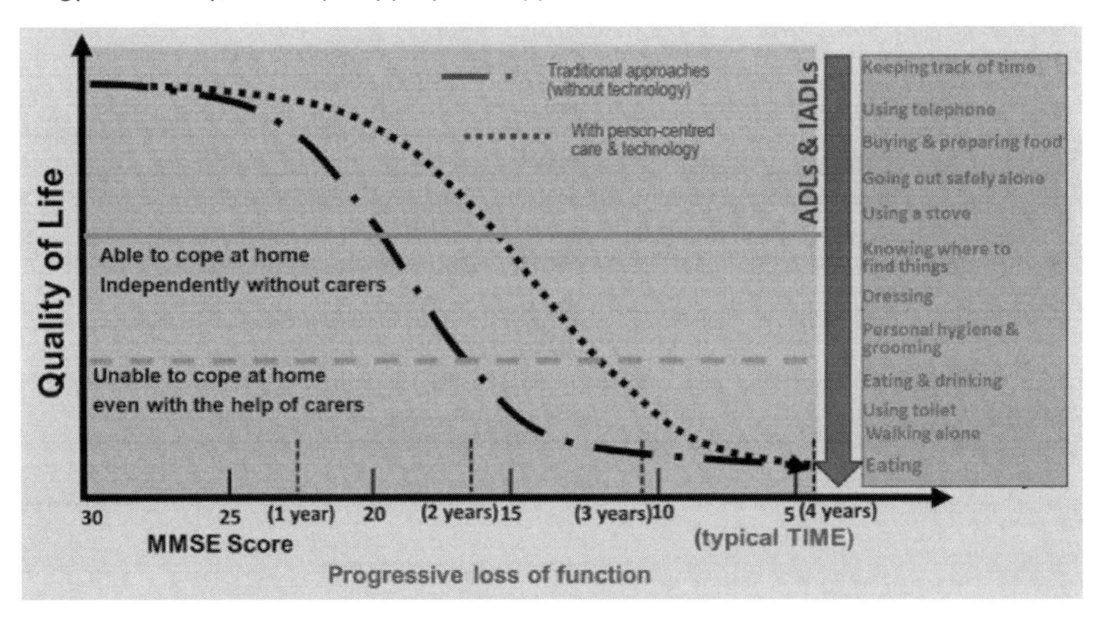

Figure 93: Cognitive and Functional Decline over Time

In practice, dementias progress in phases; this can give assessors some clues on offering solutions that are likely to benefit for a reasonable period of time. Clinically, a dementia, especially Alzheimer's Disease, may be defined as 7 stages which can take in simpler descriptions such as Early Stage. Mid-Stage and Late Stage. Table 40 defines these stages with broad descriptions of what an individual can and can't do in each stage. This table also provides some possible duration information for stages 3 to 7. The times are average and include significant variations, especially when person centred care principles are applied as in Figure 85. The total average duration of the disease can be calculated as 14 years using this approach though half of that time is spent before the Early Stage when an individual might be described as suffering from Mild Cognitive Impairment (MCI). Not everyone with

MCI develops dementia but some of the coping strategies to be described below can also be offered to them for support. Family carers can also struggle to keep their loved ones active during the many years of decline, especially when they are being cared for at home. Some specialist facilities for people with dementia have dedicated members of staff who are employed to keep residents entertained or actively involved in performing activities. Examples are discussed in Part 2 of this book. Some home-based interventions for carers are shown in Chapter 10.

Table 40: The Seven Stages of Alzheimer's Disease

Stage	Diagnosis	Description	Average Duration
1	No Dementia	The person functions normally, has no memory loss, and is mentally healthy.	**Lifetime**
2	No Dementia	This describe normal forgetfulness associated with ageing; Symptoms are not evident to family, friend or GP.	**10-20 years**
3	No Dementia	This involves increased forgetfulness, slight difficulty concentrating, decreased work performance. People may become lost occasionally or have difficulty finding the right words. A person's loved ones may begin to notice a cognitive decline.	**7 years**
4	Early-stage	People have difficulty concentrating, decreased memory of recent events, reduced ability to manage finances and to travel alone to new locations, trouble completing complex tasks properly and may be in denial about their symptoms. They may also start withdrawing from family or friends, because socialising becomes difficult. A healthcare professional can detect clear cognitive problems.	**2 years**
5	Mid-Stage	In this stage, people have major memory problems and need some assistance to complete their activities of daily living. Memory loss is more prominent and may include being unable to remember their name, where they live or their telephone number. They might also not be aware of the day, the time of day, or their current location.	**1 year**
6	Mid-Stage	Considerable assistance may be needed to carry out all daily activities. Names of close family members will be forgotten and short term memory can all but disappear though more detail of earlier life may be retained. Both urinal and faecal Incontinence can become an issue. Ability to speak declines. Personality changes, with delusions and compulsions appear, and anxiety and/or agitation may occur.	**2 years**
7	Late-Stage	Ability to speak or communicate is totally lost. Assistance is needed with all tasks including eat. Assistance with most activities. Psychomotor skills, including the ability to walk, are lost.	**2 years**

It may be evident that the most profound changes in behaviour are experienced in stages 6 and 7. Few people with dementia can be supported in their own homes during these stages unless 24/7 care is provided. These stages generally present challenges to carers and management in specialist centres for people with dementia, though there are attempts being made at introducing alternative models of support based on community living. An example worthy of mention is Hogewey, a gated village in Weesp, an industrialised suburb of Amsterdam, which is a pioneering care facility for Holland's most severe dementia sufferers.

For completeness, Figure 41 provides examples of the specific issues that are presented by people with more advanced dementias (Stages 5 to 7). Some are similar to, but usually more complex than, the problems already described in Tables 38 and 39, but are provided for completeness and as a reference guide. Unfortunately, many of the issues described in Table 41 are related to general behaviours and attitudes that are difficult to change or manage, and therefore cannot be addressed in a straightforward manner using technology. However, more generic approaches using better signage, reminders and prompts can help remove some of the factors that can cause frustrations;

these often lead to aggression and yet more challenging behaviour. Again, the need for bespoke solutions is evident because of the individual nature of the issues.

The remainder of this chapter is focused on how many, but not all, of the most common of these problems can be addressed using technology, either alone or in combination with other forms of support provided by families, friends and carers. Not all will be successful. Indeed, some approaches might initially appear to be ideal, only to be rejected after a short while due to some aspect of design or appearance. In some cases, this will simply require a particular device to be exchanged for something different, perhaps in a different colour. Until recently, this would not have been possible as there would only have been one type of device available (at least from an individual service or equipment provider).

By today, there are many different options available through different suppliers, and it may be the responsibility of the service provider, or the family, to procure the best one for the individual. A try-before-you-buy arrangement might work well, but such a facility is unlikely to be available through manufacturers and distributors, especially if a subscription is required. This opens up the possibility of introducing Technology Loan Libraries to provide equipment for short periods of time, maybe a week or 2, before purchase or long-term provision. Such facilities may operate either by service providers or by third sector organisations which could be funded by the NHS and a local authority in partnership, perhaps resembling Community Equipment services for other items of Assistive Technology. However, it is unlikely that such devices and systems could be added to existing Equipment Store inventories both because of the need for more short-term provision, and because many of the items are digital in nature and involve a subscription

Table 41: Some of the Problems Presented by People with More Advanced Dementia

Issues	
Bathing issues	Refuses to bathe or shower for a variety of reasons
Poor oral hygiene	Refuses to brush teeth or clean dentures
Won't change clothing	Doesn't appreciate that clothing is soiled or inappropriate
Poor nutrition	Fails to prepare and eat appropriate food at regular times
Hoarding, hiding	Collects and conceals random objects such as labels
Short-tempered	Shows frustration by losing control of emotions
Dresses inappropriately	Fails to match clothing to weather; loses sense of colour
Verbally abuses people	Shouts at family members and carers for no reason
Agitation, aggression	Unable to settle down; likely to strike anyone within range
Lack of security	Fails to close or lock doors and windows
Medication issues	Unable or unwilling to take medication as prescribed
Delusions	Believes that they have seen or spoken to someone unreal
Refuses help	Can't accept that they need support with any tasks
Lack of inhibition	May strip naked or make inappropriate sexual advances
Loss of continence	Fails to use toilet leading to urinary and faecal incidents
Communication problems	Can't find the words to express feelings or needs

Based on Tables 38 to 41, the issues that are most worthy of attention, and which are most likely to achieve some level of success for people with dementia are:

- Cooking – and management of electrical and gas devices
- Remembering to do things – including the time and need to perform tasks

- Managing medication – taking correct medications at the right time
- Finding items – when they are lost or stored in strange places
- Safe Walking – encouraging exercise but restricting access to dangerous locations
- Prevention of accidents in the home – especially due to water and flooding
- Security – prevention of exploitation by bogus callers and opportunist burglars

Other issues are discussed in Chapter 10 as examples of how informal carers can use technology to reduce their burden. It is also possible to improve the home environment to be more 'dementia friendly'. The most relevant issues and opportunities will be discussed in Part 2 of this book, focusing on the design of institutional facilities, and on ways that whole towns can become more supportive. However, it is apparent that the home environment can also be modified in simple ways by the use of more calming colours for walls, the use of contrasting colours for plates and crockery, and the removal of mirrors and other reflective surfaces that people with dementia find to be very confusing. Apps will also emerge that provide video instructions that can support someone with memory loss by showing sequences triggered by simple prompts or stickers. Some ideas for making the home more 'dementia friendly' are shown below.

Making Homes Dementia Friendly

Many interventions can be considered to be assistive technologies that are simple adaptations to the home rather than complex interventions that convert the property into a smart home. Fully connected homes will become an important feature in the future and will be discussed in Part 2 of this book. There are, however, many modifications to the home environment that can be performed quickly and at low cost. Some of these are discussed below and are in addition to those specific interventions that have already been discussed in this chapter and in the previous one, and are usually complementary:

- Increase the levels of lighting – this not only helps avoid falls but is important for someone with dementia as it helps with orientation and well-being. The level of natural light during the day can be increased by removing unnecessary net curtains or blinds, or by having the blinds open or close electronically by sensing the outside light levels. The number and location of light bulbs matters because it can reduce the levels of shadow. Higher wattage light bulbs can also help, as can the use of lower level LED lamps that can be attached to the bases of kitchen units or, as portable strings of lamps, used to guide the individual up or down stairs.
- Traditional telecare systems (Chapter 3) can be extended to included smoke alarms, flood detectors , heat alarms and gas detectors. These devices automatically signal when their batteries are becoming low.
- More advanced telecare systems can include alerts caused by an individual leaving the bed at night and opening outside doors and going outside without returning before long.
- Reduce unnecessary background noise such as the TV and /or radio, by providing automatic switch-off functions if they are not being listened to.
- Arranging clothing for each day in a separate wardrobe or set drawers to help the individual to dress themselves with appropriate clothes that are readily there. Figure 94 shows a prototype DRESS unit which uses a camera and lights on individual drawers to show users the order in which clothing needs to be put on.

- Electrical sockets can be replaced with lockable ones to avoid the overnight use of equipment that might cause a fire.
- Socket covers should be provided over equipment that should not be switched off, including a dispersed alarm unit, refrigerator and freezer.
- Clocks can be modified to tell the person that it's night, and time to change into night clothes. Audible messages can also be played at appropriate times.

Figure 94: The Prototype DRESS unit (left) and a fridge with a clear door (right)

- In the bathroom, special lights can illuminate the toilet so that it's easier to find; the toilet lid and seat should contrast with the colour of the pan for visual ease. Bath mats should be avoided as someone with dementia might think these are barriers they need to step over. Contrasting colours for towels, toilet paper and other items also make them easier to find.
- In the kitchen, there is a need to make contents more accessible; cupboard fronts/ doors can be removed, or a refrigerator installed which has a clear front (see right of Figure 94) so that the contents can be seen without a need to open the door.
- Hot and cold taps should be marked either with a clear colour scheme (e.g. red for hot and blue for cold or with words and signs.
- Plain plates and dishes in a solid colour can make the food stand out, and easier to find.
- Matt flooring and walls are useful as shiny surfaces may be mistaken as being wet or slippery. Single colours work better than patterns.
- Important items, such as keys, money medication and key phone numbers should be kept in the same accessible place.
- A white board or an electronic notice board may be used to provide reminders of appointment or events for the person.
- Rooms and the contents of drawers and cupboards can be labelled using clear images or words as shown in Figure 95.

Figure 95: Signage for rooms and contents of drawers and kitchen units

Technologies to Help Manage Specific Issues

a) <u>Cooking</u>

The importance of food preparation and cooking to independence, self-esteem, and general Quality of Life are well established. Cooking for oneself gives choice in what and when to eat, as well as being a facilitator of shared meals with family and friends. The freedom to manage one's own meal planning and management involves a number of separate tasks including:

- making a list of grocery, vegetables and meat requirements,
- going out to shop for them,
- putting the shopping away and storing them in the kitchen, larder, fridge or freezer,
- preparing vegetables by washing, peeling, dicing and slicing them,
- putting food items in appropriate saucepans or trays;
- timing the cooking of each item;
- stirring the food and providing seasoning or herbs during the cooking;
- plating the meal with added sauces or gravy;
- cutting up meat and eating the meal; and
- clearing away and washing up.

The above are all life skills that are important in preserving an individual's quality of life and well-being. They are supported by many devices that are assistive in nature including the cooker, electronic meal planners and cook books, tin openers, peeling machines and table top dish washers; it would be perverse to remove and make all these devices and skills redundant through removal of the cooker as a working appliance, for example. Instead, the focus should be on making the cooking experience safer, and to take measures to manage the risks of fire in the kitchen.

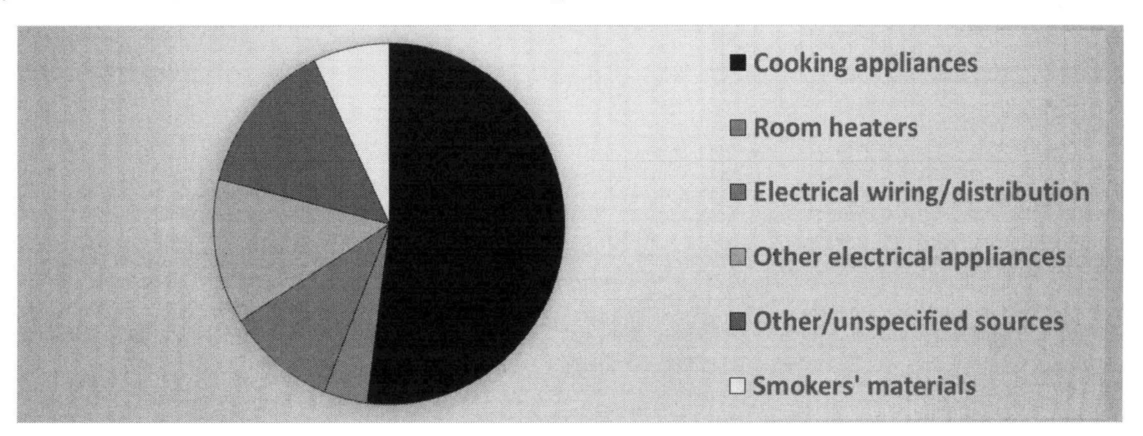

Figure 96: The Causes of Fires in the Home

Yet, the removal of the cooker, and its replacement by a microwave oven is precisely what many authorities do in order to tackle the problem of kitchen fires; this allows prepared or frozen meals to be reheated but, if the individual has not previously used this type of appliance, it would involve a need to teach new skills which may be difficult if they have a cognitive deficit. Yet, it is easy to see how easy it is for some fire incidents to be caused by individuals draping a tea towel or other inflammable items over hotplates, being distracted and allowing saucepans to boil dry or contents to burn, or wrongly estimating cooking times. In the case of electric cookers, burns can be caused when someone fails to appreciate that a hotplate can retain its heat for many minutes after being switched off. There have also been many incidents of a plastic jug kettle being placed directly onto a hot-plate, causing the kettle to melt and, unless full of water, to catch fire. Chip pan fires and others involving hot fat (from a wok for example) are a particular cause of concern both because of the potential for fat to spread quickly around the kitchen and because of the toxic fumes that cooking oils can give off when heated to high temperature before ignition.

The fact that over half of household fire were caused by cooking accidents in 2013 (see Figure 96) could be used to justify the removal of the cooker, or to place more emphasis on risk reduction through technology. Fortunately, assistive technologies and telecare are ideally suited to help manage these risks by either reducing the likelihood of an accident and/or by reducing the harmful consequences. Timers and reminder devices can be employed to warn the person of the need for vigilance, while temperature extreme detectors or devices that respond to a sudden increase in temperature above the hob can give an early warning of, for example, a saucepan that has boiled dry or a hotplate that has been left on after being used for frying or boiling food. Such temperature extreme detectors are readily available from major telecare equipment providers; they can either communicate directly and wirelessly with a telecare dispersed alarm unit (which then sends an alarm to a monitoring centre) or to the mobile phone of a relative or neighbour. They also produce an audible alert which warns the user that the temperature has increased to an abnormal level and requires immediate attention. In general, the user can respond quickly to such an alert, but if they have left the kitchen or have poor hearing, then this can become ineffective.

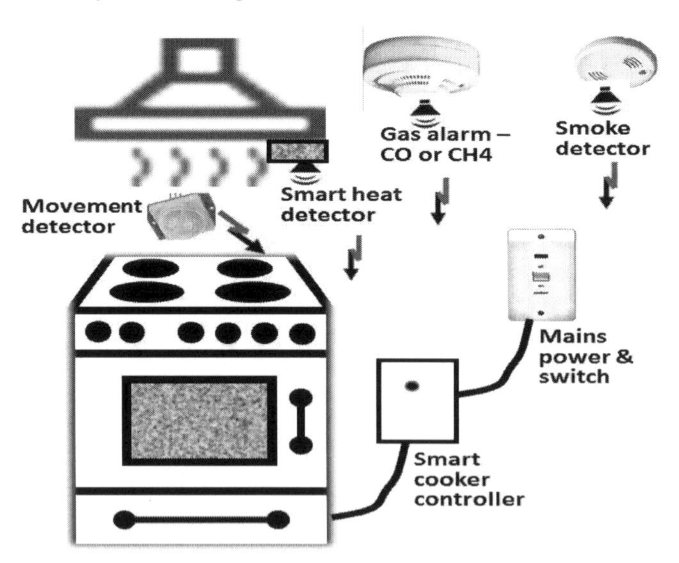

Figure 97: Schematic cooker safety arrangement using smart controller

Fortunately, smart hob temperature monitoring systems are also available from some suppliers (e.g. Innohome and STT Condigi); these are capable of automatically switching off the cooker electricity supply within a minute if a potential fire situation is detected and no action has been taken and are shown schematically in Figure 97. These systems provide radio and/or audio communication between the intelligent temperature sensor and a cooker control unit which can switch off the cooker directly or through industry standard home automation systems (or Internet of Things connectivity – see Part 2 of this book). Home automation units can be used to provide additional alerting channels, such as flashing beacons, vibrating pads and sirens, for people who have sensory impairments.

Some of these systems have the intelligence to learn their usual cooking environment and to respond accordingly depending on cooking time for example but can also be individually programmed to meet local needs. Others employ movement or activity detectors (Passive Infra-Red Detectors – PIRs) to detect the presence or absence of a person in the kitchen or at the stove so that reminders can be issued, or power withdrawn until the user returns. When used in conjunction with other safety devices such as fire blankets and appropriate fire extinguishers, the harmful consequences of a kitchen fire are greatly reduced. Unfortunately, few of the properties lived in by older people have such equipment; in many cases, it remains more appropriate for the Fire and Rescue Service to be called immediately as this can reduce the possibility of an older person being harmed trying to fight a fire. Complete digital kitchen management systems will begin to appear within a couple of years. These will monitor the status of hobs, the temperature of pans, and the times that they have been left on without supervision. The potential for such innovations will be discussed in Part 2.

An extended use of water sprinkler systems is advocated by many authorities because they are known to be associated with improved outcomes in the event of housefires. But these benefits are mainly restricted to fire in the living area or bedrooms rather than in the kitchen. Water is not necessarily the best way of tacking a kitchen fire because of the effect of water on electrical supplies and appliances, as well as the potential for causing burning fat or oil to overflow their vessels and to spread the fire quickly along other surfaces. Therefore, wet chemical extinguishers are more appropriate, but few are found outside commercial catering establishments even though small domestic versions are available.

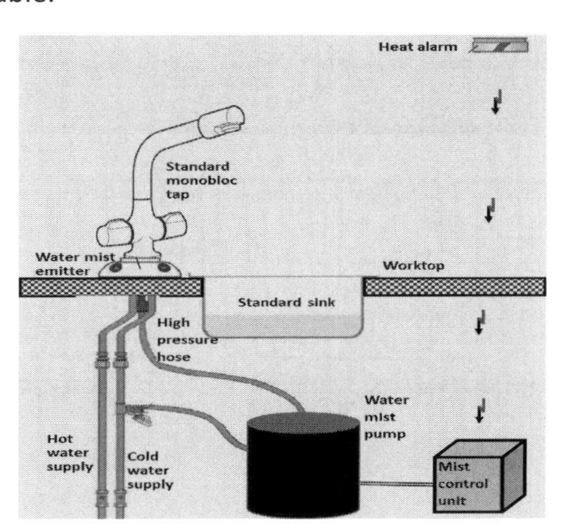

Figure 98: A High-Pressure Water Mist System Installed under the kitchen sink

An innovation to tackle house fires is the use of water mist which has previously been used only at sea and in premises where industrial plant is used. They rely on specially designed nozzles and the delivery of water through a high-pressure cylinder to create microscopic water droplets. These droplets form a fine mist which not only provides a cooling blanket but also reduces the oxygen content in the air surrounding the fire, effectively suffocating it. The volume of water needed is considerably less than is used in sprinkler systems (where it is generally unlimited) which means that it causes less water damage to absorbent material such as soft furnishings, and to surfaces that can become treacherous when wet.

Water mist is an ideal way to tackle kitchen fires because the amount of water required doesn't pose an immediate electrical shorting hazard, and also because the high-pressure action can quickly fill a room irrespective of the source of fire. Some systems are mobile and are designed to be moved between rooms where an at-risk individual spends most of their time e.g. nursing home bedrooms where patients continue to smoke despite being in an oxygen-rich environment. Others (e.g. Aquamist and Ultramist) are designed for use in a kitchen or in other dangerous areas; integration is possible with the water mist emitter being fabricated for direct insertion into the tap base as shown in Figure 98. In this case, the high-pressure system is activated automatically when the temperature reaches 57^0 C, the wireless alarm unit transmitting a radio signal to the mist control unit.

In the case of cooking with gas, there are many similar risks to contend with, including those associated with scalding, burns and fire (perhaps caused by putting a plastic jug kettle on a hob) as well as with suffocation or poisoning through the release of carbon monoxide or methane. However, the greatest concern is that the gas taps on the hob will be opened without lighting the flame either through distraction or through a lack of insight. Similarly, the flame may be extinguished by a pan boiling over, allowing unlit gas to fill the kitchen posing a severe explosion risk. This is often seen as a much greater risk than house-fires because several properties can be put at risk.

Gas explosions remain relatively rare in UK homes, but when they do occur, they can be devastating. The consequences can be death and destruction, not only of the property in which the gas leaked into the environment, but also of neighbouring properties. North Sea gas is almost pure methane and therefore has little or no smell; additives are used to make it easier to detect. It is, however, highly combustible and explosive. It is little surprise that it is not used in many housing schemes designed for older people even though the cost of heating and cooking with gas is considerably lower than with electricity.

In practice, there is little evidence to suggest that gas explosions are more likely to be caused by the reduced cognitive abilities of older gas customers than by accidental gas leaks from pipes or faulty appliances or, indeed, by poor installation by illegal amateur fitters. However, such is the fear of explosion or suffocation that common practice is to respond rapidly and, perhaps, inappropriately:

1. ***Replace the gas cooker with an electric cooker*** – an option which is rarely popular because of the considerable differences in operation
2. ***Use a microwave oven to reheat prepared or frozen meals*** – an approach which removes the need to perform many of the life skills that are associated with independence
3. ***Install a methane alarm to provide an alert if gas is present in high concentration*** – but with no guarantee that the user will hear or respond to the alert (when connected through a telecare system this leaves the call handler having to take rapid and decisive action)
4. ***Control the gas supply using a manual valve operated by a carer*** – this prevents the individual from cooking when alone but imposes significant responsibilities on people coming into the home every day

5. ***Use an electronic valve to control the gas supply*** – this could be operated by a timer device or by a linked electronic gas detector; or
6. ***Provide a new gas cooker with safety features*** - often not considered because of a perception that older people cannot become familiar with a new and expensive appliance.

The risk of gas release posed by older gas cookers has been recognised by the industry for many years and has resulted in the use of flame safety detection devices (FSDs) in all current models. Indeed, it is illegal to install a gas cooker (new or previously owned) without such devices in a block of flats even if the cooker itself has been tested and confirmed to be safe. FSDs effectively switch off the gas supply to individual burners if a probe located at the burner does not exceed ambient temperature. This prevents the release of dangerous volumes of gas irrespective of the position of the gas control knobs. Thus, no gas is emitted if a hob burner fails to ignite nor if a lit hob burner is extinguished by water boiling over from a saucepan for example.

In practice, people with cognitive impairments are capable of learning new skills with appropriate training and can be taught to operate many new household appliances, such as TVs, radios, fridges, freezers, vacuum cleaners and cookers, especially if the functionality is similar, and there are not too many additional features that they don't need. Many new cookers are available priced in the range £200 to £300 and with a standard layout of 4 burners controlled by knobs arranged horizontally at the front of the appliance; these are similar in design to cookers that were on the market during the 1960s. It means that replacing the cooker becomes a viable option in some cases.

If the individual will not be able to cope with a new cooker, then an intelligent arrangement of gas detector and electronic valve can be retrofitted to ensure that the system is safe. These systems have appropriate bayonet fittings that allow them to be quickly installed in-line on the gas supply to the cooker. Some control systems can also switch off the gas supply on detecting alerts from other safety devices such as smoke detectors and carbon monoxide sensors similar to the approach shown in Figure 97 for an electric cooker. However, the cost of such arrangement, and their installation by a qualified Gas Safe engineer, can be prohibitive, especially if the solution is only temporary.

Remembering to do things at the right time

In principle, timely reminders or prompts can help everyone to remember to perform specific tasks that might range from getting up appropriately in the morning through to taking their medications during the day. The importance of such tasks depends on the individual, their informal support networks, and their respective lifestyles; in general, it may be simple to offer one such reminder per day, and to gauge how successful it might be. But when a person needs to be reminded to perform multiple tasks on many different occasions during the day, there are many more opportunities for confusion. There is therefore a need to ensure that prompts or reminders are articulated in a clear way that overcomes the likelihood of the wrong task being initiated in response to a specific reminder. It follows that user interfaces need to be personalised to match the preferences and limitations of the user. There is also a need to ensure that temporal orientation issues are managed by ready access to the time whether through clocks or alternative time announcing devices.

Figure 99: Day and Date Clocks that are Useful for People with Cognitive Issues

People with dementia can often lose track of time, which leads to poor sleep patterns, missed appointments, and a loss of daily routines. Some people need help to know which day it is, and this may be a problem with them maintaining the routines that they require e.g. collect pension on a Thursday. Others may be prone to confuse day with night especially during the long days of the winter, and if their sleep patterns have been disturbed. Figure 99 shows a variety of clocks that present time, date and day information in different formats. Analogue clocks remain the easiest to read and to understand for most older people. Unfortunately, regular clocks don't differentiate between am and pm, so additional information is added in the form of words of pictures to indicate the time of day. In some cases, a third hand is added to show the day of the week. Digital devices such as tablets can be programmed to display information in more flexible ways and in a variety of languages. They offer a means of displaying time, day and information such as morning, afternoon, evening and night, though few handle the transition at midnight in ways that avoid confusion.

Alarms are often additional functions of clocks and can be used to wake people up in the morning, irrespective of their level of cognition. Multiple alarms are not usually offered on clocks, so reminder devices can be used. These range from alarm watches that were originally designed as medication alerts, through to fully programmable systems that can provide speech or visual messages, as in some Echo devices from Amazon. Some examples are shown in Figure 100. Their limitation is likely to be the need for them to be programmed for individual needs, and then reprogrammed to accommodate new reminders as they arise. They will be considered in more detail in Chapter 10. Medication management is an especially difficult issue for people with cognitive problems and is considered separately below.

Figure 100: Examples of Visual Reminder Devices Which May Include Audio Messages

b) <u>Managing medication</u>

The role played by medication in helping people to recover from illness, injury or surgery is well-established. In the case of acute illnesses, the appropriate choice of medication can quickly overcome the symptoms and help to defeat the disease resulting in a cure. However, many chronic diseases from diabetes through to arthritis cannot yet be cured in most cases, and certainly not through medication alone. In such examples, medication is needed to manage the symptoms, but needs to be taken for many years. This raises the possibility (or perhaps probability) that adherence to the prescribed medication therapy will decline over time, limiting control of symptoms. Indeed, if the decline in health over subsequent years is slow, people may not associate cause with effect. Few medications are yet available to manage cognitive impairment, though efforts continue to find a drug (or combinations of drugs) that can delay if not reverse some of the progressive symptoms such as memory loss.

It follows that adherence to medication prescriptions will be a particular challenge for people with dementia; not only is there an increased chance that some doses will be missed, there are also likely to be practical issues associated with removing medication from their packaging and reordering them on a regular basis. Today, about 10 million people in the UK are aged 65 or older, a number that will increase significantly in the future because of demographic trends and increasing life expectancy (see Figure 2 in Chapter 1). However, because many types of chronic disease are the result of lifestyle choices, including the effects of smoking, alcohol, obesity and a lack of exercise, more than half this population will suffer from not a single disorder but from multiple comorbidities. Each disease may require specific medications, resulting in a need for polypharmacy as shown in Figure 101 with nearly 3 in 5 men and women aged over 85 requiring 5 or more different medications every day. Managing these prescriptions is a significant problem, and an issue that leads to poor outcomes, and health and social care system inefficiencies. Special attention is needed to support people with cognitive impairments throughout all disease phases, and for all aspects of management.

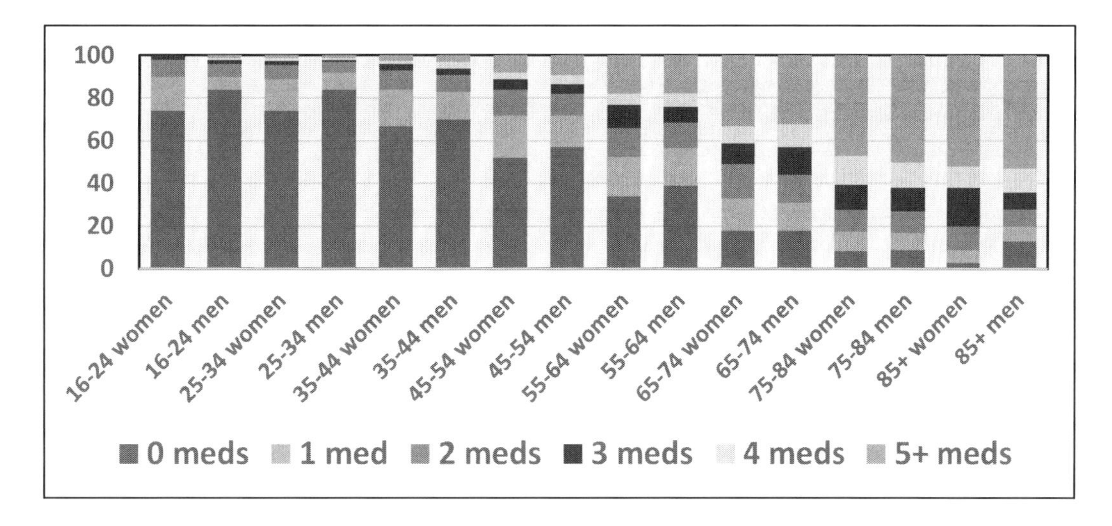

Figure 101: Age Dependency of Polypharmacy

The World Health Organisation introduced a model based on 5 factors - social/economic, therapy-related, patient-related, condition-related, and health system related – to explain medication adherence. Philips Healthcare considered emotional and perceptual influences alongside situational factors to yield a more holistic impression of why an individual is more or less likely to follow their medication regime. It may be evident that some people cannot be helped, because their hearts aren't in it, while others may need more practical support and encouragement, some of which can be provided by the connected and assistive technology options described below. Many people who have dementia might form a special group for whom these factors don't apply, but others will combine all these effects.

People who struggle to take one or more different medications at different but set times during the day, may benefit from their different pills, ointments or medicines being presented to them in arrangements that simplify their administration. There are effectively two approaches – pharmacy-prepared blister-packs (as shown on the left of Figure 102, and a range of different medication boxes (such as the others shown in the same figure) which are effectively reusable versions of blister packs. Most follow a weekly matrix arrangement which allows for up to 4 doses a day, but when more doses or medications are required, a second pack or box can be provided.

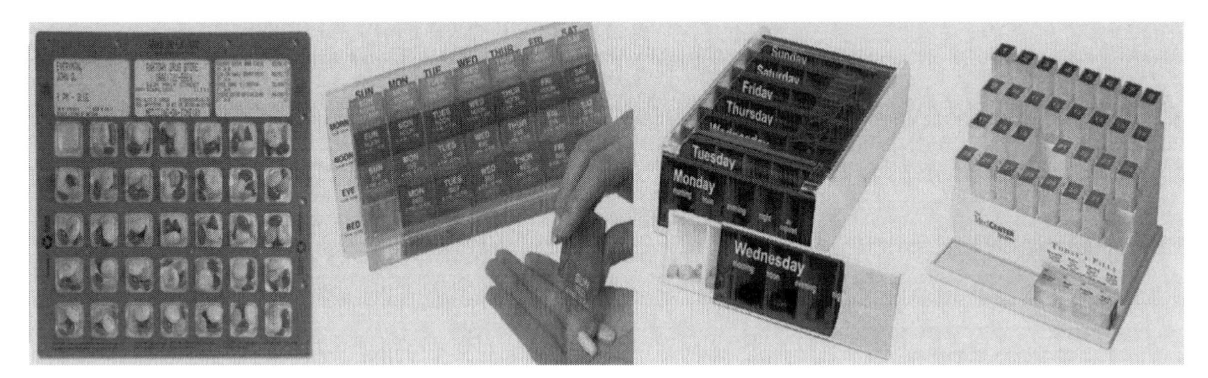

Figure102: Blister Pack and Medication Arrangements for Improved Organisation

These arrangements can be prescribed by the GP for patients in the community. They are prepared (and often delivered) by pharmacies and therefore comply with all relevant regulations, including the avoidance of secondary dispensing, though there is a cost to the NHS for their preparation. Adherence can be checked by monitoring how many pills remain at the end of the day, though some people will take their pills out of their box and throw them away rather than swallow them. This is their choice but represents an enormous waste of money. Supervision is often the chosen solution.

Many people struggle to take medication due to swallowing problems or issues associated with their lack of dexterity or their low vision (poor eyesight). A range of low-tech devices is shown in Figure 103; these help to deal with common problems ranging from releasing pills from manufacturer's blister packs (see left hand image) through to self-administration devices for eye drops (see right hand image). Other devices shown in this figure are a pill crusher, a beaker to help with swallowing pills, a special flavoured spray to make swallowing easier, and a spoon that enables larger and measured doses of medicine to be taken.

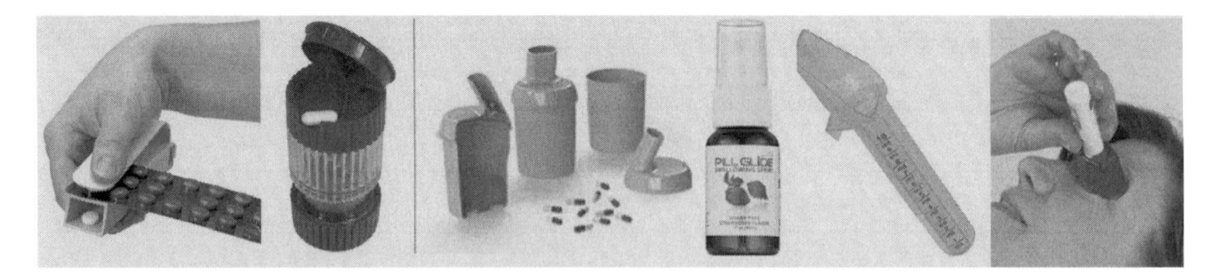

Figure 103: Low Tech and Low-Cost Devices to Make Medication Administration Easier

Some people, especially those with dementia and who live alone, will fail to medicate successfully on a continuous basis despite having a simplified administration regime. They need further prompting and, in some cases, external monitoring, to enable their level of adherence to be quantified. Figure

94 shows 2 simple unconnected alarm options (on the left). The other three options can all be connected either through a telecare alarm centre. The Pivotell carousel (centre of Figure 94) can deliver doses of medication only during the programmed time slots. It is compatible with leading telecare services and provides a wireless alert to a monitoring centre through a dispersed alarm unit, or to a remote family carer, enabling a telephone reminder to be offered before the medication window closes. It is the most popular system used in the UK, but generally requires the medication doses to be preloaded by the pharmacist (unless a family member is prepared to take the responsibility for secondary dispensing).

Next to the Pivotell device is the larger smart MD2 linked medication dispensing device used exclusively in the Philips Lifeline telecare device. It contains 60 preloaded cups each with a medication dose for a particular time. It is not currently available in the UK. Finally, the Medido system (on the right of Figure 104) uses pre-packaged medication pouches that are prepared at a central facility where robotic packaging systems produce a roll of medication that fits into an Internet-linked electronic reminder and dispensing system. This arrangement is becoming popular in mainland Europe and can be cost-effective through economies of scale, though the role of the community pharmacist changes significantly.

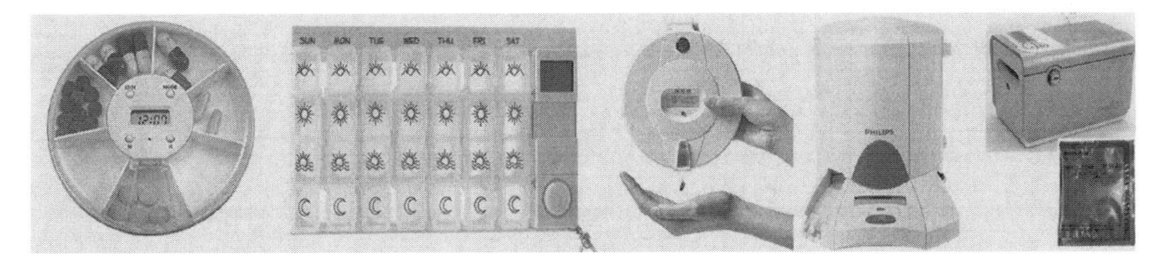

Figure 104: Pill Organisers with Alarms or Connected Functionality

Despite all these devices and innovations, it is unlikely that current technology will succeed in supporting all people with cognitive impairments. However, most individuals will be able to benefit from at least one type of device and will therefore improve their medication management to some degree, especially during the early stages of the disease. More advanced systems using Internet of Things principles and digital technologies may be more successful and will be described in Part 2.

c) <u>Finding misplaced items</u>

Everyone misplaces important items occasionally, whether it's a purse, a set of keys, a bill or the reading glasses. People with dementia usually have poor short-term memory, and therefore are more likely to forget where they might have left everyday objects. However, they might also hoard things and hide them away too, especially if they have delusions about people stealing their things. It can be very frustrating for carers and for family members especially. Fortunately, Bluetooth tags and modules in smartphones have enabled devices to be tethered so that both the tag and the phone can provide an alert either if they are separated by beyond a certain distance, or to enable them to be found quickly if lost within the range of Bluetooth (which is typically about 10 metres indoors). This can be a low-cost approach using small, low-powered tags that can readily be attached to or put inside the valuables and which have batteries that will last 12 months.

Unfortunately, they require an app and a smartphone to operate. People with dementia are amongst the least likely to own and successfully use a smartphone, so this approach may not help an individual to live independently. Alternative approaches such as the MENGGOOD Wireless Key and Phone Finder, and the Loc8tor Lite's wireless system (shown in Figure 105) may be more robust and universally useful. The latter has a much greater range, enabling lost objects to be found when they are in another room. Most lower cost items have more limited use.

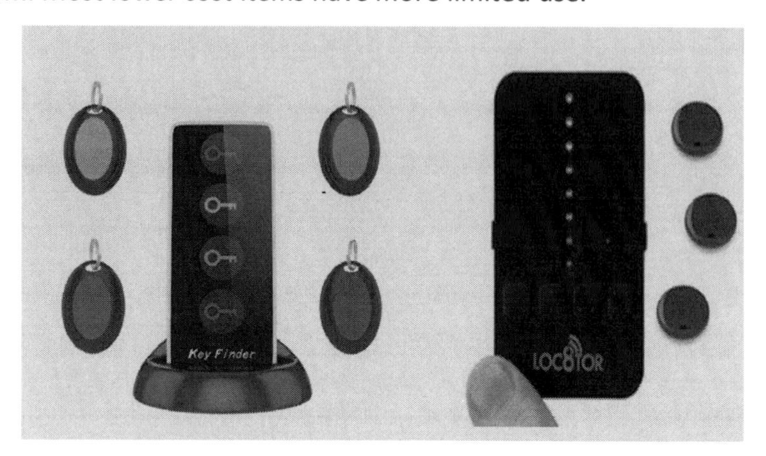

Figure 105: Low cost object locators using r.f. tags

Both devices can be used to find up to 4 different tags, using their respective detectors. The former uses colour coding while the latter has different keys for each tag. The Loc8tor detector is credit card size. This makes it easy to carry, but also easy to lose! The locator can be used to show the radio signal strength from a tag when activated and can therefore be used to find the direction to look. The Loc8tor Lite costs about £60 with replacement tags available for £20. The MENGGOOD device is about £20.

d) <u>Safe walking</u>

Many people with dementia go through a phase of restlessness, especially in the evenings or at night. In the USA it is often known as Sundowning. They go out with a defined purpose which either relates to not realising the time (see section 2) or a desire to go to a location which was important to them perhaps many years ago. This could be to visit somewhere that they worked, or perhaps the house where they used to live. In the dark, and unfamiliar with structural changes, new roads and buildings, it is easy for them to become lost. This can lead to great distress and the potential for panic; it could lead to accidents on roads, rivers or other waterways, as well as incidents of elderly abuse by people without conscience.

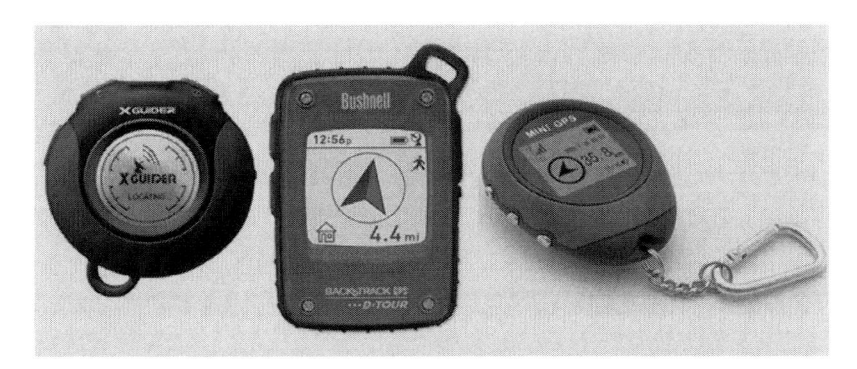

Figure 106: GPS Backtrackers with no connection requirement

In the early stages of dementia, an individual may recognise the need to return home, but may not be able to find their own way without help. GPS backtracker devices (such as those shown in Figure 106) can be useful because they are effectively digital compasses that point the user in the direction of their programmed home location, also indicating how away far it is. Some devices can be programmed with several safe locations, but this can cause confusion. The use of these devices is particularly successful when an individual has the insight to enable them to learn how to use the technology, though the original programming of the home location can be performed by a family member of by an installer from a telecare service organisation. The devices are entirely self-contained, requiring no SIM card nor mobile phone connection. Users therefore feel genuinely independent, empowering them to go out at any time knowing that they will be able to find their way home without having to ask anyone.

Figure 107: Locations around the home and outside that may be used as monitoring zones

Access to the great outdoors, and exercise generally, is beneficial to the general health and well-being, of people with dementia as well as to their cognitive capabilities. This is probably due to increases in blood flow to the brain enabling more oxygen to be made available. The challenge is often one of supporting this freedom without increasing the risk of accident to such an extent that it might appear to be reckless. In practice, the risks are present everywhere, both inside the home and outside it. Within a property there will be safe spaces or rooms where an individual can be left on their own without great worry. These might include the living room, the bedrooms and the bathroom. But there might also be unsafe area, such as the kitchen, the attic or the basement. Access to these areas may be restricted or an alarm needs to be raised if the individual enters these spaces. Figure 97 extends these principles to the home, the garden, and various locations in the wider world.

Each of the areas defined in Figure 107 can be regarded as location monitoring zones that are defined as being safe or unsafe. When the individual leaves a safe zone and/or enters an unsafe zone, this may be the trigger for an alarm. The earliest property exit devices used within telecare systems could be programmed (or activated by a carer) to raise an alarm if the monitoring individual opened the front door at an inappropriate time, and then left the property without returning within a defined period. This became a successful element in telecare systems. In some cases, a monitoring centre could speak to the individual as soon as they opened the exit door, advising them to go back inside. Distraction technologies might also be used. This could be a pile of magazines left on the inside of the door, or it could be an electronic messaging device such as a Memominder, or an illuminated message board. These could offer simple messages such as "Go back to bed" or "Wait until the milkman has been". Examples of the devices are shown in Figure 108. Some of these can offer different messages when someone is entering the property.

Figure 108: Devices to provide advice to an individual planning to make an exit

When someone lives alone, if they don't take any notice of warning devices, and leave the property, it may be a challenge to locate them. They are not in immediate danger, so it may be premature to launch a full-scale alert, especially if they have more than 10 minutes head-start, and if they are mobile, or close to a bus route or railway station. However, the sooner that their absence is known, then the risk of them coming can be reduced, providing that the family or a responder service is available to look for them and take them back home. In practice, responder services can struggle to find them unless they have information on where they are likely to be heading, and they have a recent photograph (and a description of their clothing) to help identify them. If and when they are found, convincing them to return home may not be trivial, and there remains the possibility that they might go out again within a short period of time. A complete and integrated service might include the provision of places of safety where such people may spend the night in a warm, comfortable but safe environment. This could be in a staffed facility such as a residential care home, a hospital (not acute) or an Extracare scheme. A&E departments are inappropriate, as is the local police station!

There has, fortunately, been some useful development in established shared information to help the police to deal with some of these issues. The Herbert Protocol is a national scheme being introduced by many police forces in partnership with other agencies which encourages carers to compile useful information which could be used in the event of a vulnerable person going missing. Carers, family members and friends can complete a form recording all details, such as medication required, mobile numbers, places previously located, a photograph etc. The Herbert Protocol initiative is named after George Herbert, a War veteran of the Normandy landings, who lived with dementia.

Perhaps the innovation that offers the greatest opportunity for success in rapidly locating people who have gone out (usually at night) and who need to be found before they come to harm are miniature GPS devices. Not only do they have satellite receivers to enable them to calculate their exact position, but they also have mobile communication capability, enabling this information to be sent on request (by a remote carer or monitoring organisation) or automatically if they move outside a programmed safe zone. The challenge is often to find a form factor that is suitable for the user, and also to make sure that it is kept charged and taken with the user every time they go out. Since first being introduced in the early 2000s, their functionality, aesthetics and battery lives have been increased considerably; users can be offered significant choice, but this may be limited by their service providers and by the capability of a monitoring centre that coordinates alerts. Local authorities, and their commissioned service providers and partners (including NHS trusts), usually restrict access to a small number of devices. Individuals may choose to buy their own devices but need to be reminded that they may not be interoperable with other elements of their telecare service.

Figure 109: Watch-based GPS Locators and Alarm Devices

Devices that are worn on the wrist are often the most acceptable type of device for men, partly because they are easy to access, but mainly because they appear natural rather than stigmatising; many can display the time in either digital or analogue form. Unfortunately, they remain relatively thick and the plastic appearance of some models makes them unattractive to most ladies. Figure 109 shows a selection of models that are currently available. Most cost upwards of £200 but also need a contract to cover connectivity and access to monitoring data. Lower-cost devices often have poorer user interfaces and are less accurate and sensitive.

Alternative form factors can resemble a mobile phone so that they can easily be carried in a pocket or in a handbag, or they can be attached to a key-ring or worn around the neck. In each case, the challenge is to ensure that the device is with the individual when they leave their home, and that it is fully charged and available for use. Some of the more popular devices are shown in Figure 110.

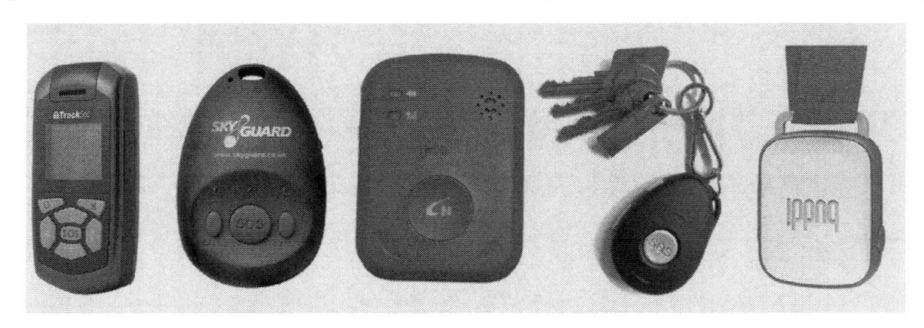

Figure 110: A Variety of GPS locator devices

If these devices are also unsuitable for an individual user, other options are available as examples shown in Figure 111. They are also wearable devices that can be worn naturally as a belt or in a pair of trainers or within as insoles in shoes; but they are not perfect as an individual is likely to have many belts and trousers, as well as many pairs of shoes. There is no way of ensuring that the individuals wear the clothing expected, especially during the night when they might be wearing pyjamas and slippers. The more innovative approaches employ contactless charging options which can greatly increase their usability, and their likelihood of being deployed successfully for extended periods of time. The alarm functions are designed to alert a family member but can be directed to a monitoring centre. However, some devices can only be reprogrammed using the manufacturer or distributor's own (or designated) centre. This can be a limitation for seamless service.

Figure 111: More innovative examples of GPS locators

Most, if not all, GPS devices can be programmed to alarm if the user moves outside a safe zone, though the shape, size, symmetry and time-dependence vary significantly between the devices. The accuracy of the devices also varies according to design and the potential for them to be screened from full satellite coverage by high buildings, tunnels and other obstacles. Thus, the assessment should ensure that the device that appears to be most appropriate from a perspective of appearance is also capable of providing the required functionality – as well as being available for prescription or monitoring in the user's area of the country. Mobile phone reception is a requirement for all alarm types, though there are a small number of devices available that use satellite-based communication systems. These are used mainly by mountaineers and by sailors.

In the future, a GPS-enabled chip may be embedded within an individual's body, perhaps body-heat activated with voice recognition and the ability to track vital signs. This would prove useful for many medical conditions, including heart disease, Parkinson's Disease, Huntington's Disease, Pick's Disease, Creutzfeldt-Jakob Disease, and Wernicke-Korsakoff Syndrome.

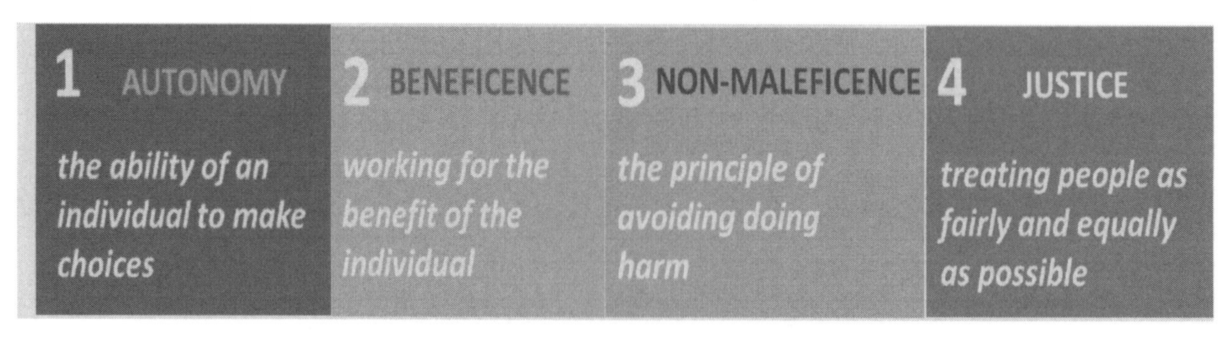

Figure 112: A Four Point Ethical Framework for Using Technology Support Applications

Finally, it is apparent that the use of personal electronic tags, such as GPS locators, poses many ethical questions, some of which are described in Table 42. These relate to the whole composition of the technology service prescription but especially to the role devices and systems that might involve a monitoring of personal information, some of which might be assumed to be private. If the assessor is confident that an item (or items) of technology is genuinely in the best interest of the service user, and its use is limited and proportionate, there remain few people who continue to object to its use. Many service commissioners are advocating the use of an ethical framework that could be used by anyone with a cognitive or learning disability. This would overcome the issue of considering each and every item of technology before a full prescription can be offered. A code of practice based on the principles shown in Figure 112 could be produced at a local level and following a general debate amongst stakeholders. Appropriate technologies might then be proposed without a need to refer to debate each and every item prior to offering choices to the individual and their families.

Table 42: Questions to be Asked Before Using Technology Support Applications

No.	Question
1	Is the service user and/or his/her family involved in designing the care package?
2	Does the service user genuinely understand the capability of the technology?
3	Can the technology really help the person to live better and/or independently?
4	Who is able to connect to the system and receive data and/or alerts?
5	Are the circumstances for data access defined and restricted to maintain privacy?
6	Does the technology impact on the appearance of the home environment?
7	Can the technology be personalised to reflect changing needs?
8	Who benefits the most from the use of the technology?
9	Is there a clear cost-benefits case to support the use of the technology?
10	If the application is provided only for a trial period, what happens afterwards?
11	Is the service user capable of providing informer consent?

The questions shown in Table 42 might also be relevant to other applications of technology, especially those which use applications such as GPS that can provide information on location that the user may not also want. GPS trackers and alarm devices are also being used for lone worker protection, especially in mental health services, and by victims of domestic abuse. In the former case, devices are part of the duty of care that employers have towards their staff. In the latter case, the devices are often provided by (and funded by) the police.

e) <u>Accident prevention</u>

Accidents should, by definition, be random events. However, they always have a cause, and this is often related to an individual (or individuals) not performing a proper risk assessment, not appreciating the hazardous nature of their actions, not being aware of some hazards, or failing to take heed of safety advice that has been offered. Unfortunately, people with a cognitive impairment are more likely to be the victims of accidents for all the above reasons. Technology can help to reduce the likelihood and/or impact of such adverse events if introduced sufficiently early, and in ways that are acceptable.

Figure 113: Low-tech approaches to reducing the risk of electrical accidents

Accidents relating to cooking and to falls have already been described in this chapter, and strategies to minimise the risk and harm have been discussed. There remain a number of common accident types and risks that can also be managed using low-tech approaches that are more accessible because they are usually easy to implement, though some do require an electrician to install. Some of these relate to the use of electrical appliances, including high power household devices such as electric fires, curling tongs, kettles and hair dryers. Figure 113 shows examples of these approaches that can help to restrict access to power-points in locations that are difficult to reach, or which are out of sight, and which might then lead to the devices to be left on. They also include devices or simple labels that can advise the user not to switch off or remove a power plug. These can help avoid fridges and freezers being switched off at night, leading to their contents perishing and also the risk of food poisoning. Mains-operated telecare devices, including dispersed alarm units, need also to be left on at all times, even though their batteries are likely to keep them operational for over 24 hours, during which time they will have sent a 'loss of mains power' alarm to the monitoring centre. Simple timer devices are very popular with Occupational Therapists. Integrated intelligent sockets are available that can be programmed to operate for specific lengths of time.

One of the types of accident that are most feared by landlords are those that involve water. Exposure to hot water, in a bath or a shower, can lead to scalding, while overflows of water due to blocked toilets, overflowing baths/basins, and plumbing leaks can lead to flooding causing structure damage to properties beneath the one in which the flood started. Water can lead to electrical shorts affecting heating supplies, lighting and communication systems, all of which increase the risk of accidents.

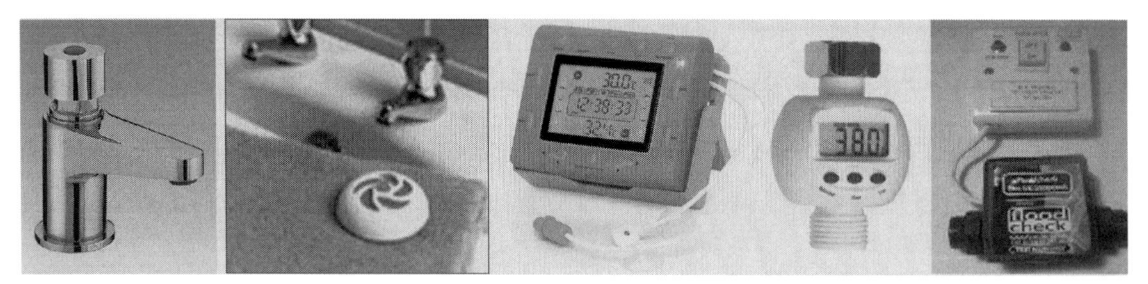

Figure 114: Technology applications for avoiding floods and scalds in the home

Some of the most effective technologies for reducing the risk of such accidents are the simplest, and include the removal of plugs in basins, sinks and baths or their replacement with spring activated plugs (Magiplugs) that compress when the water level (and pressure) reach a defined depth. Taps that provide water for only a short period of time are also useful in wash basins. They are low cost and often used in public wash rooms. Electronic devices can provide alarms for temperature of showers or baths and can also offer alerts when the water level reaches a certain depth. The most sophisticated devices provide automatic switch-off of the water supply when water continues to flow for too long, or when it is sensed in the overflow system. It is fortunate that devices such as Magiplugs are low cost because they are sometimes removed and thrown away by service users who don't know what they are. Examples of these devices are shown in Figure 114.

f) Security concerns

People with dementia may be more likely than others to be victims of bogus callers, who target older people in an attempt to perform a distraction burglary, gaining entry to their home, often with an accomplice, pretending to be from a utility company. While one asks to be shown the power, gas or telephone supply, the other takes the opportunity to search the property for valuables. A number of measures are available to limit the opportunities for such crime, including the use of alert buttons placed on the back of the door. These allow the resident to operate the alarm remotely and silently enabling their monitoring centre to listen in to the conversation and to contact the organisation that the callers claim to represent, in the first instance. If the callers appear to be bogus, then the police can be called, or the monitoring centre can warn them directly that they should exit the property immediately. An example is shown in Figure 115 together with other low-cost devices that can be used to protect the property at the front door; these include peep-holes viewers, intercom devices and a door chain which sprays a mist onto an intruder who might be trying to force their way into the property. It should be added that people with a cognitive impairment are also more likely to leave a window or a door unlocked (or perhaps open) when they are not at home, or during the night when they are at home. This possibility can be managed by linking a door/window status sensor and movement detectors to a telecare system and programming the dispersed alarm unit to provide an alert to the monitoring centre.

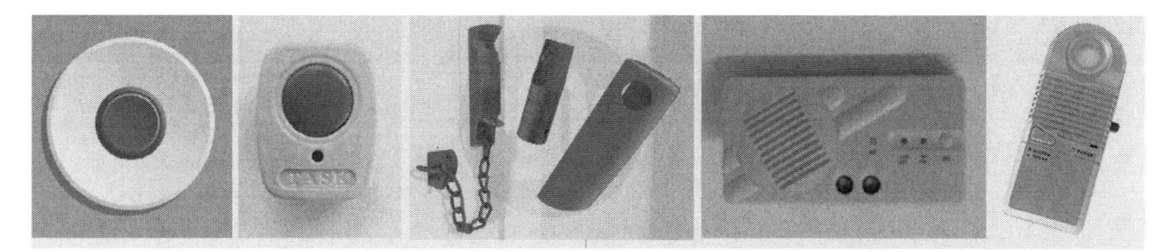

Figure 115: Bogus caller buttons and Low-Cost Units to Improve Home Security

Balancing Risk Management with Other Aspects of Person-Centred Care

The examples shown in the previous section can play an important role in helping people with dementia to retain at least some of their independence by managing the risks that would otherwise harm them or damage their property. However, on their own, they do little to support a higher quality of life and encourage activities and other entertainments that are should be central to person centred care. Tom Kitwood, at the University of Bradford during the 1990s, championed improvements of assessment to achieve these ambitions. Table 43 shows 4 of the needs identified by Kitson, and which can be satisfied, in part, using technology.

Table 43: Some of Tom Kitson's Person Centred Care Needs of People with Dementia

Need for Inclusion – to feel part of a group is essential so that a person doesn't retreat into a world of isolation.
Need for communication – to be able to convey desires even when normal speech may be lost
Need for Attachment – without the reassurance of family and other social bonds, people cannot function well.
Need for Comfort – am inability to continue performing normal activities makes it important to gain comfort from close engagement.

It should be noted that these needs tie in well with the Hierarchy of Needs as developed by Maslow for the whole population (see chapter 6). This helps to identify the interventions that can support the higher-level needs such as belonging, self-esteem and self-actualisation. For example, Alternative and Augmentative Communication aids (see chapter 5) can produce synthesised (or recorded) speech to improve communication, whilst pet and doll therapies can both provide a mechanism for bonding and provide comfort. The Paro Harp seal is the most well-known example of electronic pets, but some success has also been achieved with electronic monkeys. Indeed, Figure 116 shows a range of electronic companion puppies that are orders of magnitude less expensive (~ £30 each compared with several thousand pounds for Paro) but which can provide comfort to some dementia sufferers. These opportunities will be addressed in more detail in Part 2 of this book where future role of technology in residential and nursing homes will be considered.

Figure 116: Low-cost companion puppies currently available

In practice, services provided for dementia sufferers across the country vary both in quality and in scope. Most local authorities offer, through their telecare services, a limited range of connected equipment such as a dispersed alarm unit, portable trigger, a programmable door exit alarm, bed occupancy sensor, and environmental sensor that can quickly detect incidents such as flood, fire, inflammable gas release, and extreme temperatures caused by hot unattended cooker hob. Few can offer access to platforms that feature benefits such as those shown in Table 44 that My Home Helper. Me Maxi or ReMe can provide. Their use is therefore restricted to private payers and to families who can help their loved ones to realise the benefits to ensuring that they are available to share experiences with them in their own homes. Such systems will benefit from a digital transformation of services (see Part 2) but are also idea for supporting the family carer (see Chapter 10). There are costs associated with both the telecare and the digital platform solutions, but these remain small compared with the cost of residential or nursing care.

Table 44: Activities that Can be Supported by Digital Platform Systems

Music	Diary	Entertainment	Communication	Life Threads
Radio	Carers	News	Video calls	School days
Song favourites	Therapy	Messages	Group chats	Family pictures
Karaoke	Appointments	TV	Reminders	Work experience
Quizzes	Food recipes	Films	Voice control	Big days
Dance	Schedules	Adverts	Messaging	Favourite foods

People with Learning Disabilities (LD)

The health of people with learning disabilities is generally poor compared with the general population, as described in Table 9 in Chapter 4. Significant improvement in their care and medical attention has increased life expectancy by many years, though it remains rather less than others. Some experts consider both dementia and conditions such as Down Syndrome to be intellectual disabilities, giving rise to similar impairments such as a lack of reasoning and impaired communication. It follows that some of the issues that arise in the case of people with dementia might also be relevant to people with learning disabilities. This includes diagnostic overshadowing:

1. A person with LD may be unable to articulate pain, discomfort or a general feeling of being unwell – this leads to late presentation and poorer outcomes
2. Presenting symptoms may be associated with “learning disability” rather treatable causes
3. Physical problems such as a deterioration of hearing or sight may not be addresses
4. Psychological problems such as depression, anxiety, dementia or psychosis may not be recognised
5. Social problems such as a change in carer, bereavement or abuse may be ignored

But there are some important similarities but differences also, between people with dementia and those with learning disabilities. These are described in Table 45.

Table 45: Capabilities and Issues Affecting People with Dementia or with Learning Disabilities

People with Dementia	Adults with Learning Disabilities
Struggle to deal with new concepts or devices	Capable of learning new skills
Condition worsens over time as condition progresses	Little deterioration in coping ability over time until/unless symptoms of dementia appear
May lack insight	May lack insight
Short term memory is poor	Independence compromised by over-zealous protection from hate crime
At risk of forgetting sequences, faces and places	At risk of accidents due to lack of practice at domestic tasks and activities
Independence compromised by likelihood of accidents affecting them and others	Able to accept support from personal technologies such as smart phones & apps
Reminiscence therapies using computers can be extremely beneficial and enjoyable	Capable of performing many activities of daily living if given the opportunity
Able to perform many self-help tasks on occasions with prompts	Can adapt well to the use of new devices & alarms in their homes

An increasing number of specialists in neurological disorders are concluding that having an intellectual or learning disabilities is a strong risk factor for developing dementia over time. Dementia commonly occurs in elderly people with intellectual disability, especially those with Down's syndrome. Figure mmm (courtesy of Steve Barnard, HfT) shows estimates by Sadly-Ann Cooper of the age dependence of dementia in the general population, people with learning disability and those with Down's Syndrome. It shows that the dementia curve for people with a learning disability is shifted to lower ages by approximately 10 years for those with a learning disability, and also that the prevalence may be twice as great. The effect is yet more pronounced for people with Down's Syndrome; more than half have dementia before by the time they reach the age of 70.

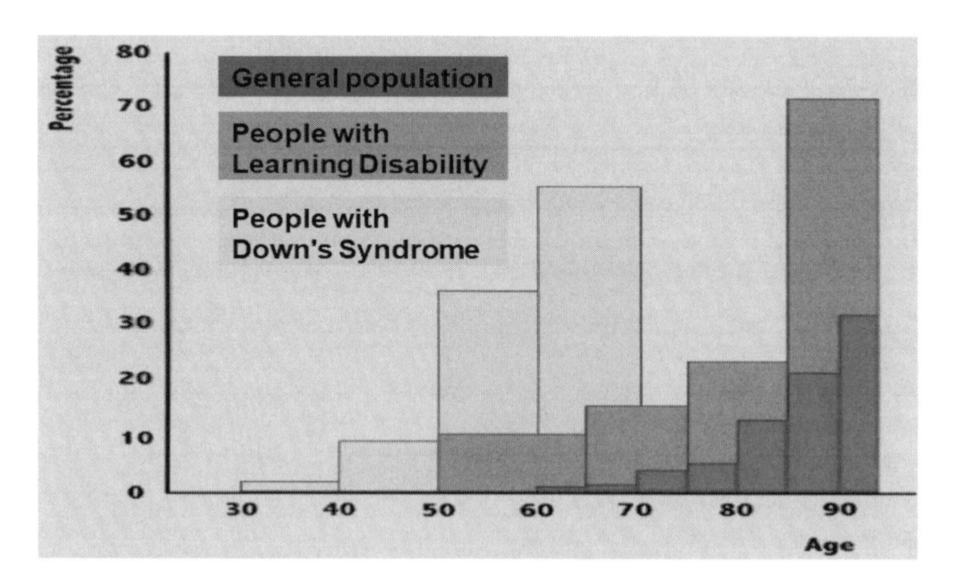

Figure 117: Age dependence of dementia for 3 population groups

Support Arrangements

Many people with learning disabilities live with their families, but others are supported in the community in grouped homes or residential homes. However, there have been significant changes in support services for people with LD in recent years, these changes are likely to be accelerated by the factors described in Table 46. In particular, there have been major shifts from residential care to the development of supported living and other approaches to housing and support provision, (including own tenancies and ownership), which provide for greater independence and choice for service users.

Table 46: Factors influencing the funding landscape for Supporting People with LD

Factor	Explanation
Regulation and Inspection	The Care Quality Commission (and equivalent inspectorates in developed nations) are under pressure to increase regulation and inspection to prevent failings; these measures are likely to increase costs for service providers
Wage and Cost Inflation	Energy, transport and housing costs continue to increase at rates more quickly than inflation; wages will also increase over time
Mobility and Sensory Impairments	People with LD are more likely to have a physical disability, a visual impairment and poor hearing than the general population; these problems will be exacerbated as people with a learning disability age
Welfare Reform	The process should be cost neutral, but many people are losing benefits, causing pressure on support services; older parents may also be affected
Housing Shortages	A reduction in building activity fuels an increase in rental charges making it difficult to find suitable properties for independent living or for small groups; social housing grants to build properties have also been reduced
Mental Health	Efficiency targets imposed on the NHS are reducing the ability of mental health services to provide lower level support; integrated services may reduce inefficiencies but only if the role of technology is recognised
Personalisation	Individuals take advantage of opportunities for defining their own priorities, causing service providers to lose economies of scale. People need expert advice to make informed choices on potential of assistive technologies and telecare
Chronic disease	As the life expectancy of people with LD increases, more will suffer from long term conditions; the management of their health will increase the cost of medication, and medical care without new applications of technology

Supported living enables individuals to have their own tenancies, with the flexibility to move to a new house, or remain in the same house, but change their support provider if that is their wish. Supported living care provision lies outside the scope of the National Minimum Care Standards introduced in 2002. This means that properties may be smaller and more like other forms of general needs housing. Shared accommodation would be possible similar to that occupied by students where a shared bathroom would be more likely and en-suite facilities more unusual. This could reduce costs considerably. However, individual local authorities may specify their own requirements for new supported living and this might include a desire for self-contained flats with separate front door, bathroom and kitchen. Some grant support is necessary to make individual flats affordable and also desirable both from the perspective of the tenant and the housing provider who has fewer problems in managing voids.

The impact of these changes on providers has been considerable, especially if they had invested in the purchase of larger properties for residential care purposes. The potential for offering economies of scale has all but disappeared and they are left with the cost of either splitting properties into 2 or 3 smaller units suitable for sharing by a maximum of 4 tenants, or of planning a more radical remodelling of properties into a number of single tenancy flats or apartments. The cost of conversion can be considerable, especially for the latter option, but can produce a clustering effect which can support a more flexible staff support option, especially when assistive technology and telecare options are included. Indeed, the shared homes option for people with higher support needs might only be financially viable if they also make more use of technology to reduce the need for wakeful or sleep-in support. Dedicated one to one support is strongly advocated by commissioners on the basis that it can provide more 'personalised' and effective support, enabling the individual interests of the service user to be followed separately from others. In such cases, staff need to be able to work with little supervision but in an environment where their safety is protected through lone worker support strategies. Fortunately, the same telecare technologies that can protect service users can also be deployed to support their carers, especially if low level telecarers are employed to respond to emergencies and to provide a low level of shared support to several service users who live in adjacent (within perhaps 200 metres of each other) locations.

Despite clear evidence for success by organisations such as HFT, the uptake of telecare and AT services has remained disappointing, probably due to a lack of awareness of the potential for improved outcomes for all stakeholders including health and social care professionals. There remains considerable work to be done to catch up with the lead provided by older person services including those for people with dementia. However, there are numerous examples of where an increasingly wide range of technological devices and techniques, often combined, have been used successfully to increase the ability of service users to live independently, as well as to be able to monitor their activities, actions, reactions and risks in the home remotely. These enable them to live safely, whilst not needing so much on-site support, thereby reducing both unnecessary staff intrusion and staff support costs. Some applications have enabled people to have more confidence that they can operate unsupported at times whilst reassuring them (and their families) that telecarers will be alerted if problems arise. Examples of success for those living with their own families are provided in Chapter 10. Part 2 of this book will describe the use of more digital technologies that can be used successfully by people with learning disabilities who are living independently in the community, demonstrating that, as a group, they are different to others with cognitive impairments in that they can learn to use new technologies and applications.

Other benefits include an increased amount 'home alone' time where desired, giving service users opportunities to read, listen to music or to catch up on sleep. Personal interactive platforms such as tablet devices, and smartphones together with cool interface devices such as wrist worn communicators, head-up displays and touch sensitive surfaces are all likely to ensure that younger people with learning disabilities will be able to embrace AT in ways that defy the older generations, and that they will become adept at using digital media to exploit the full range of benefits. Specific applications using a new generation of 'apps' will also advance the role of enabling technologies. These can be embraced by people with LD.

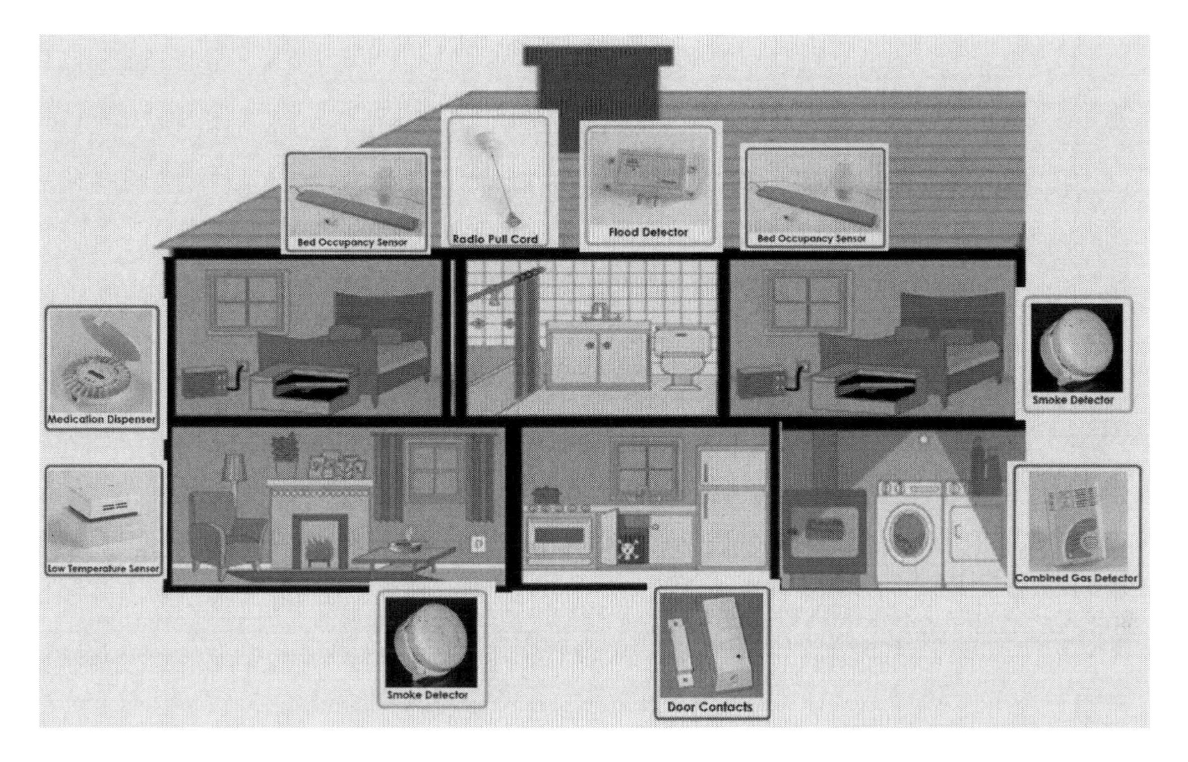

Figure 118: **Smart sensors in a telecare package that supports independence in a tenancy**

Many apps support self-care and can be used by the entire populations; others are being developed specifically to support people with LD to organise schedules, to provide reminders, to offer context-aware speech, and to help with travel using public transport and safe walking. Other apps provide emergency contact and communication options which help to manage personal security issues while GPS enabled smartphones will literally guide their users on journeys to and from home. New applications, such as the Endeavor series, are emerging that will offer video prompts and reminders in the form of complete sequences recorded in the individual's own home using their own appliances and utensils. They will enable users to be able complete complex kitchen tasks ranging from making a cup of tea through to preparing a meal. Of course, they will be relevant to everyone with a cognitive impairment but are likely first to be deployed by people with a learning disability because of their ability to learn to use digital equipment relatively quickly but effectively. Digital transformation will again be fundamental, a theme that will be continued extensively in Part 2.

Back in the analogue world, people with independent tenancies may benefit from a spectrum of telecare devices (as shown in Figure 104). These should be subject to a full and comprehensive assessment to identify the issues that are likely to compromise the ability of an individual to live independently. Telecare systems can be deployed to detect environmental problems such as temperature extremes, flooding and gas emission, and can also manage the risk of being targeted by bogus callers and other criminals. These systems can be enhanced through external cameras and by activity monitoring that can identify changes in lifestyle that could indicate a failure to cope or signs that the individual is becoming socially isolated.

Applications of technology relating to the detection of epileptic seizures during the night by local responders are presented in Chapter 10. It is likely that they, together with other digital platforms and applications will be integrated into home environments so that they combine with other smart elements, reminder systems, scheduling prompts etc to create a scaffolding of support that goes way beyond risk management; intelligent dementia systems will become digital companions that can guide and encourage people to live more normally and by increasingly overcoming the challenges of cognitive impairment.

Concluding Remarks

The challenges posed by poor short-term memory and a lack of reasoning, fundamental of all major neurological conditions, pose a threat to the independence and quality of life of millions of sufferers across the world. Their progressive nature of dementias, and the failure of the pharmaceutical industry to develop drugs that can noticeably slow down, or reverse, these effects must inevitably lead to more use of assistive technologies to support individuals and their families.

Over the past decade, there has been an explosion in the number of devices that have been designed to help manage particular issues. Most are standalone, or non-optimised add-ons to existing telecare systems. They have not yet reached their potential, partly due to a limited assessment process that struggles to identify what would really make a different to their lives, but mainly due to a lack of knowledge of what is available on the market. Individualisation, a major plank of person-centred care, depends on all aspects of support being offered at the right time and in the right way. This means having an intimate knowledge of the individual's preferences and having the patience to try lots of different devices to find the one that works best for that person. This is an inefficient process and is far removed from the standardisation approach which existing telecare services employ as they try to reduce their equipment inventories in order to reduce cost and increase efficiency.

Perhaps the greatest need is to increase the knowledge of assessors, commissioners, families and other support organisations, such as charities, of the range of equipment available, and their success in addressing identified issues. The number of different devices increases so quickly that few people can keep up. In the new digital world, products may have a half-life of less than 2 years, so they are already being superseded before they have been able to prove themselves. Manufacturers cannot be blamed for producing their products in relatively small numbers, and for the implication that this has on the retail price. It is not surprising that those which are sold for a low up-front cost then try to recover cost with higher monthly subscriptions. There must be a role for products to be sold directly to the public through retail operators such as Telmenow and Unforgettable, but there might also be a role for loan libraries which can fill in the gap between private provision and the state. They could be operated by the 3rd sector provided that they have the support of expert occupational therapists.

Chapter 10

Caring for the carers with plesiocare

Introduction

Whilst the focus of providing care and support for older and vulnerable people has been on those commissioned formally by the NHS or by local authorities, it remains the case that more than one in eight members of the public is an informal carer, usually caring for a family member or a friend. They provide a wide range of functions as shown in Figure 119, It may be apparent that those who provide more than 20 hours of care a week include more care functions including personal tasks and medication management.

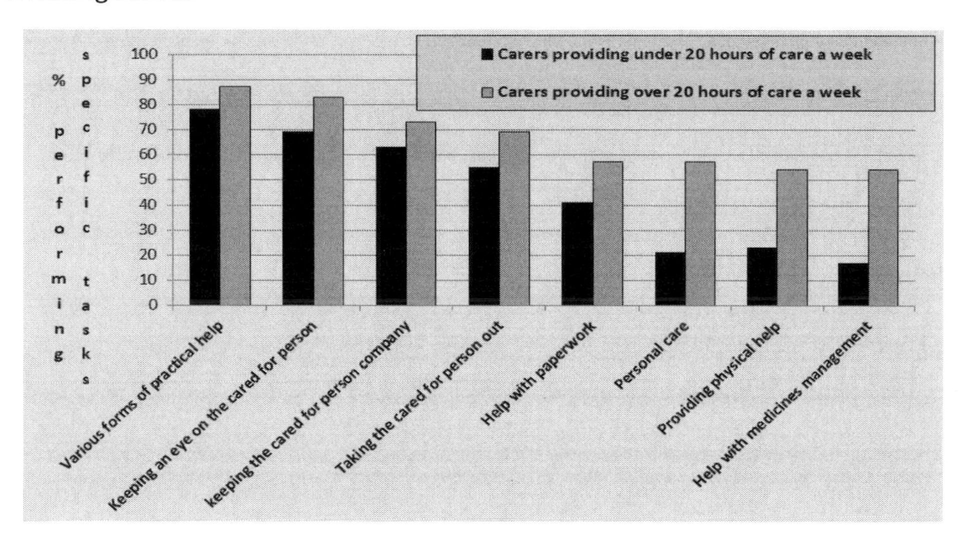

Figure 119: Tasks Performed by Two Groups of Carers

Furthermore, Figure 120 shows that in more than two-thirds of cases informal care is provided for between 1 and 10 years though it Is not unusual for carers to provide more than 20 hours of support for over 20 years. It may be apparent that the number of hours of care increases over the years as the condition of the cared for person declines through age and the progression of underlying illness or disability. Combining observations of Figure 119 with those of Figure 120, it may be concluded that informal carers perform many tasks that might otherwise have to be provided through formal care agencies. This is often (and perhaps probably) paid for by the public purse with some contributions from the individuals themselves, as well as leading to premature use of residential care, and a loss of autonomy and quality of life for the individuals in many cases.

The value of this informal care is enormous and has been compared to the entire cost of running the NHS. It follows that the support of informal carers is essential to reduce the demand on formal care services, and that telecare and assistive technologies can play an important role if introduced appropriately and at a stage where both individuals and their family members can appreciate and accept the benefits. Targeting AT and telecare services at informal carers is therefore an approach which has proved increasingly attractive to local authorities and has been shown to reduce the carer burden in many cases, whilst enabling some of the thousands of informal carers to carry on working when they might otherwise have had to retire with financial loss to themselves and their families, and also to government through loss of taxation income.

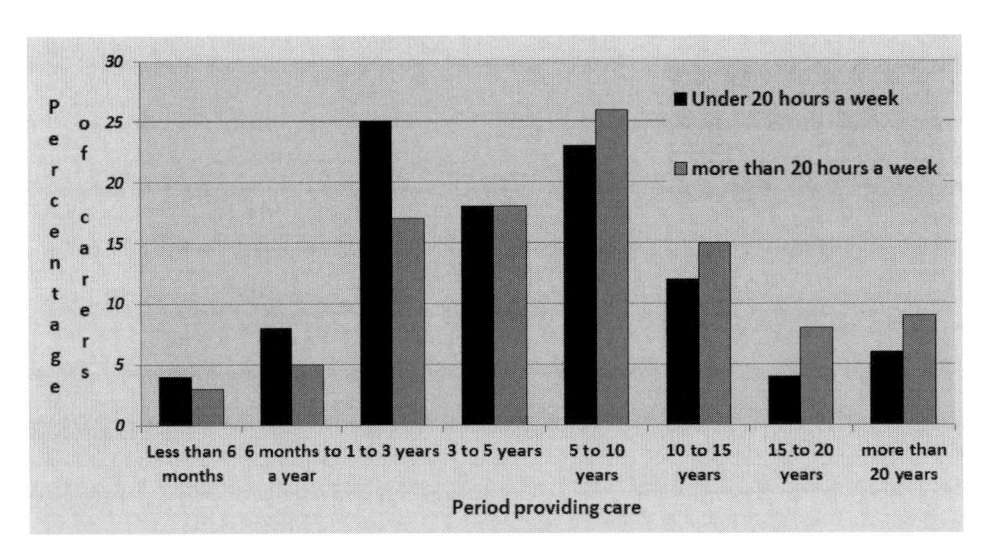

Figure 120: Periods over which informal care is provided

Models of Support

The type of telecare and AT support system described in Chapter 3 could, of course, be used to support anyone who is vulnerable when they are in their own homes, and especially if they live alone. It uses a 24-hour monitoring centre that can alert a family member (or the emergency services) in the event of a problem, and access the property using a key held in a key-safe. But if the family member (or a full-time carer) lives with them, and is close to them all the time, day and night, there is no need to involve a monitoring centre. If their main carer leaves them alone for some of the time, perhaps to go to work or to go shopping, they will still be the most appropriate person to receive an alert in the event of an incident. Again, there may be no role for a monitoring centre unless, for example, the informal carer needs to go away on business or needs respite. Alerts can be delivered using a pager-type device.

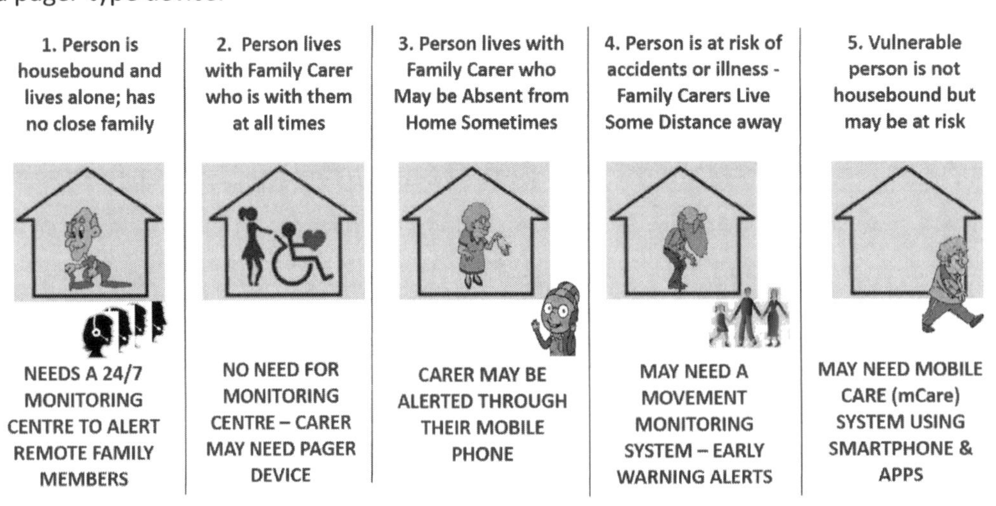

Figure 121: 5 Different Models of Remote Support for Vulnerable People

More generally, people who are not at high risk of having an accident, and who are able to live independently though usually at home for much of the day (and night), may need a continuous monitoring system that includes programmed knowledge of their lifestyle, and will the raise an alert only if a behavioural rule is broken. The technology to perform this level of monitoring was described in Chapter 4. Finally, a vulnerable person who is mobile and able to spend much of their time away from the home, but who is capable of carrying and using a mobile phone, may need a mobile care approach (m-Care). This involves a new range of mobile, often wearable, sensor devices that link wirelessly to the smartphone to detect a range of potential emergency situation including falls, periods of inactivity, convulsive seizures and becoming lost. Their use is to become increasingly important as older people embrace the benefits of the 4th wave of technology support. Five different care scenarios and models of telemonitoring are shown in Figure 107. It follows that assessors need to consider the optimum type of telemonitoring to meet the needs of individuals, as the equipment, the cost and the intrusion levels can be markedly different.

Once again, the availability of equipment for support models other than the first shown in Figure 107 will be incomplete. It will depend on the service provider and their willingness to offer products and services that do not use a 24-hour monitoring centre as the first point of contact for all emergencies. As more families need to consider supporting their vulnerable members using technology, there will be demands for access to other equipment, including those that require mobile connections, and those that need to connect to family members within the home. Thus, the former will lead to a greater use of digital solutions, the latter may involve lower cost options where communication might only be needed between rooms in the same property. This may suit three groups of people:

a) Older people living with their spouse or partner
b) An older person living with their children
c) Disabled children living with their parents; and
d) A disabled person living with their full-time carers.

Plesiocare Technology

Telecare services can be successful for informal carers when they live with the cared for person and are therefore close by to respond to any emergency. Because they are now local rather than distant, the technology to support them should be known as ***plesiocare*** rather than ***telecare*** from the Greek plesio meaning local. We should therefore consider the technology available for plesiocare compared with that used for telecare.

Figure 122 shows the conventional telecare approach to monitoring a situation in the home (upper part of diagram), together with the equivalent plesiocare arrangement. Both include one or more sensor. In the former case, the sensors connect to, and must be compatible with, a dedicated telecare radio transmitter module; this operates at specific frequencies in the UHF band (around 869MHz) or in the VHF band (around 169MHz) where only low power devices for social alarm applications are allowed. This ensures that opportunities for cross-talk are minimised but also limits the scope for mass production, resulting in unit sale prices of over £50 per module. The sensors sold in this way must therefore satisfy the most general and standard telecare applications to ensure cost effective supply. It means that they may not be suitable for bespoke applications unless a universal sensor transmitter module is added (costing in the order of £100). In the same way, the radio receivers must reject all frequencies outside the required band and are therefore Class 1 designs. High quality Dispersed Alarm Units which employ Class 1 designs may cost up to £100, or more if additional features are included; they are considerably more expensive than simpler designs that accept a wider range of frequencies. The telecare system then needs access to the Public Switched Telephone Network (PSTN) which means a line rental as well as a contract with a monitoring centre to open a channel to the home through a land-line telephone. This approach is robust and appropriate to a number of cases where an individual is at considerable risk and where they live alone as it guarantees a successful link from sensor to the carer's telephone in all but the most extreme circumstances. It also offers a remote identification of the source of the alert and provides an escalation route in the event that the carer is unable to answer the telephone, as well as an audit trail for review. It might however lead to an over-prescription of sensor devices simply because the alarm infrastructure is already available through the dispersed alarm unit.

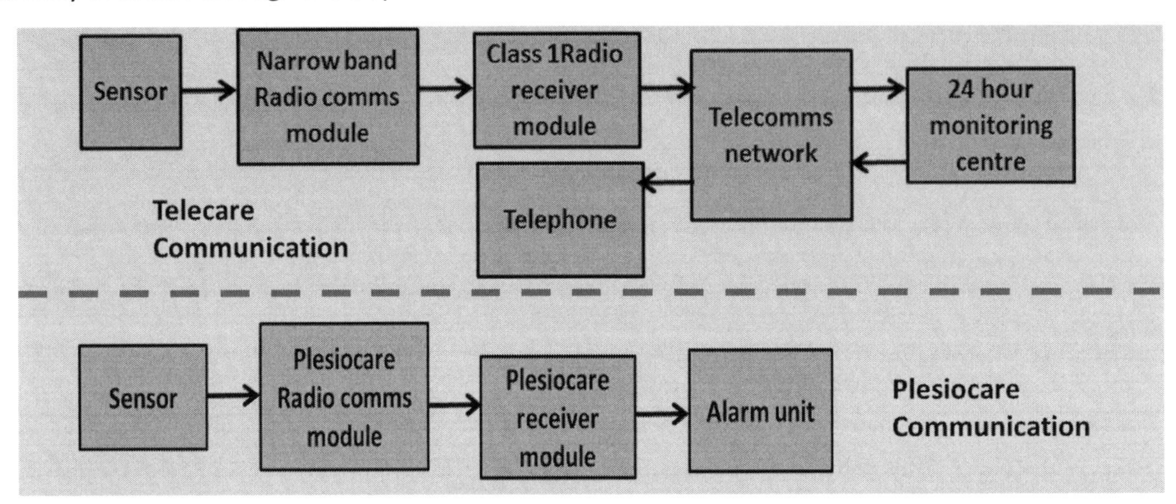

Figure 122: Telecare (upper part) and Plesiocare (lower part) Communication Systems

The Plesiocare arrangement is far simpler, and has fewer elements, because alerts do not leave the property. Plesiocare radio transmitter and receivers could be as simple, and as low cost, as a radio door bell. However, this would not provide a sufficiently reliable approach for most care and other mission critical applications as it could suffer from interference or saturation from other systems (including car alarms and remote controls); these operate at frequencies used by generic security and alarm units. Fortunately, there are reliable radio systems such as Safelink that have designed specifically for high risk medical alarm applications and which use super-heterodyne receivers that are highly selective and sensitive giving good immunity from interference and long range within the home environment. SafeLink uses a balanced modulation called Manchester Encoding, which optimises signal detection. The speed of data transfer is also very low which increases the system immunity to other radio sources and general radio noise. Thus, a SafeLink communication channel is suitable for use with all the sensors that might be used with a telecare system plus a number of personal care sensors or medical devices that may be described below. The reduced number of elements also implies an improved overall reliability at a lower overall cost – and no subscriptions.

It should be added that a telecare system can be changed into a plesiocare system by replacing the displaced alarm unit with a local pager unit that can be left in a central location or carried about by the carer. Examples are shown in Figure 123. Each needs to be recharged on a daily basis. This approach is often used in supported housing arrangements where there may be several vulnerable people being monitored using a range of different sensors. In such cases, members of staff will need to identify both the individual and the nature of an alert before responding as they may need to summon additional help to deal with some types of emergency. The portable receiver devices are considerably more expensive than the dispersed alarm units that they replace.

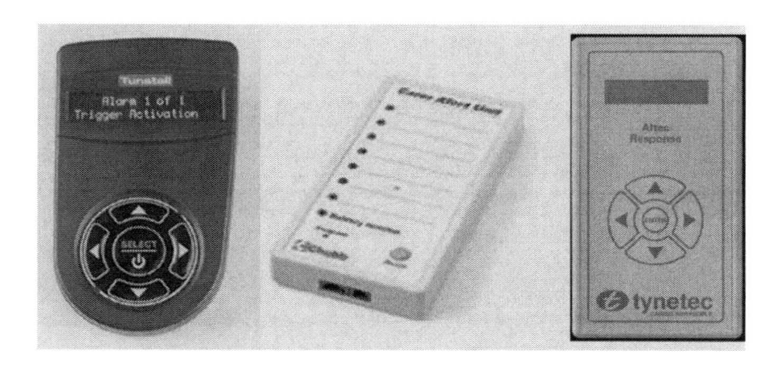

Figure 123: Portable Telecare Alarm Receivers

The use of plesiocare transmission systems enables prescribers to focus on identifying the most appropriate sensing device to meet the exact user conditions. This means that an individual who is at risk of falling if they try to get up out of bed and try to walk without carer assistance, and who sleeps in an electric profiling bed (perhaps with an alternating pressure mattress), can be monitored effectively through combinations of sensors. approaches These would not normally be available using the conventional bed pressure sensors used by major telecare equipment vendors. This personalised approach to technology can be offered more simply and at much lower cost using a plesiocare approach. Thus, plesiocare may replace telecare in many applications involving people with learning disabilities as well as those where an informal carer is present in the same property.

This approach can lead to improved outcomes and lower costs, but the evidence suggests that this may not always be the case because individual assessments are often incomplete or of a poor quality. It is no surprise that the prescription process is also limited, especially when professionals have no knowledge of the complete range of technologies available and of an extended inventory of suitable products. It is clearly of benefit to avoid arrangements that may restrict access to products available through a single manufacturer or source. It follows that commissioners are increasingly aiming to avoid the restrictions on purchasing imposed by old-fashioned framework agreements and to seek supply consortia that can provide a much wider range of devices, with choices where possible.

Applications for Plesiocare

The major uses for system that need plesiocare involve situations where an immediate response is needed to avoid an emergency. This includes potentially risky activities such as:

- trying to get out of a chair or bed without help,
- moving into dangerous areas of the house such as stairs or kitchen without supervision, and
- exiting the property alone perhaps onto a busy road.

The sensor requirements for each of these activities have been described in Chapters 8 and 9. The carer simply needs a receiver that will alert them of the nature of the emergency so that they can act quickly and retrieve their loved one in a timely manner.

Some plesiocare systems use voice messaging receivers as a friendlier approach to alerting a family carer of an incident ranging from a fall through to opening an exit door. Some of the devices are shown in Figure 124. Of particular relevance is the Voice Alert 6, a US security system that can be used with outdoor or indoor detectors. The PIR devices can be put at various locations around the house or grounds enabling a sequence of pre-recorded messages to be played. For example: "Mum is moving around the bedroom"; "Mum is on the landing"; Mum if on the stairs"; " Mum is in the hall"; and "Mum is outside". This enables the carer to choose if and when to intervene. The disadvantage of voice alerts is that they may generate a message that can be heard by visitors, and which may not be discrete e.g. "Mum has wet herself". Other situations, not covered already in this book, can be equally demanding on a carer. A number of solutions are available as described below.

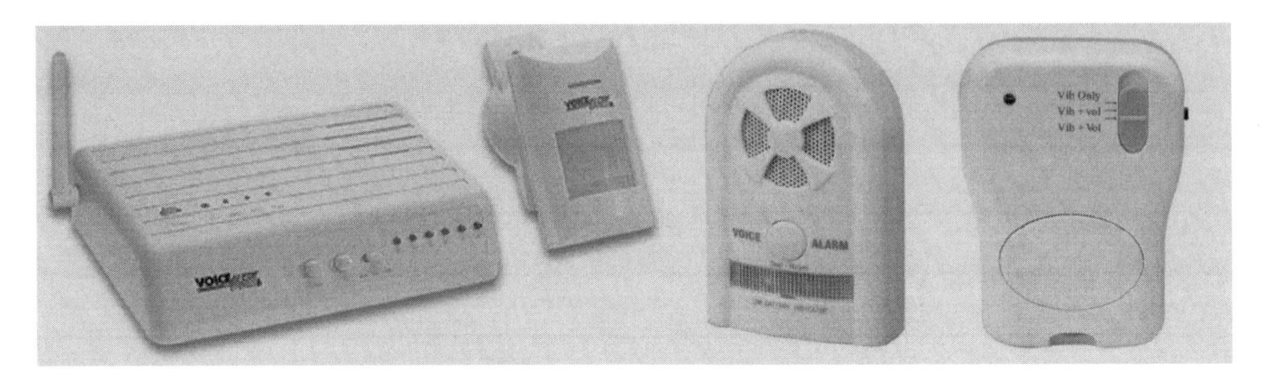

Figure 124: Plesiocare systems with Voice Alerts

Enuresis

This refers to the inability of an individual to control urination. It can also be used for children who have yet to learn bladder control, especially overnight. The term is usually limited to describing people old enough to be expected to exercise such control. Involuntary urination is also known as urinary incontinence (UI) which can be acute, and caused by uncontrolled diabetes, or infections.
However, most continence issues in older people are chronic, and can be characterised by the 5 types described in Table 47. Their impact can be considerable on the quality of life of the individual who will often have a negative view of their overall health status. It can also increase the likelihood of skin viability issues, and the need for careful management of pressure areas (see below). Some researchers have found that emotional well-being, particularly anxiety, frustration, and embarrassment, are common in older people who have had to move in with their children for care during the final years of their lives.

The effects of UI also have a considerable negative impact on family carers. It is not only the physical burden associated with the extra laundry created by having to change the bed and clothing frequently, but also issues created by the role change. Sleep duration and quality, finances, intimacy, and social isolation can also be impacted. There might also be negative emotions such as embarrassment and an on-going debate with a spouse about a care home placement. This creates a wrestling with conscience relating to a potentially distressing outcome for the older person and an often-long-term source of guilt for the carer.

Table 47: Common Types of Chronic UI in Older People

Type of UI	Characteristics of Involuntary Leakage
Urge	Associated with sudden strong need to void perhaps due to overactive bladder
Stress	Caused by sudden actions - coughing, sneezing, laughing or physical exercise
Mixed	A combination of urge and stress UI symptoms
Overflow	Loss of bladder muscle strength causing incomplete emptying of bladder – often a result of surgery such as the removal of prostate, or bowel resection
Functional	Cognitive, functional, dexterity or mobility difficulties that make use of toilet difficult

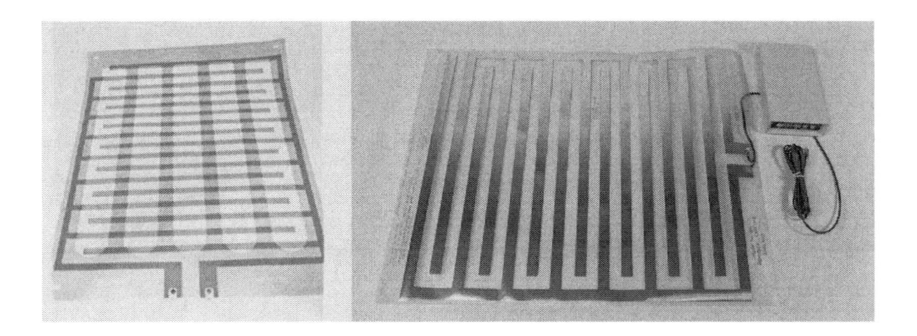

Figure 125: Examples of Enuresis Detection Sheets for Use on Beds

Fortunately, there are a number of different approaches to detecting incidents or enuresis at an early stage. During the night, bed-wetting can be detected using large plastic sheets with a grid of conducting electrodes as shown in Figure 125. These devices are placed beneath the top sheet and above any absorbed layers (such as washable 'Kylie' sheets). Although they respond quickly to moisture, the alarm often sounds after the incident has already led to the nightclothes and the mattress becoming wet and needing to be changed.

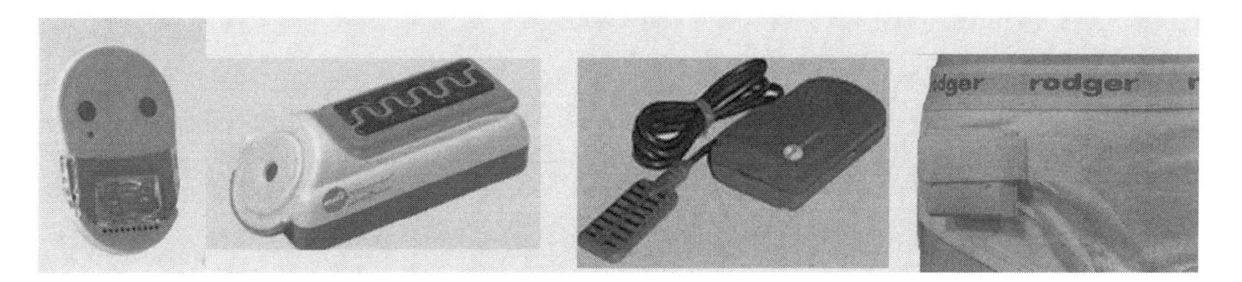

Figure 126: Miniature enuresis detectors and alarms

Miniature sensors are also available to detect the moisture at an earlier stage and may prevent the spread of liquid to bedclothes. The examples shown in Figure 126 are small and can be attached to or worn within underwear either during the day or at night. The Simavita device (extreme left) can transmit the change in moisture nature to a carer using the mobile phone network. Others require the carer to detect the alarm sounds and to respond appropriately. Scientists are developing smart underwear that can send a text messages to warn about protection leaks. The technology uses embedded circuits made of conductive thread. Singapore's National University Hospital is using wearable technology from Dutch company, Lifesense Group to assess an app-based programme for supporting people with stress incontinence to strengthen their pelvic floor and regain bladder control.

Epilepsy

Up to 38% of people with a learning disability live with their parents for all or most of their lives. Nearly 1% of UK adults have some form of epilepsy including about 20% of all people (including children) who have a learning or intellectual disability. Throughout their lives many are likely to experience "Tonic Clonic" (Grand-Mal) seizures. These can lead to injury from a fall or from subsequent reactions to the event. Prolonged seizures can be life-threatening and need to be detected quickly and reliably. These seizures are perhaps the most acutely dangerous of all the issues that parents may face when caring for a child with epilepsy. The fear that they might suffer a seizure during the night can force parents to sleep with their children, leading to poor sleep quality and other disruptions to household routines.

There are many different types of seizure to consider including:

- Generalised: Tonic-clonic;
- Simple: Partial;
- Complex: Partial;
- Tonic;
- Atonic;
- Myoclonic;
- Ictal Tachycardia; and
- Ictal Bradycardia

Some are more difficult to recognise than others, but the convulsions that characterise Tonic-clonic seizures are relatively easy to detect using appropriate sensor technologies, leading to a range of alarm systems that are commercially available. The latest devices employ worn wrist-bands and use companion apps and Internet of Things approaches to management. They were introduced in Chapter 7 and will be discussed further in Part 2 of this book.

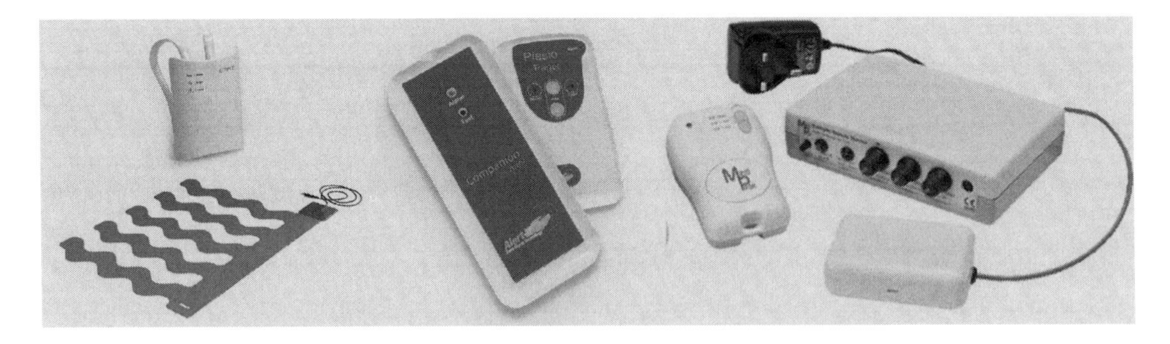

Figure 127: Emfit, Alert-It Mini and Medpage Bed Seizure Alarms

The challenge for seizure detection is knowing the type of seizure, their frequency, intensity and duration. A professional assessment usually leads to both a better selection of monitoring system, and the setting of sensitivity controls that avoid false alarms during restless sleep, for example. Three of the most popular systems are shown in Figure 127, each being compatible with plesiocare arrangements that provide a wireless link to the family carers, who are likely to be in the next bedroom. Other devices provide sensors that can detect excess saliva and noises that may be consistent with some forms of seizure. The correct use of technology can greatly improve management and outcomes.

Others issues that family carers might need to contend with in people who have epilepsy might include any of the following:

- Challenging and Disruptive Behaviour
- Cognition and Insight deficits
- Attention and concentration disorders
- Central coherence deficits
- Executive function problems
- Intellectual ability and memory disfunction
- Functional and Daily Living Skills deficits
- Eating and Feeding problems; and
- Motor Sensory challenges.

For some people, challenging behaviour is wide-ranging and refers to anything an individual may do which is inappropriate. It can include problems as diverse as refusing food, staying awake all night, wetting the bed, removing clothes in public or flicking fingers. The term refers only to those behaviours which are likely to cause significant harm to the individual or to his or her carers. It is restricted to problems such as aggression, self-harm, throwing tantrums or wandering off unsupervised. These are all complex cases that cannot be managed by single sensor solutions. Complete connected (or smart) homes might offer support. Such options are considered in Part 2.

Pressure sore/ulcer prevention

It's estimated that nearly 0.5 million people in the UK will develop at least one pressure ulcer in any given year. They are debilitating and distressing, and can quickly lead to infection, poor quality of life, and premature death. The people who are most likely to be affected are those with an underlying health condition e.g. around 5% of people who are admitted to hospital with a sudden illness will develop a pressure ulcer. People over 75 years old are particularly vulnerable, as they are more likely to have mobility problems and ageing skin. Physically disabled people who spend long period of the day (and night) sitting of lying down are also at enhanced risk. Those most at risk can be provided with pressure relief mattresses in which the air is alternated between cells so that pressure is relieved at points of contact with the bed or chair such as the heels and the base of the back. These devices are known to be effective over long periods of time.

The Waterlow pressure ulcer risk assessment/prevention policy tool has been used for many years in the U.K. to measure the risk using a simple scoring system. Points are awarded for:

- increasing frailty,
- skin type and appearance,
- being older,
- level of nutrition,
- degree of continence; and
- lack of mobility

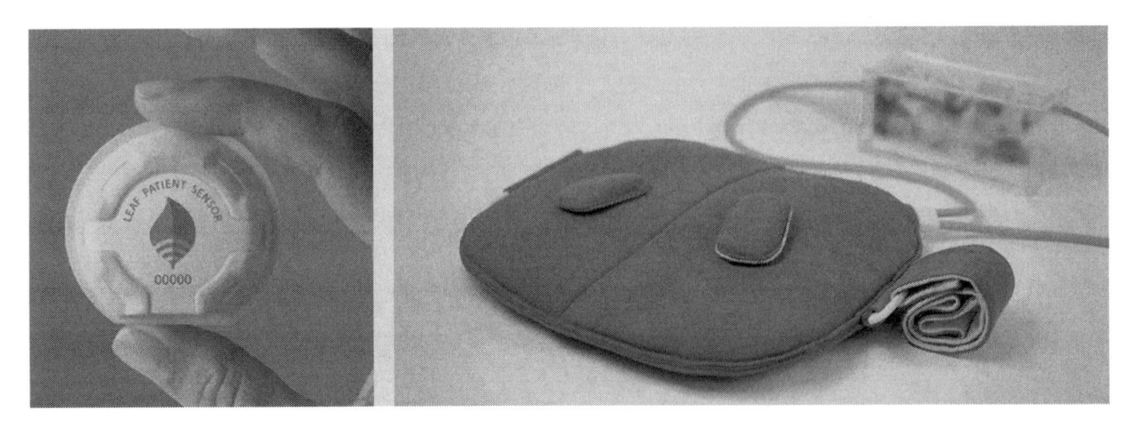

Figure 128: A sensor to remind carers to turn someone, and an automatic turning device

High scores indicate a high risk, indicating a need to make appropriate arrangements to minimise the risk of sore development, and is appropriate for both institutional care and for care in the home environment. The NHS Classic Safety Thermometer is another tool that can be used to identify patients who are at risk of pressure ulcer development. An important part of care by the family and by 24-hour nursing staff, is the need for regular turning, so that the pressure is not concentrated on any part of the body for longer than a certain length of time. The Leaf Patient Sensor System shown in Figure 114 provides a reminder through a pager system if the person should be turned. It is worn on the skin on a flat surface below the clavicle to measure position and movement. It is used successfully in many U.S. hospitals and is being used increasingly to support care in the home.

Also shown in Figure 128 is the Flipod arrangement consisting of 2 air bags, magnets and a pump assembly. The bag is positioned at the back of the patient where the pump simply pumps air into the bag, allowing the body of the patient to be adjusted. Flipod isn't only for lifting the patient's body to the side after prolonged static positions; it also features pulsating bags that allow for dynamic muscle stimulation, which helps scoliotic patients since it's able to relieve various pressure points in a patient's back. The significance of automatic turning devices of this type is that they are both automatic in operation, and thus capable of working on a 24-hour basis, but they also take away from the informal carer the need for strenuous physical effort.

Hypoglycaemia

According to the charity, Diabetes UK, more people than ever have diabetes partly because the population is ageing, and partly because of increasing levels of obesity and sedentary lifestyles. If nothing changes, more than five million people will have diabetes in the UK by 2025. Around 90% of these will have Type 2 diabetes while the remainder have Type 1 diabetes. The latter are more likely to need to inject with insulin several times a day in order to keep their blood glucose levels down and under control, though many older people also need insulin when oral medication fails to manage their blood glucose levels. The need for insulin can change irregularly during the day and the night, sometime causing blood sugar/levels to fall too low. A low blood sugar level is called hypoglycaemia or a "hypo" resulting in different symptoms for different people. These may change over time, making management very difficult for carers.

If not treated in a timely manner, a number of symptoms may appear; these include:

- general weakness
- blurred vision
- difficulty concentrating
- confusion
- unusual behaviour, slurred speech or clumsiness
- feeling sleepy
- experiencing fits or seizures; or
- collapsing or passing out

These effects can be dangerous if the individual has hypos while sleeping. This is especially true for vulnerable people who are unable to take responsive action without help. A carer needs to be made aware of the hypo so that they can administer a sugary liquid or sweet very quickly. The devices shown in Figure 115 are effectively cold sweat alarms that are worn on the wrist. They produce an audible alarm in many people with diabetes when they detect the physiological effects that are likely to accompany a hypoglycaemic incident. They are not medical devices but can be very useful aids to self-care. Their current limitation is that they don't link into a plesiocare system and can't therefore inform an informal carer directly. However, they can be used with a simple intercom system (such as a baby monitor); more advanced sensor systems that detects the sounds and automatically identifies them as alarm triggers might also be employed to alert the carer.

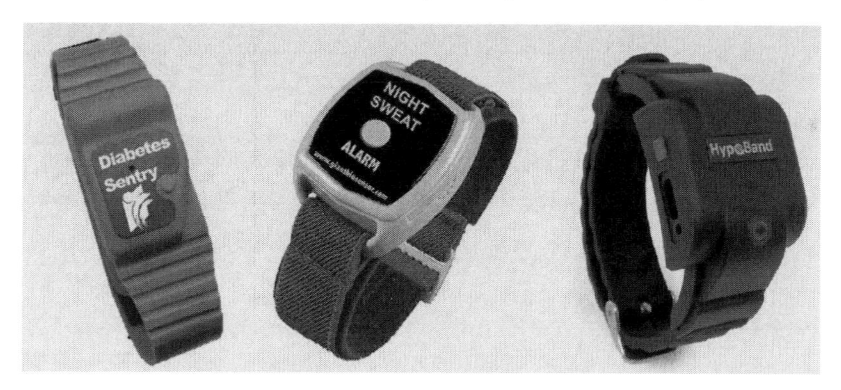

Figure 129: Types of Hypoglycaemic Event Alerts that Work for Some People

Autistic Spectrum Disorders (ASD)

The range of conditions and syndromes that affect an individual's behaviour and lifestyle is called Autistic Spectrum Disorders. Their prevalence is about 1%, and they appear to be rather more common in males than in females. There's no "cure" for ASD (nor is it a disease), but speech and language therapy, occupational therapy, educational support, plus other interventions, including technology support can help with their management. Some are aimed at the individuals, but many can help parents and other informal carers to reduce the burden of care that they might experience.

People with ASD tend to have problems with social interaction and with the development of hobbies and interests. Communication can also be a problem, especially in the use of non-verbal behaviours such as body language and gestures to interact with others. They might not use eye contact and, at times, ignore people, irrespective of if they know them. Children with ASD tend to play alone, while adults have difficulty in making friends and in becoming a member of a team or to interact with fellow employees in an organisation. All groups like to stick with a known regime and are reluctant to try new things. They are a challenge to their parents and families and need special attention in order that they can be supported in their development, learning and employment.

Figure 130: Alarm devices to locate a child who has wandered away from their carer

One of the behaviours that causing parents and family carers most concern is that a child with autism will wander off in shops or in busy places. The child would not understand the risks, nor the distress caused to their carers, and would be unlikely to ask for help. A range of alarm devices has been used with some success, and without resorting to the use of GPS trackers that would not work indoors (such as in a large supermarket or shopping centre). Three devices are shown in Figure 130 each in two parts, one for the child and one for the carer. The Helix Child Alert (shown on the left of Figure 116) uses a fixed 10 metres range while the Teddy Bear child alarm and the Bosieboo Out of Distance monitor operate at line of sight distances of up to 50 metres.

A number of apps have been developed in recent years to support the development of children with an autistic spectrum disorder, including the There are dozens of resources that list examples of apps including coordinated efforts in some countries to use the Padogogy Wheel approach to provide such information. One of the best known is ***Apps for Students with Autism Spectrum Disorders v.4*** shown in Figure 131 by Mark Coppin. It offers 10 or more apps, in 6 different domains – Communication, Social Skills, Need for Sameness or Routine, Sensory Sensitivity, Difficulty with traditional Learning Methods, and Behaviour.

The wheel extends outwards from a core of common learning characteristics through Common Learning Traits to App Categories, before providing some examples of apps that can provide an appropriate starting point. Other apps can then be deployed to build on developing skills or for personalisation through the learning process. Dedicated apps that can be used for self-care and for improving quality of life of older people who have autism and other issues that limit their well-being will be described in Part 2 of this book.

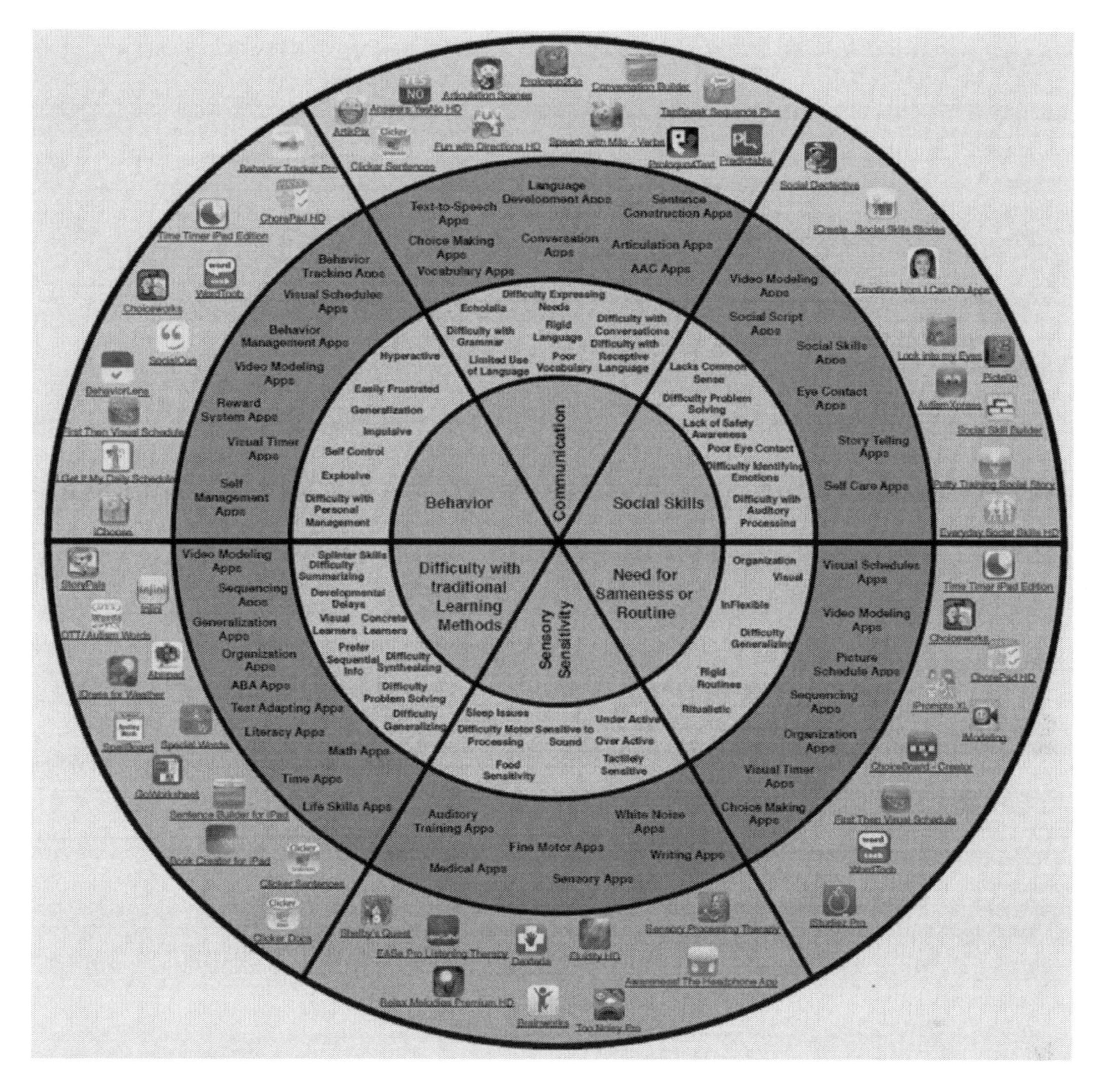

Figure 131: Apps for Students with Autism Spectrum Disorders (version 4 – 2016)

Caring in the Home for People with Dementia

Many of the technology support options described in Chapters 8 and 9 are for people who live alone for all (or much of) the time. They are also relevant to someone who has a cognitive impairment but lives with an informal carer. The communication technology is plesiocare, but the sensors are often similar to those used for telecare. Some of the care challenges that have been described previously in this chapter will also be applicable to people whose primary diagnosis is dementia; they can also suffer from enuresis which is a common symptom of dementia as it develops. The solutions offered may be essential to managing their care.

However, family carers also have to deal with the added burden of keeping their loved one occupied and entertained, both to avoid depression, and to suppress the possibility of them going out at inappropriate time when there is nobody to accompany them, or not sleeping properly during the night, and losing track of time. The Café culture approach shown in Table 48 is a useful way of designing the home environment to be more supportive of the needs of someone with dementia.

Table 48: The Café Culture for the Home Environment

Acronym	Description	Examples
Calm	Moving slowly to avoid surprises and accidents	Use of soft music, aromatherapy and singing. Choosing reds and yellows rather than greens and blues
Appropriate	Using visual signs and cues rather than verbal signage	Photos on doors of bedroom, bathroom etc. Good signposting below usual level
Familiar	Keeping things homely rather than institutional	Analogue rather than digital clocks; glass-fronted fridge
Ensuring Safety and Stimulation	Remove hazards to promote use of garden and open spaces	Replace steps with ramps; create paths that lead back to the door; create interesting views from accessible windows

More generally, many activities will stimulate the mind, and will prolong the period where the individual fees better in themselves. These can build on existing interests whenever possible, sometimes reminding the individual of their own past or childhood. Unfortunately, many older people (including those with a cognitive impairment) find technology difficult. Touchscreen devices are changing this, as they are tactile, versatile and very simple to use and the apps make it easy to play accessible games, which help memory and interaction. The potential is greatest when an informal carer helps them to access and use these devices and applications. The need for volunteers (and the entire community) to help the carer may be apparent and is discussed further in Part 2.

Cooking and domestic tasks

The carer will, on occasions, need to perform some tasks for their loved one, especially if preparing involves potentially dangerous tasks such as using a sharp knife for dicing, slicing and peeling. It may be easier to supervise these actions, providing prompts when the next action is not apparent. Apps are now available that allow such actions to be recorded using video, and then played back as required. This is relatively easy provided that the recipes selected are relatively simple, and if the carer has received appropriate training in developing the content. Once more, some of these new digital approaches will be described in Part 2.

Puzzles and games

Any kind of pastime that keeps the mind active is beneficial and can also be a great way of spending time with other people. Although performing puzzles, such as crosswords and Sudoku, have been individual pursuits, they can be continued with support from a carer as a means of very informal support and encouragement. Boards games can be equally satisfying. The Activity and Reminiscence Year book us a comprehensive source of original material with a different theme for each week which can then be personalised with local events and individual anniversaries.

A number of electronic games are also available that offer both visual and speech outputs to encourage engagement. Many companies and support organisations make games and puzzles specifically for people older people and for those with dementia. Some examples of puzzles and quizzes are shown in Figure 132. Jigsaws are considered to be particularly useful because the number of pieces can be varied from 100 down to a dozen or less. They resemble the puzzles used for children, but the scenes can be more adult. Some organisations will create a jigsaw from a supplied image, enabling a memory to be reinforced.

Figure 132: Jigsaws and puzzles to exercise the mind of someone with dementia

Technology for Entertainment

There are many virtual puzzles that can be displayed on a tablet computing device. Tablet computers have many advantages over desktop computers for an informal carer. They are smaller and portable, enabling their use in the lounge, on a sofa, rather than on a desk. They also use touchscreen technology to enable users to use the fingers for control and selection, rather than using a mouse or keyboard. It is possible therefore to both search the Internet for YouTube videos, to communicate face-to-face with relatives and friends, for running apps that run games, and other past-times that are suitable for someone with a cognitive impairment. Figure 133 show some popular examples for the iPad, including several that allows virtual activities such as growing flowers, making pottery and fishing, to be performed from the comfort of the living room and without making a mess. Tablet devices also enable books (as well as newspapers and magazines) to be read out aloud, again easing the burden on the carer.

Figure 133: Screenshots from popular apps for carers to use with dementia sufferers

Some of the more sophisticated apps can be used for multiple entertainment and support purposes. For example, RemindMeCare provides a number of specific activities as shown in Table 49. MindMate is a three-fold app that offer games, exercise and TV and music activities to stimulate the brain and to offer gentle reminders.

Table 49: Activities that are enabled through RemindMeCare (ReMe)

Activity	Description
Image therapy	Uses pictures and videos to explore experiences to help with recall and to support reassurance and engagement
Calming strategies	Uses familiar content to reassure, and then helps find out what works
Music reminiscence	Supports the creation of a playlist of musical favourites that is easily accessible by the carer for entertainment or therapy
Family resources	Allows easy uploading of content that can then be used alongside a calendar to explore key dates and events such as birthdays and Christmas
Group sessions	Allows the carer to present favourite songs, music and pictures to present to other family members based on the loved one's memories

MyHomeHelper, shown in Figure 134, is a support system based on a tablet computer that aims to support the individual with a cognitive impairment (see chapter 9), but which can be equally successful for use by the carer. In addition to providing a clock and visual reminders (similar to standalone products and reminders also discussed in Chapter 9), it can provide news updates in the same way as personal electronic assistants, such as Amazon's Echo Screen and Show, can offer, because it is linked to the Internet and to news feeds. It can also provide a calendar of events and offers many of the functions that would otherwise require separate apps.

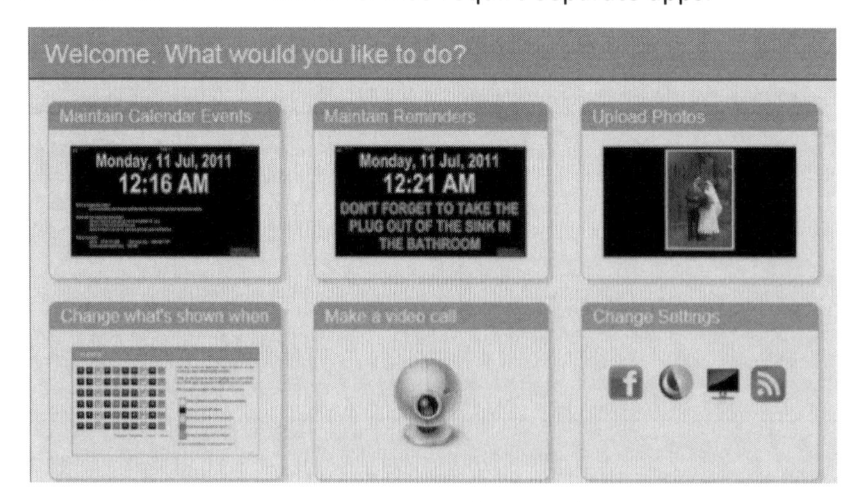

Figure 134: The MyHomeHelper Welcome Screen

Life story and reminiscence

The importance of knowing the individual was brought out in Chapter 6 as a means of improving assessment processes. The family carer is in a unique position to help their loved one to develop the story of their life whether as a scrapbook, a photo album or an enhanced diary. Apps and digital technologies provide a medium to record and remember details of one's life, including experiences, values and beliefs and how they might have changed over the course of the years. The work is best performed as a joint activity that can lead to reminiscence sessions during which things from the past can be talked about using the prompts provided by music, pictures, films and objects. When complete, it can paint a more relevant picture of the individual that they can look through themselves as a reminder of what has mattered to them previously, and as an indication of their experiences and achievements.

One excellence resource to support the Life Story development is ***Book of You*** which is an online resource which brings families, friends and carers together to share an individual's key moments in a simple way making storytelling easy, fun and beneficial. It uses words, pictures, music and film to help show who a person was and who they are now. In particular, it can become the means to conduct on-going reminiscence sessions which have many benefits including:

- Encouraging and enhancing communication
- Improve mood and well-being
- Stimulate memories and emotions
- Being tailored to an individual and their stage of memory loss
- Serve as a legacy for the family including future generations
- Maintaining or improving a sense of personal identity; and
- Providing the knowledge of the person which is essential in providing person-centred care

Reminiscence therapy has become an important component of care in specialist facilities and homes for people with dementia. It will be considered in more depth in Part 2 of the book, along with the role of dolls and electronic pets in provide providing people with a means of attachment to something physical that can be held and stroked.

Carer Support Networks

The challenges facing a carer can lead to a need for improved time management, and prioritisation of both household and social activities. Although hobbies, such as reading, swimming, walking, going to the cinema, being a member of a choir, and going out to meet friends may not be considered as essential, they are central to the carer's well-being, self-esteem and independence. It is important that time is set aside for the carer's social life. This is where other family members and friends have a role to play. Unfortunately, when families are geographically scattered around the country, this is more difficult. But there are local charities that can provide sitting services, and which can also take the loved one out occasionally for a car ride or a visit to a community centre. More volunteering roles are considered in Part 2, along with town or community initiatives that can make areas more age-friendly or dementia-friendly.

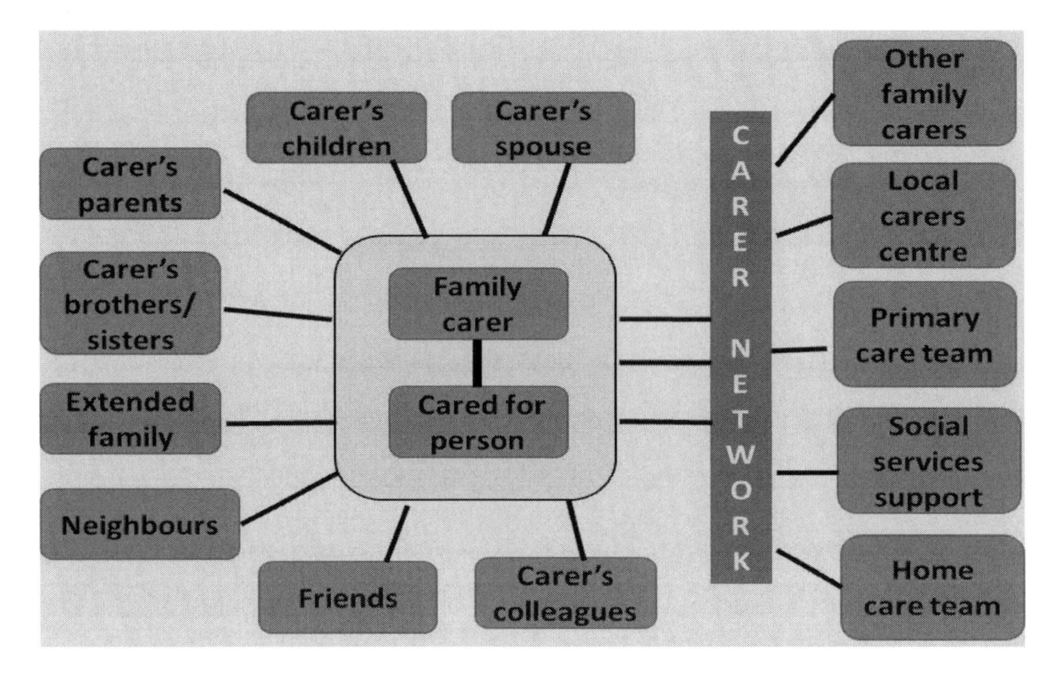

Figure 135: A Carers Support Network Based on Telecommunications

The challenge remains to mobilise the resources and to coordinate the care function. Fortunately, technology can help in this respect, as well as providing additional support from other agencies in the event of an emergency. Figure 135 shows a network of support that has been designed to meet this need. The methods that are available involve websites and hardware that enable the sharing of photographs, tasks, stories and information. These can be as simple as an app which can be run on a smartphone or tablet device, through to TV set-top boxes, controllers and cameras that can be used to extend telephone conversations through the use of video (5th wave applications). Although many of these applications focus on the needs of family carers who are supporting someone who has dementia, their application is more extensive.

Carers need to be afforded the means to have holidays and rest time away from their caring responsibilities which would otherwise be continuous and, potentially, overpowering. Practical Digital Respite facilities might include hotels, chalets or perhaps caravans that have been suitably adapted to ensure that they can be used by families who have a disabled member. However, respite can be virtual by giving a carer the opportunity to used digital technologies to share their burden with an extended family group and with other members of the community. This is the basis for a number of apps that have been developed specifically to benefit carers by enabling them to seek help from others and to engage with them in developing shared care plans. Some examples of apps are described in Table 47, which also shows whether they are free to use or part of a subscription service. The trend is to move from individual apps into digital platform applications on which several apps may be launched. These device-agnostic approaches allow carers and their family members to use their own devices and to build digital toolboxes that will support their caring duties.

Table 50: Apps and Platforms that Can Support Family Carers in Need for Respite

Name of App	Features	URL	Condition Focus	Cost etc.
Oscar Senior	A secure platform, with large prompts, photo-sharing, video communications	https://login.oscarsenior.com	Older people	Free
Jointly	Developed through Carer Charities – offers group communication, task lists, calendars, profiles, medications, contact list etc	https://www.jointlyapp.com/#welcome	All	£2.99
Rallyround	Offers response service, and allows group support for tasks, discussions and volunteers, reminders & alerts in funded localities	https://www.rallyroundme.com/welcome	All	LAs or CCGs licence
Caring Ties	Designed for family carers – offers Reminders, Note taking, Information sharing, Medication list and prompts	https://www.facebook.com/CaringTies	All	Free - web based
Lotsa Helping Hands	Provides a care calendar through a supporting community including requests for support such as with meals, rides, or visits	http://lotsahelpinghands.com/	All	Free
Nourish	Enables circles of care to connect easily to keep families informed about eating, health and status	www.nourishcare.co.uk/familu-care/	All	Not yet priced
Tyze	Communicate with family, friends and helpers, Schedule appointments & events on a shared calendar, photos, updates	http://tyze.com/	All	Free
CareZone	Creates patient profiles, medication info and schedule reminders, share to-do lists, journal entries, photo albums, important files, and updates	https://carezone.com/home	All	Free
Konnektis	Platform for providing complete picture of needs, from nutrition and hydration through to sleep and missed care visits	http://www.konnektis.com/	All	Free
Magic window	Views of beautiful landscapes and simulation of passage of time through time-lapse photography	http://www.jetsoncreative.com/apps/	House-bound	£7.99 - iTunes

The apps and platforms described in Table 50 are designed for people who already have smartphones and/or tablet devices and laptops. Despite a big increase in the number of older people who have such devices of their own, there remain many who either cannot afford the investment, or who prefer to have separate devices for their caring responsibilities. ***MyHomeHelper*** is an example of such a device and has been described previously in this chapter. Three other devices for carers began as dedicated tablet devices but are now available with different options – ***MimoConnect***, ***MeMaxi***, and ***Respexi***. The tablet version of all three devices are shown in Figure 136.

Figure 136: The MimoConnect, MeMaxi, and Respexi Tablet Carer Support Systems

MimoConnect addresses social isolation by creating a simple family social channel. A simple touch screen for the elder and family and friends can be involved in downloading memories and familiar days from the past. They can talk, exchange family videos and photos, download eBooks and Music memories. All on a private family network. Talking timed messages can be posted on a daily reminder sheet. Mum or Grandad can now hear a familiar voice reminding them about something special or just something that needs doing.

Memaxi is a communications and care network that is created for someone requiring long-term assistance or care. Users can access Memaxi via a specialised web portal or through the complimentary app which has two sides to it. Firstly, the display mode, which is intended to be used by the one who needs assistance and, secondly, the management mode which is used by family, friends and health and social care professionals. The Memaxi display usually runs on a dedicated tablet device to create a fixed point of reference in the home for everything relating to the daily support plan, rotas or schedules.

Respexi is a fully supported care service, providing the tablet user and their family, friends and professional care groups with the capability to stay in touch, prevent isolation, support independence and offer peace of mind to all. The respexiTM service sets reminders and alerts family / carers of any that are missed by the user; offers simple radio, video calling and browsing of photos sent in real time over a broadband connection with easy access to on device assistance and a professional support help desk. It has been created by Sight and Sound, a company with an established history of supporting people who have vision or hearing disabilities. The device is therefore ideal for someone who has a visual impairment due to its ability to provide offer display changes and colour preferences.

Final Comments

It is apparent that many organisations have been making considerable efforts to support informal carers through technology. This isn't surprising in view of the huge level of care that family members provide, and which therefore saves millions (or perhaps billions) of pounds every year for both the NHS and for local authorities. Yet, the processes that are employed in assessment are poorly developed as far as offering technological solutions, equipment and support. Indeed, few local authorities are able to provide plesiocare items, and are far more likely to offer more expensive telecare equipment.

Ideally, service providers, both in-house teams and outsourced organisations, should be involved in coordinating equipment provision, and in ensuring that carers are prescribed the most appropriate equipment and service, including maintenance and support. The alternative is that carers will purchase their own devices, which may be sub-optimal, and which might offer only a partial solution. Table 51 provides a summary of some solutions that can be used by family members locally or remotely to support their loved ones.

Table 51: Technologies that Can Support the Monitoring of Family Members by a Remote Carer

Technology type	Description	Examples	Benefits	Limitations
Emergency communication	Means of raising an alarm rapidly	OwnFone mPERS smartphone apps	Allows user to easily inform carers of accidents	Device may need to replace successful older technology
Welfare checks	Automated check-up communication	"are you ok?" calls Text messages Proactive calling Acticheck band	Requires proactive response; offers a preventive approach	User may choose not to be honest Requires correct telecommunications
Medication systems	Automated reminders and alarms	Pivotell dispenser Cadex watch alarm	Can improve medication adherence	Devices may need to be loaded and/or programmed
Movement trackers	Systems showing room occupancy	Just Checking Canary TextCare	Show clearly where person spends their time	Provides accurate data for person living alone
Activity tracker systems	Monitors of domestic tasks	Lively 3rings Alcove	Allows monitoring of specific tasks	Activities cannot be shown on single screen
Location alerts and trackers	Mobile GPS trackers and alarm device	Bushnell backtracker; Doro Secure 480	Provide accurate location information	Need to be carried and kept charged
Vital signs monitoring	Medical sensors and data collection	Pulse oximeter MedWand	Supports self-management of chronic disease	Need clinical input for interpretation of comorbid diseases
Video camera systems	Remote monitoring of rooms or entrance	Nest Cam Digital Peephole Door Viewer	Gives carers peace of mind regarding who is present	Ethical issues regarding intrusion and privacy

People who devote their lives to the care and support of a loved one can often feel lonely and socially isolated because they lose contact with their friends and aren't able to continue with their own social agendas. This is a source of loneliness in some people – a topic that will be considered in depth in Chapter 11.

Chapter 11

Social isolation, loneliness and rurality

Introduction

The subject of loneliness has attracted considerable interest for over a decade both in the UK and in other developed countries; it could be an unintended consequence of greater life expectancy, of losing a lifelong partner and being left alone for many years, or indeed of divorce and separation in later life. This has led to the publication of several reports and manifestoes that have reviewed the evidence and concluded that the impact can often be devastating if feelings of loneliness persist over extended periods of time. The effects can be both physical and psychological; they may be comparable in scale to other public health issues such as obesity and smoking. This is due to the complex interplay of health and contextual factors that appear to put some people, and especially those who are older, at risk of loneliness. Particular factors that are relevant include a relocation to a care home, the loss of a loved one who can share confidences and feelings, and spending much (too much) of the time alone and with nobody to talk to.

Table 52 The scale and impact of the loneliness problem in the UK

Numbers or Percentages of Population Affected	Source
>9 million adults say that they are often or always lonely	British Red Cross & Co-op
3.6 million people aged 65+ have TV as their main form of company	Age UK
50% of disabled people are lonely on any given day	Sense
>1 in 10 men say they are lonely, but would not admit it	Royal Voluntary Service
75% of GPs say they see between 1 and 5 people a day who attend their surgery mainly because they are lonely	Campaign to End Loneliness
38% of people with dementia say they lost friends after diagnosis	Alzheimer's UK
> 1 in 3 people aged 75+ say that feelings of loneliness are out of their control.	Independent Age

Loneliness is a subjective feeling of unwanted isolation and a lack of companionship. The evidence suggests that some groups of people are more at risk than others as highlighted in Table 52 which is compiled from research by 3rd sector organisations. The groups at higher risk include:

- Physically disabled people, and those with mobility issues;
- People with sensory/speech/communication impairments, or mental health issues;
- People who have long term conditions or who are in poor health [9];
- Those who live alone, especially following a bereavement of a partner, a relative or friend;
- People suffering from dementia or other cognitive impairments [8];
- People who live in deprived urban environments or in rural areas;
- Those who live in, rather than close to, the home of an adult child;
- Tenants or owner occupiers of sub-standard homes;

- The oldest people (i.e. those aged 85+) who have outlived most of their friends;
- Those who have migrated to another area or community following retirement;
- People with low incomes or few assets, including common consumer goods;
- Those who are socially (or digitally) isolated with little access to information sources;
- Family or other long-term informal carers;
- People from some (but not all) minority ethnic communities;
- Gay men and lesbians or people with transgender sexual preferences; and
- People who have a self-sufficient personality.

The above factors have statistical significance, especially with regard to social isolation, which is a more objective term, and which is influenced by social networks, contacts and support mechanisms. But they are not directly responsible for an individual feeling lonely, which is also dependent on their attitude and need for social contact, intimacy and self-worth. It should be emphasised that people can feel lonely when they are surrounded by people, while others who are socially isolated (in the sense that they receive few if any visitors) may not feel lonely. Interventions aimed at reducing loneliness must therefore be clear that they are not simply addressing issues of connectedness without understanding why people feel lonely.

People can also be socially isolated due to geography, and by the physical distance between them, their neighbours, their families and other support groups. In practice, people choose to live in rural areas where there are different challenges in establishing viable communities, housing choices, and networks of communication. This can impact the success of interventions, and of the potential for using technology as a means of reducing the impact. These issues are discussed in more details at the end of this chapter.

Why Are People Lonely?

According to the English Longitudinal Study of Ageing, there are 4 key elements to loneliness in older age which are described in Table 53. They provide a starting point for strategies to overcome some degree of loneliness in older people, and which can be addressed through community efforts rather than through pharmaceutical interventions. However, as described above, items 1 and 2 may be causes of social isolation while items 3 and 4 are more directly associated with loneliness if the individual understands or feels that the effects of the issue(s) are not to do with their physical condition or location. A lack of companionship can be due to having few family members or having them distant geographically. People who are able to spend time with their family appear to be less lonely than those who socialise only with strangers in community groups. Item 4 could also mean that they have few hobbies or interests to keep them occupied or have forgotten how to spend quality time alone.

It follows that strategies to relieve loneliness and social isolation need to address one or more of the feelings described in Table 53. In practice, this can be achieved using the general approaches described in Table 54, each modified to deal with an individual's preferences and psychology.

However, most older people develop their own coping strategies, and these are very individual to them, their lifestyle, and their circumstances.

Table 53: Four Elements of Loneliness in Older People

Number	Feelings That Support Loneliness	Explanation
1	Isolated from others	Physical distance to their friends, a lack of transport or not having the financial means to travel to where they want to go
2	Left out	Unable to be included in social events and meetings often due to sensory impairments or mobility issues
3	Lack of companionship	Being alone from friends and loved ones for most of the day, or for periods of many days,
4	Out of tune with most people	Having their own preferences on conversation, music, and entertainment that are different to the majority

Many general interventions target easy to find people who may suffer from short-term loneliness but fail to find those with long term loneliness issues because understanding which of the factors shown in Table 49 is often not apparent.

Table 54: General Approaches to Overcoming Causes of Loneliness

General Approach	Examples and reasoning
Changing philosophy and lifestyle	Chronic loneliness can increase sensitivity to hostility and negative ideas in social conversations and situations. People need to change their outlook and be more positive when interpreting the words of others, and to become more confident in looking for opportunities to socialise over an extended period of time.
Social training	Over the years, some people will have lost confidence in their ability to communicate with acquaintances or strangers, especially if their former spouse or partner had previously been a dominant figure. They need to be encouraged to speak to people face-to-face or on the phone, to make small talk, feel happy with silence, and learn positive non-verbal communication techniques.
Professional or focused support	People who have lost a loved one may be traumatised and/or left with such a sense of bereavement that they can shut themselves away for an extended period of time, rejecting attempts at friendship and sympathy. They can benefit from professional counselling or from the support of faith-based or community groups. Elderly people who have downsized or relocated can also benefit from focused interventions to help them engage with a new community.
More social interaction opportunities	Increasing opportunities to meet other people can enable a lonely person to connect with others who may have shared interests, backgrounds or experiences. This form of match-making can be successful if they are not obviously contrived, and if individual feel that they are able to make choices. Most opportunities involve the creation of group events.

The general approaches shown in Table 54 are more theoretical rather than practical. They may be further divided into 6 different types of intervention shown in Figure 137, which may be combined within strategies aimed at particular groups. The role of technology within these groups has not been evident in most studies possibly due to a lack of technical knowledge of the researchers, or perhaps a view that technology can be the cause of loneliness rather than part of the solution. It is apparent that changes in the way that we live, including fewer roles for individuals to perform menial delivery and collection jobs, can increase social isolation. Digital transformation can accelerate such trends (see Part 2 of this book), but short contact periods with people performing their professional roles, whether as carers or as tradesmen and women, are not substitutes for the close personal relationships that are needed to avoid loneliness. A good carer may be happy to have a chat while performing domiciliary tasks, but it would be wrong to assume that they should be genuine friends irrespective of how much empathy they offer, and how "caring" they appear to be. The reality is that they are paid a relatively low wage and are expected to complete a defined list of tasks within a few minutes.

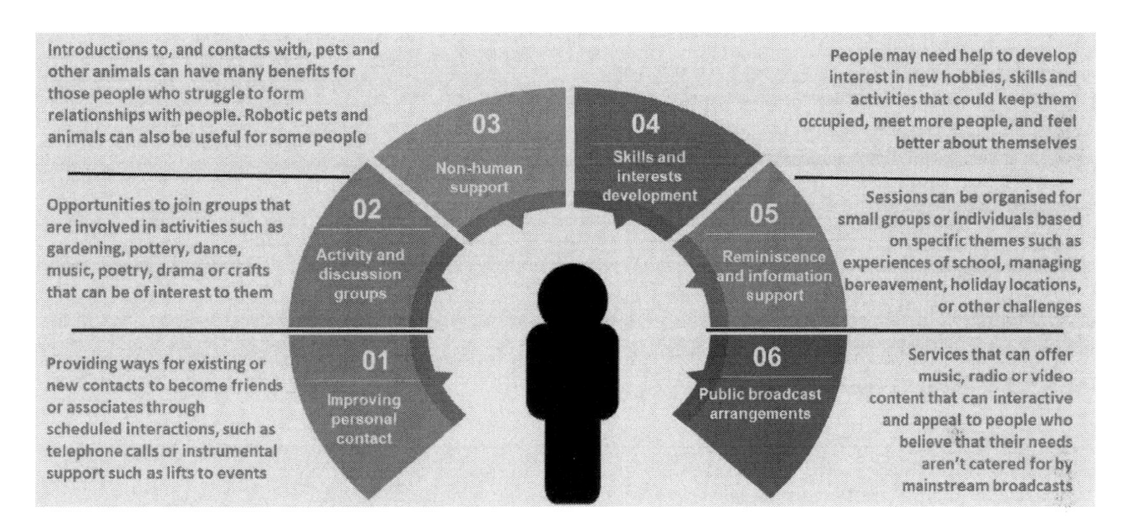

Figure 137: Interventions for Loneliness by Type

Practical Approaches to Intervention

The importance of personalised approaches to all aspects of social care are recognised in legislation and in the assessments required by the 2014 New Care Act and similar legislation in the devolved nations. They may require more time and effort to complete to obtain a better picture of an individual's ambitions, strengths and family support; the benefits appear downstream in terms of improved outcomes, especially when interventions occur early and are of a more preventive nature. Good assessments made at this time recognise the issues that prevent people from achieving their life goals. This should enable a new focus on tackling the underlying causes of problems such as loneliness and should lead to offering people opportunities to learn new coping strategies including self-help techniques focusing on occupation, purpose, learning and gaining confidence.

Table 51 describes some general approaches that may be used for the benefit of individuals, and the resources needed to make them successful. Most of these activities are private but may lead to conversations and chance meetings with others who share similar interests. Thus, the acquaintances do not appear to be contrived. These activities have been employed with some success by some people for many years, though the evidence on how effective they can be is weak, mainly because they are difficult to research on large heterogeneous populations. Many can be enhanced, or may be made more efficient, by utilising information and communication technologies mainly to offer access to information. Opportunities for extending the scope of activities by using more digital technologies will be discussed below.

Table 55: Individual Activities with Benefits and Limitations

Activity	Benefits	Resources Needed	Limitations
Gardening	Encourages fresh-air and some exercise;	Window box or allotment with individual areas	Facility needs to be close by
Caring for a pet	Many emotional, security and exercise benefits through dog ownership;	A facility for exercising pets that is safe and accessible; dog training/walking schemes	Type/size of pet need to be appropriate for space available
Lifelong learning	Provides purpose and motivation to improve self-esteem;	Access to library, and computing resources; mentors and digital buddies	Targets need to be achievable
Brain games	Mental stimulation through mind exercises or games such as bridge or chess	Access to digital hardware and software;	Level needs to be appropriate to experience/ability
Building a family tree	An individual form of reminiscence that can help find members of the family;	Access to Internet through appropriate digital devices; mentors to find websites	May accentuate lack of support
Bereavement counselling	Provides support at times of greatest need	Empathetic listeners; advice on financial matters	Cannot compensate for physical loss
Hairdressing	Encourages a positive body image;	Local salon or hairdresser who can visit the home	cost may be prohibitive for some
Volunteering	Provides a sense of purpose and encourages people to appreciate their strengths;	Access to volunteering noticeboards, and help with travel where appropriate	May highlight physical deficits
Fishing	Enables people who enjoy their own company to engage in a pastime that offers solitude and purpose;	Appropriate fishing tackle and access to bait or skills to manufacture flies; help with travel to fishing sites.	Appeals to a minority of people, mainly men
Exercise	Encourages weight loss and body toning;	Gym with equipment and a personal trainer, or devices for home loan	Older people may be unfamiliar with devices

The link between social isolation and loneliness is apparent in many, but not all, cases. It depends very much on whether an individual is an intentional loner (or introvert), or an unintentional loner who has lost their social connectedness due to an experience in their life. This might be, for example, a mental disorder or a reaction to the loss of a loved one or being physically relocated as a result of housing redevelopments. The unintentional loners may benefit most from increasing their social activities, especially if they can be engaged with and become part of groups with whom they might share some common interests and values. There have been many approaches aimed primarily at these people, though generally with limited targeting; this prevented their effectiveness from being researched in such a way that the benefits can be gauged definitively, and evidence provided for future interventions. The resources needed are generally low-cost and are available to the widest groups of people.

The approaches shown in Table 55 are relevant once individuals have been identified, and full assessments performed to determine why they are lonely, and what might make a difference to them. Unfortunately, some of the loneliest people are not known to formal agencies nor to community groups. Table 56 shows some of the interventions that are employed to identify and support those who might benefit from group approaches in the community. In addition, some of the activities shown in Table 55 can be extended for groups. In each case, activities need to be organised and led by individuals and/or organisations who can provide an empathetic approach. They must therefore be more formal in their approach so that interventions can be reported, along with their relative success, which may require further discussions in order that strong evidence can be collected.

Table 56: Community Actions to Support Unintentional Loners

Action	Description	Comments
Identification of those at-risk using delivery personnel	Use of people delivering the post or milk, window cleaners etc. to find socially and digitally isolated people	Many older people see nobody from day to day except workmen/women
Community/neighbourhood support and information/news groups	Neighbourhood Watch and similar trusted coordinators to introduce new tenants to meetings and initiatives	Good neighbours can be a lifeline to people who are lonely or isolated
Improve community and public transport	Dial-a-ride schemes, scheduled volunteer buses, bus route changes	Transport initiatives are vital to those with no car
Support faith-based organisations	Help people to re-establish spiritual connections by service attendance	Lonely people find comfort through faith-led activities
Establish local special interest or language groups	Consider hobbies such as crosswords and those that appeal to minorities	Men in sheds groups are proving to be popular
Cooking/baking classes	Opportunities for lessons and for sharing their recipes/skills and knowledge of nutrition	Sessions can lead to social events offering tea/cakes; Improves nourishment
Luncheon clubs	Provide people with limited mobility the chance to dine out with others	Many hotels support events during winter
Dance, music or cinema groups	Sessions in community to attract groups with focus on 1950s to 1970s	Reminiscing, & films as conversation tools
Inter-generational groups	Introduction to schoolchildren to encourage story-telling and learning	Gives older people chance to "borrow" grandchildren

Using Technology to Help Facilitate Interventions

There are several opportunities to use technology directly to help overcome feelings of loneliness and social isolation. These are discussed below. However, technology can also have an immediate impact in making existing initiatives of the type described in Tables 55 and 56 more effective. In each case, there is a need for more digital awareness in the community, and of improved learning by volunteers and health, social care and housing professionals and third sector employees. When these are in place many of the benefits of digital transformation are enabled, and this apply to all six intervention types shown in Figure 122. The process of supporting digital inclusion itself can be beneficial to those who go through a formal programme, especially if they can they help others in a similar position to engage.

Table 57: Technology Application to Facilitate Activities to Help Overcome Loneliness

Activity	Intervention Type	Role of Technology
Clubs, societies and meetings	2, 4, 5	Electronic bulletin boards, websites and recommendations; Display monitors in community settings
Personalised information	1, 2, 5	Access to social media sites where relevant information can be curated and presented in simple formats;
Identification of needs and interests	2, 4	Use of digital profiling and stratification tools to identify people who might benefit most from proactive interventions at key times in their lives
Personal trainer; exercise tracker	4, 5	Wearable devices to monitor exercise and to provide motivation through feedback and messaging advice
Transport and travel	2, 4, 5	Enable people making a journey to publicise it and find a passenger; Can be used as evidence to plan new routes.
Brain training puzzles and competition	2, 4, 5	Offering an unlimited source of materials suitable for any audience; Allowing people to compete on line with like-minded individuals anywhere
Searching for music, records or people	2, 5, 6	Using the Internet as a resource to trace old friends, music, poetry and other items that can be used to stimulate memory or family connections
Visits or visits to far-off locations	2, 5, 6	Enabling virtual visits through Google Street View or technologies that give access to real-time web-cams
Learning a new language	2, 5	Methods such as those developed by Babel can make the learning of a new language simpler, and more successful

Several different uses of technology for enhancing or extending activities to help overcome feelings of loneliness are described in Table 57. This also separates these uses according to the type of intervention (Figure 137). It may be apparent that the list is dominated by opportunities to give lonely people involvement in leisurely activities where they will meet with other people who may share new and existing interests. The challenge is to match the activities with the individuals, and to provide the resources at locations where they are able to gather without having to travel too far, especially if public transport runs at inconvenient times. These are effectively opportunities for greater community involvement, often utilising the resources and volunteers that charities are able to mobilise

Unfortunately, no scheme will cater for everyone's needs, even if they do help to overcome social isolation. Technology needs therefore to not only help facilitate existing ideas but also to help people who are more concerned about the quality of their relationships with a small number of people than with the number of interactions with everyone. Problems may arise due to physical factors such as lack of mobility, or location with respect to their loved ones.

New Opportunities for Support Using Technology

1. <u>Video and distant communication</u>

It is no longer necessary to have expensive and bespoke equipment in the home to enable older people to communicate safely with relatives and friends using both audio and video channels. Some approaches, such as Komp, *Mindings*, and US products (examples shown in Figure 138) use dedicated touch-screen tablet devices. These offer simplified user-interfaces with, in the case of Komp, a single switch control, to avoid confusion.

Figure 138: Examples of Tablet Devices to Simplify Video Interaction

Other approaches make use of an individual's own TV to provide a large screen suitable for people with poor vision, and to avoid the need for additional (and perhaps costly) hardware. Systems such as *V-connect, MyLinqs* and *Speakset* (shown in Figure 139) are provided specifically for this purpose and use wireless interfaces to enable the system to be used as a secure means of communicating with NHS professionals or with others in their circle of contacts. Unfortunately, these devices have tended to cost significantly more than the large screen TVs with which they can work, though the same is true for many of the specialist tablet devices described previously. It is not surprising therefore that they have struggled to succeed on a commercial level despite good user experiences.

Touch screen facility can now be retrofitted to TVs to allow them to mimic the operation of tablet devices (https://www.ssidisplays.com/touch-screen-overlay-lcd-tv/). Perhaps new TVs in the future will offer such an interface as standard.

Figure 139: Video communication systems based on user's own TV

However, *Skype* has proved to be an effective and free way of linking computers, tablets and smartphones using the Internet. A professional version provides added security and has been used by NHS organisations to initiate links with patients in their homes or in care homes. However, the greatest impact is being made by people using smartphones to provide messaging, audio or video communication between individuals or groups using apps such as *WhatsApp, Facetime, Google Hangouts, Viber* or *Tango*. These may provide the basis of more advanced and interactive opportunities in the future using additional digital technologies as will be described in Part 2.

There are few who would currently suggest that virtual visits and meetings can fully replace face-to-face interactions, including those provided by services such as day care centres. These are being shut down both because of the costs involved (including transport and provision of food) and because there is little evidence of their financial value, as determined by the willingness of people to pay a market price for the service. However, this could be the result of many people not fully appreciating the benefits to them, and therefore refusing to pay any charges that are applied by social services or by a charity. The examples provided below do not provide alternatives to day care centres, but they do offer ways of using technology to tackle some of the other issues that are faced by people who feel lonely.

2. Sensory testing and enhancements

Poor hearing or eyesight lead to a loss of opportunities to appreciate shows or events and can make the individual self-conscious in joining in conversations of groups. Vanity and stigma may force individuals to resist (or delay) the fitting of hearing aids, while the price of the latest digital versions might also put off people with limited financial means from having a hearing test. Hardware devices, as assistive technologies were discussed in Chapter 5 but may be surpassed by fully digital solutions that will be described in Part 2 of this book. Fortunately, apps such as uHear, uSound, Mimi Hearing Test, and Hearing check offer free triage arrangements using no hardware other than a smartphone, while AUD1, Hear Coach, EarMachine and BioAid offer assisted listening, sound processing and amplification as appropriate. They are used with ear-phones which means that the use of the apps appears to be no different to someone listening to music through a MP3 player. Wireless earbuds can also help users to use amplification without having large devices in their ears or connecting wires. Telephones with high or variable amplification also simplify remote conversations as described previously.

Sight amplifiers have not been developed in the same way as hearing aids, but *the* ***eSight Vision system*** and the *NuEyes smart glasses* can both restore some sight to people with serious vision loss, while the *Relúmĭno app and glasses*, use image processing software to enhance vision by redefining shapes and making outlines more visually prominent, reducing blurriness. Apps aren't as effective for enhancing vision as they are for overcoming hearing loss, but the camera on a smartphone can allow people with little or no vision to be confident that they will not become lost, and when using beacon or text reading apps, they can interpret written signs, thus enabling them to go out and attend more social events with more confidence.

3. AAC (Augmentative and Alternative Communication)

Conditions such as Aphasia can reduce an individual's ability to communicate especially by speech, thus isolating them from many people. Stroke is the most common cause of aphasia, but it can arise from head injuries that have caused damage to the language processing centres of the brain. Some people on the autistic spectrum can also have difficulties in expressing themselves verbally, causing them to be excluded from social events. There have been hardware devices that can generate speech through the pressing of switches, but such approaches are both limiting and expensive as described in Chapter 5.

The use of smartphones and tablet devices has enabled people with speech issues to use bespoke apps to replace their own voices and to gain confidence in speaking in public and social situations. These have developed rapidly in recent years and there are now many excellent examples of apps that can help adults who have difficulty expressing the wants and needs that form part of daily living. *APP2Speak*, *Proloquo2Go*, *iCommunicate*, *iComm*, *My Talk Tools Mobile*, *Look2Learn*, *Voice4U* and *iConverse*, are all apps that operate on Apple devices (such as iPads and iPhones). a wheel of apps for Android devices produced by Edinburgh University that offers solutions in many different areas of applications, many of which can be used to give confidence to the user.

4. Continence

Urinary frequency or urgency affects many people, especially older women (who have had children many years ago), and men with enlarged prostates. It can be embarrassing and a reason that prevents them going out far from home in case they are taken short. This can reduce opportunities for socialising and is a stigma that they are unable to discuss even with friends. Fortunately, modern incontinence pads and briefs are much smaller and lighter than the ones that were used for bed-bound patients 20 years ago. They are now discrete and an example of how new materials and good engineering design have been used to minimise the effects of bladder leakage. There are also new catheter designs for Intermittent Self-Catheterisation (ISC) which are reusable for up to a day's use.

Chapter 10 described Enuresis alarms that can be configured to provide more discrete alerting to an individual to avoid any stigma associated with an audible alarm, prompting an individual to find a toilet. There are also apps available to locate public toilets enabling people who want to participate in rambling events to go out with more confidence. Abena Nova is a sanitary pad product developed through MediSensor which has built-in sensors which register when the product is wet and send an alert to carers' smartphones or tablets. Plesiocare systems described in Chapter 10 can also be used to support carers from becoming isolated by worrying what happens if they go out.

5. <u>Virtual meetings – Telepresence (Use of Virtual and Augmented Reality)</u>

Poor mobility, whether due to giving up driving due to health or eyesight issues, or to a lack of suitable travel options or companions, can be a major factor in reducing opportunities for social interaction and enjoyment. TV companies, such as Sky and BT Sports are already competing to provide viewers with an immersive experience, offering different views and options to enable them to improve their experience and to achieve a feeling as close as possible to being present at the event. Some churches and crematorium facilities have also been equipped with cameras and microphones that can stream services to those who can follow using the Internet.

These approaches can overcome a sense of being left out that can affect some people, but virtual presence can only start to replace the sense of exclusion if it can allow the user to interact with the distant players in the same way as they might do if they were in the local audience. Some people may be satisfied with being able to shout, but it is the feeling that the shouting can be heard by others that creates the necessary virtual environment.

Figure 140: Examples of Virtual (Oculus), Augmented, and Mixed Reality (Magic Leap)

The microphone arrays that have made Digital Assistants such as the Amazon Echo and the Google Home so successful as a home system interface, together with surround sound systems for use in a lounge, are preparing the way for more immersive virtual environments. These can be enhanced by the use of virtual reality headsets, by augmented reality systems and by mixed reality hardware and spectacles (see the examples in Figure 140). Virtual Reality (VR) is a powerful and innovative way to travel using nothing more than computing power and the optical display in a headset. The advanced version has motion tracking enabling the user to look around and discover a virtual space as if they are actually immersed in it.

Augmented reality systems present a live direct (or sometimes indirect) view of a physical, real-world environment whose elements are "augmented" by computer-generated images or by sensory inputs such as sound, video, graphics, haptics or location-specific data. It means that an actual and recognised environment can be mixed with artificial elements to create scenarios that wouldn't otherwise exist. Applications could be created that allowed users to accompany friends or physical guides to explore places or situations that would not be possible without the technology. Virtual team might thus be created. An augmented reality image may be displayed on a portable device such as a tablet or a smartphone, included the Bayern Munich selfie opportunity shown in the centre of Figure 140. Mixed reality takes augmented reality to another level offering the potential to improve the overall experience, and to provide social interaction through digital connections and communication with friends or colleagues using a dedicated headset of the type shown on the right of Figure 140.

SyncVR harnesses a user-centered design, well-developed software anVR content to enrich the lives of older people. It is a Virtual Reality (VR) system that allows elderly people to undergo unique experiences alongside others if appropriate. For example, several older people can immerse into a different place (in a different time) together and are triggered to connect over things they see and hear; this makes the connections they make in real life after the VR experience even more special than the VR experience itself. Moreover, a growing volume of content allows the use of unique VR content from global partners, so that lonely people can choose to visit inspiring places from city museums to wildlife parks.

6. Devices for Virtual Intimacy

Many lonely people miss the loss of intimacy as well as companionship when a loved one dies or when a long-term relationship ends. A lack of physical contact can lead to decreased well-being and a sense of hopelessness. Friendly, platonic touching, such as hand holding or hugging, from friends and family can lower stress and help to overcome loneliness even in people who might not previously have considered themselves to be tactile.

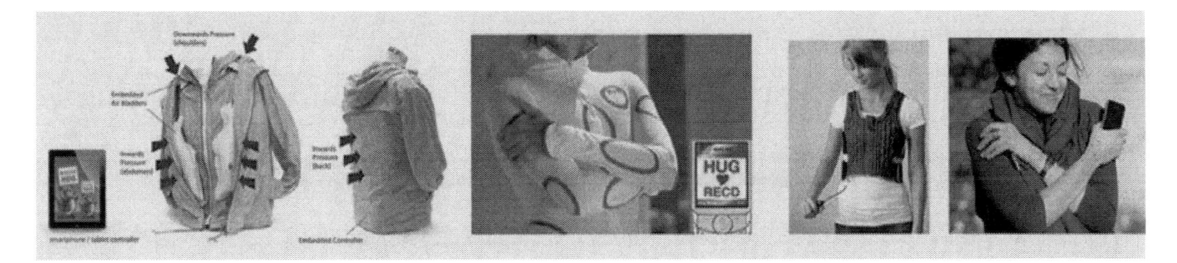

Figure 141: T-Jacket, Hug-short, the Squease and Like-A-Hug for Remote Hugging

Various wearable devices have been proposed to support social care and these can be extended to provided touch interfaces. Figure 141 shows examples of wearable devices that are designed to enable people to share hugs or squeezes at a distance. The technologies combine sensors and actuators and have primary applications in providing comfort to people on the Autistic Spectrum who have difficulties processing sensory information, and for creating a personalised form of remote physical interaction for children with ADHD. These devices have yet to be used for older people who seek a form of touching from their relatives who may live at a distance.

There are also portable devices that can simulate more direct contact such as kissing or feeling a heartbeat at a distance. The devices shown in Figure 142 are, from left to right, the Kissenger which provides "robopigs" as a physical interface, Pillow Talk, which consists of a ring worn on the finger and a panel inside a pillow, and the Taion which transmits a squeeze to the partner device to simulate heart rate and pressure via a mobile phone. The Dutch company, Somnox, is also developing a sleep robot which can be cuddled, and which can offer music, guided meditations or breathing rhythms. It will also contain sensors to provide feedback and sleep tracking.

 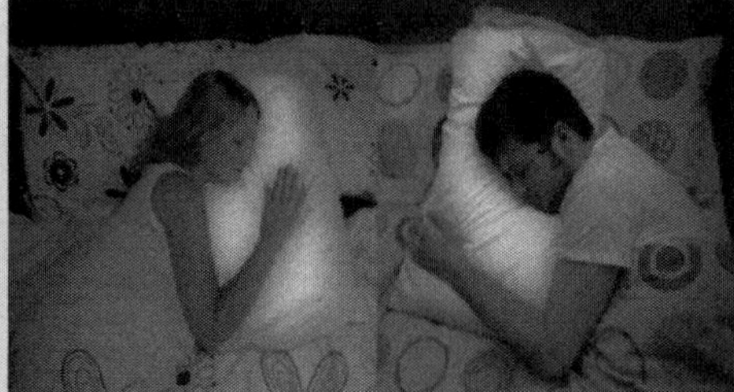 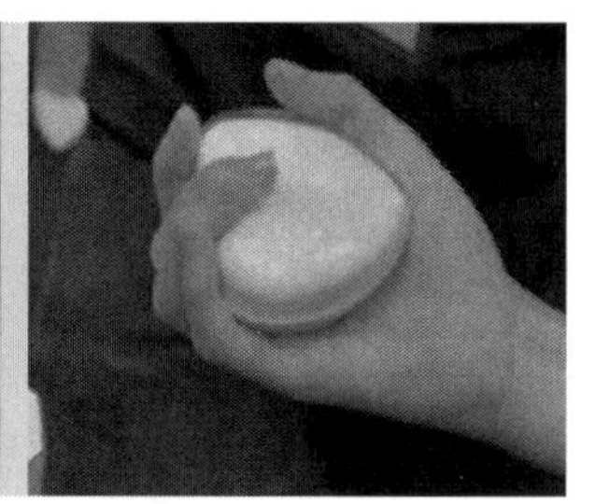

Figure 142: Portable Devices for Simulating Contact Remotely

7. Robotic and Virtual pets

The benefits of owning and caring for a cat or a dog are well established, especially for people who live alone. In particular, dogs require exercise and regular feeding, and therefore can help their owners to establish a routine that involves taking walks where they may interact with other dog-walkers. However, mobility issues, and a lack of access to open spaces, can limit an individual's ability to own a pet and to realise all the benefits. Table 58 compares the pros and cons of real and virtual pets. Figure 143 shows Paro the seal, and MiRo, a biomimetic robotic dog, together with a robotic cat specifically designed to be held and stroked by older people. Their main application is people with dementia, but they have more general uses in overcoming the need for companionship.

Numerous studies have demonstrated the value of Paro, the robotic seal, especially in nursing homes for people with dementia, where residents include some of the loneliest people in society. These applications will be discussed at more length in Part 2. However, outside such environments, robotic pets may occupy people for relatively short periods, especially if they have novelty features (such as an ability to repeat what is said to them) there is little or no evidence that a robotic pet can offer tangible benefits to compare with those of actual pets.

Table 58: Pros and Cons of Robotic Pets Versus Real Ones

Real Pets		Robotic Pets	
Advantages	**Disadvantages**	**Advantages**	**Disadvantages**
Soft, furry and easy to stroke and pet	Can bite or become a trip hazard	Can memorise faces and personality traits	Novelty can wear thin after a few weeks
Each can have an individual personality	Vet and food bills	Can provide visual or spoken information	Limited ability to read a situation
Provide security through barking	Can shed hair	No vet or food bills	Needs regular maintenance
Can be taught new tricks and habits	Can cause allergic reactions	Inherently safe and easy to hold and pet	Internet connection needed
Helps establish eating and activity routines	Need to go out in inclement weather	Doesn't need exercise or attention	Useful lifespan may be short
Encourages exercise and socialising	Needs attention for waste disposal	Can be programmed or repaired	High initial cost of ownership

American toy manufacturing company Hasbro and Brown University have teamed up to create affordable assistive Artificial Intelligence (AI) that is expected to benefit ageing populations all around the world. The "Affordable Robot Intelligence for Elderly Support" (ARIES) is a three-year collaboration which recently received a $1million grant from the National Science Foundation in America. The project is focused on helping elderly people remain independent using a "smart robotic companion capable of assisting older people with simple but sometimes challenging everyday tasks". The funding provided is to go towards a series of user studies that will help determine how ARIES will impact the lives of older adults.

Figure 143: Examples of Robotic Pets for Older People

8. <u>Companion Robots and Chatbots</u>

Companion robots are autonomous robots *that have been designed fundamentally to interact and communicate with humans (or other autonomous physical agents) rather than to perform specific support functions such sweeping the floor or mowing the lawn. They operate by following* social *behaviours and rules attached to their role by their designers/programmers.* Thus, they can show emotion, have the capability to hold conversations at an adult level, understand their social partners at a psychological level, form social relationships, and act on natural communication cues. This gives them a form of personality and social capability that can make them meaningful companions to lonely people and to those who need intellectual stimulation or motivation.

Their physical presence, their humanoid appearance, their offline capability, and their ability to move about and to show emotion through head or eye (camera) movement differentiates them from social assistants such as the Amazon Echo and the Google Home products that have become popular interfaces for information and for electronic device operation. They are therefore inherently more acceptable from a data security and intrusion perspective and are therefore more likely to become trusted companions. The two leading examples are Pepper, by the French company, Aldebaran, and the Jibo product from the USA, both shown in Figure 144 alongside Alfie, a prototype device from the University of London as part of the ENRICHME project, and Billy-Billy, an interactive flowerpot which aims to make the lives of older people more comfortable by using its temperature and light sensors to help them look after their house-plants. More sophisticated social robots will be discussed in Part 2 of this book when more autonomous digital assistants are introduced as part of the digital transformation of care that is already underway.

Figure 144: Pepper, Jibo, Alfie and Billy-Billy Companion Robots

Southend-on-Sea Borough council has become the first local authority to buy and use Pepper under an academic licence. Pepper will not be used to carry out any direct or personal one to one care, but may be used in a range of settings, including in residential care homes, and sheltered housing schemes as an information and advice point in relevant. It might also be used for certain social care tasks, such as running a reminiscence group or helping stroke victims with physiotherapy, freeing up staff for other tasks and becoming an exemplar for digital transformation of care services.

The limitation of robotic devices is their local memory while personal digital assistants are reliant on broadband to enable them to access the full databases of information that enables them to answer questions and provide information. On the other hand, the simplest chatbots operate over a telephone line and therefore need considerably less bandwidth with all the processing, including the Artificial Intelligence and data storage, being provided at a distance. *A* chatbot *is a computer program which conducts a conversation (sometimes by texting or typing) that is designed to simulate how a human would behave as a conversational partner.* They can pass the Turing test thus potentially fooling the person that has engaged them in conversation into thinking that they are an actual human being. Table 59 gives examples.

Table 59: Some Popular Chat-bots

Chatbot name	***Description***
ELIZA	*One of the first chatbots and capable of providing person centered therapy*
Eviebot	*In use for over a decade, and capable of speaking 5 different languages*
Cleverbot	*Uses context of conversation and its archive of previous chats to respond*
Mitsuku	*Won award for most human-like AI in both 2013 and in 2017*
Answer	*A customer service platform for creating personal, productive relationships*
Kiyana	*The chatbot persona is of a flirt who likes to sing and talk about cats*
Nicki	*Answers questions about the dangers of smoking but also engages in chatter*
Botster	*An open-source AI agent for intellectual, useful and entertaining conversations*
Yhaken	*Engages with users through natural language at 2 discreet levels*
Vashenka	*Chats about her interests in mushrooms, purple stuff, science and knowledge*

Chatbots are being used increasingly by telephone call centres to provide unlimited capacity to answer customer queries. This includes the use of Amelia, a cognitive agent who can take on a number of service desk role and communicates in natural language with customers for Enfield Council. Many operate through the web computer as avatars offering the opportunity to present a visual image of a person whose lips can move as words are being said. There are some ethical issues to address before chatbots may be used without restriction to provide companionship and interaction with older, vulnerable, lonely people,

Care.coach, a US company, has succeeded in combining an avatar-based bot system with a team of health professionals to combine the benefits of artificial intelligence and an attractive interface (on a tablet or smartphone) in such a way that it can improve health through better engagement, psychosocial support, scheduled reminders and automated alerts. They have succeeded in providing case studies and academic research to validate outcomes across the care spectrum.

9. Online Dating

Some people feel that the level of companionship that they crave can only be provided within a stable, long-term relationship irrespective of their age and the factors that may have caused them to become lonely. If they fear being left alone and feel lonely for much of the time as a result of their limited mobility, or any of the other risk factors for social isolation described previously, they may be tempted to use Internet (or mobile device) based dating. These Internet sites enable strangers to find and introduce themselves to new personal connections, often with the aim of developing personal or romantic relationships (though many want nothing more than companionship providing that it is someone that they can respect and share time with).

The language used by many of the sites tends to focus on love and on soul mates rather than on new friendships, but they share some principles as far as simplicity of operation is concerned, and their claim that their matching algorithms are superior to those of other dating services. In practice, none of these claims has been robustly tested and this poses a threat to their long-term future if they are advertised with exaggerated outcomes. Several different English language dating sites or apps are described in Table 60.

The success, or otherwise, of these sites is not necessarily the number of people who are "matched" but by the overall experience of members in being connected to people that they would not otherwise have met. Each claim to have algorithms that have been optimised for the older demographic, but few have promoted scientific testing of their scoring systems, and on the significance of the photographs compared with other features such as accent, educational achievements, sporting and cultural interests, and other hobbies. Academics have applied machine learning to compare how well initial attraction is the product of two people's self-reported traits and preferences. But theoretical models were unable to predict relationship variance using any combination of traits and preferences reported before they actually met.

It is not therefore surprising that a scattergun approach is employed, with most sites using profiling questions that are similar to those used in social media sites. Social care assessment might well in the future follow such profiling methodology in obtaining standard form input data for determining future care needs and preferences.

Table 60: Friendship Matching Web-sites and Apps

Dating Site/App	Description
Plenty of Fish	Broad age range with free messaging
Elite Singles	Nearly one in three members is aged over 50
eHarmony	Leading UK site for all ages but with special advice for older people
Senior People Meet	Has > 1.5 million members worldwide and has geographic search tool
Stitch	Focuses on travel companions and for introductions to people who share an interest in a particular hobby
AgelessFish	Long established US site focusing on over-50s
OlderDatingOnline	Offers a wide range of features to its members
Mature Dating uk	Offers tips and directory of UK success story
Ourtime	Ensures that profiles of over-50s on the site are authentic and genuine.
Our Time	Only accepts paying members aged 50+ – ideal for finding travel companions
OkCupid	Allows use of 2 profiles, one for photos and the other for information
SeniorMatch	Nearly 4 million verified members, none under 40
SeniorsMeet	An international senior singles website serving older people and mature singles from the USA, Canada, UK, Europe and Australia
OlderHearts	Not only s dating site but also a large social network for older singles

10. Apps and Websites to tackle Loneliness and to Fill in Time

In addition to apps that help individuals to find other people who are in search of companionship, such as ***Huggle, Hug!, Pocket Boyfriend*** and ***Hey! Vina,*** there are others that provide support more directly either by linking directly or virtually with a therapist. These include ***Happy***, ***Talkspace***, **Back Talk** and ***PRIORI***. Watching Cute Girl is a popular app that has recordings of a Japanese woman who is watching, giving the impression that an individual is not actually alone when they are eating dinner or watching the TV.

There are also many other apps and websites that can offer distraction or occupation. These might enable an individual to forget about feelings of loneliness, through self-entertainment. Table 61 describes some leading titles that might appeal to people who have specific interests. None of these sites currently offer techniques such as cognitive behavioural therapy to change the way that people think; if they are more positive about themselves then they may harbour fewer thoughts of loneliness, and perhaps place more value in the time that they have alone, when they can watch TV programmes of their own choosing, listen to their own favourite music, or enjoy their own selection of food.

Table 61: Apps and Websites to Provide Occupation for Older People

App	Description or Purpose
Lumosity	This is a brain-training app that features cognitive research tasks that have been transformed into games that should appeal to older users
Playlist for Life	The app works in association with the Spotify online music service to allow the user to listen to the music immediately, enabling them to associate a song with memories or moods and create different playlists for different occasions
Replay Sporting Memories	Allows users to access sporting memories from years gone by and to share their own with others so that anyone can. The app makes these memories instantly accessible, and it is easy for a user to add and save a collection of memories to replay later, reminisce with others, or create their very own vintage newspaper, Sporting Pink.
Pocket Physio	The app features videos and an easy-to-use guide for older people recovering from orthopaedic surgery or those wanting to build up strength with appropriate physiotherapy exercises
Bookshare	This is the world's largest library of accessible eBooks for people who can't read standard print, and features apps such as *Go Read*, and *Capti Narrator*
Which? Elderly-Care	Features a Care Services Directory and information from finance through to organising care for older people
RetireMove	An independent website to offer news on new schemes that might interest people who are looking for a new home in later life
TheyWorkForYou	Enables people with an interest in politics to keep track of issues and the actions of their elected representatives

The EU Mobile Age project aims to develop apps and other digital services to help deal with complex issues including loneliness and social isolation. It includes a UK arm led by Lancaster University researchers who are working with groups in the South Lakeland area, and are focusing on providing information that will help older people to go out; this includes the terrain, the weather and the accessibility of toilets. It is an example of how authorities are appreciating the potential of apps and digital resources to support older people across Europe. The problem that faces researchers across the developed world remains the individual nature of loneliness, and the current failure to find universal solutions that will offer support to individuals irrespective of their age, personal circumstances and preferences.

Social Isolation and Loneliness in a Rural Environment

Rurality is considered by some to be a subjective term, even though it is well-defined in individual countries using a specific index. The Rural Urban Classification is an example of an Official Statistic used to distinguish rural and urban areas. The Classification defines areas as rural if they are outside settlements, or towns, with more than 10,000 resident population. Similarly, the Cloke index uses 16 different ways of drawing conclusions that result in areas of England and Wales being categorised using five criteria:

- extreme rural,

- intermediate rural,
- intermediate non-rural;
- extreme non-rural; and
- urban.

This suggests that rurality is not simply about geography but implies positive lifestyle choices that residents make. This might, in turn, be relevant to aspects of mental health. People who live in rural environments might therefore be considered to be more resilient to challenges such as social isolation and loneliness.

In larger countries, including Australia, Canada, Brazil and the USA, the scale is significantly different, and population density and factors such as temperature, rainfall, and the terrain are relevant in the way that they support opportunities for farming and how rural an area might then be considered. Several different classification systems have been developed to define remoteness and rurality in Australia. These tend to define in terms of the size of a community, distance from population centres, and access to services. Figure 145 demonstrates the large variations that exist by comparing Australia, almost a continent in its own right, with Wales, one of the smallest nations by both population and area within the British Isles (including Ireland but excluding the Orkneys and Shetlands).

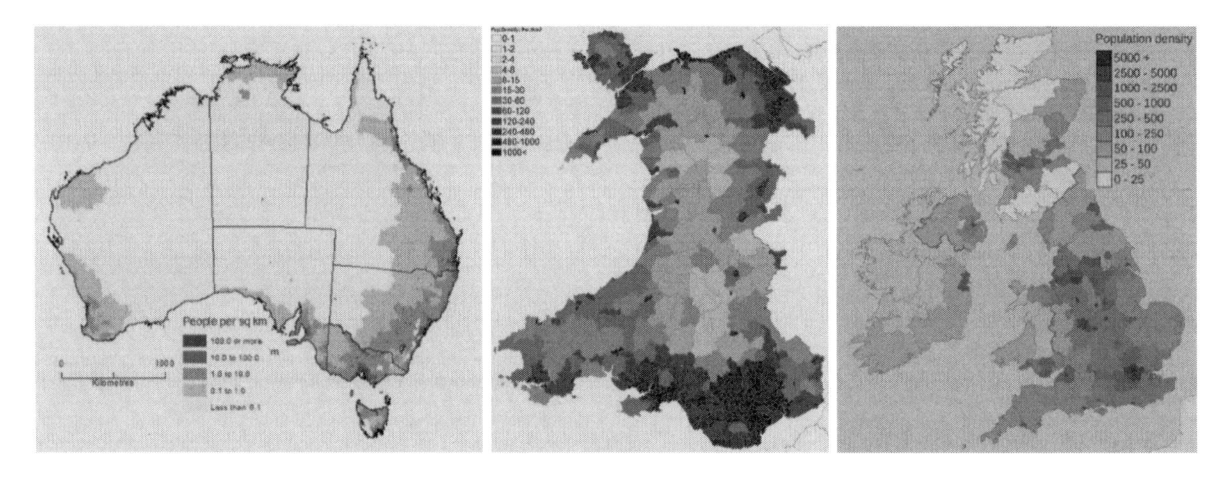

Figure 145: Population Density across Australia, Wales and the British Isles

In the case of Australia, the scales range from 0.1 to 1 person per sq. km. up to 100 persons and more, a category which is relevant only to areas around the major cities of Sydney, Brisbane, Melbourne, Adelaide and Perth. By comparison, Wales, a small nation of 3.2 million people has most of its population collected in towns and cities but has more than half of its area populated by fewer than 15 people per square kilometre but nowhere with fewer than 2 people per kilometre. This should be viewed in the context of the British Isles which shows that there neither England nor Wales have areas that are as rural as large areas of the Highlands of Scotland as well as the Borders. It follows that rurality exists in most countries, but the scale depends on the countries themselves.

Table 62: Use of Technology to Support Isolated Older People in Rural Locations

Initiative	Description
Satellite broadband connections to a village	High speed internet to areas with poor mobile reception that are too far from telephone exchange for broadband
Wireless Wideband Area Network to a community	Low bandwidth long range transmissions enabling Internet of Things connectivity
Internet/library van	Mobile access to Internet access and support
Transport network	Enhanced dial-a-ride opportunities
Volunteer-based proactive telephone calling service	Non-emergency telecare telephone service where the call handlers are socially isolated older people
Virtual worship	Broadcasts of services from churches, chapels, mosques, and synagogues including funerals and weddings
Digital neighbour scheme	WhatsApp or similar grouping of people in locality each with responsibility for checking up on their neighbours
Share a meal bank	Storage and distribution of spare meals cooked by neighbours
Respite network	Volunteers to take out disabled people, or to sit with them at home so that their family carers can go to social event
Assistive technology library	A loan pool of electronic assistive devices such as reminders and alarms that are not in the inventory of local authority
Mobile day centre	Bus to collect isolated people and to provide them with digital entertainments on board or at a local hotel
Community postmen	Country-wide roll-out of Jersey Post initiative

The significance of rurality in many regions, is that the term can be used to determine the level of economic support available across a spectrum of activities whether to build a house or to set up a business. In the USA, remoteness is considered differently for healthcare than for other government organisations and applications. This acknowledges the significance of rurality on the healthcare agenda and on its importance in maintaining the well-being of citizens. It requires the careful planning and introduction of measures that can mitigate the effects of rurality, both with respect to receiving health and social care services, and in ensuring that the lack of population doesn't necessarily mean that there are fewer people who can become involved in providing care for older and disabled people.

The solution lies in the community, and the availability of initiatives to build capacity to support the lifestyles of inhabitants without forcing them from their homes and into conclaves where support can be provided more efficiently. Table 62 describes some approaches that may help provide some of the glue needed to keep such communities together, ensuring that people can remain in their own homes for as long as possible. If they eventually have to move into care institutions or Extracare schemes, these need to be smaller and more intimate to reflect the status of small communities.

Concluding Remarks

It is apparent that loneliness is a very complex condition, which is neither restricted to older and disabled people not to those who are socially isolated, whether due to where they live or the lifestyles that they have chosen to live. It can also hit at any time. There can therefore be no single 'golden bullet' that will transform the lives of people who are lonely. The need is for a raft of different measures that need to be made available to meet the need on an individual basis. It follows that the matching of the intervention to the individual will be crucial for success – and that means having a full and comprehensive assessment of the individual in place, which is the aim of the assessment processes described in Chapter 6. This must focus on what the individual wants, but must also probe deeper into their preferences, so that a (digital) assessment algorithm or artificial intelligence can be applied with sufficient data to understand and learn by example.

Although this chapter has focused on the role that technology can play, in most cases the solutions are not the technologies themselves, but the ways that technology can facilitate interventions by helping to remove obstacles. Indeed, the examples shown in Figure 58 could be provided anywhere but only if the communities are connected digitally using every available channel from the old telephone through to satellite-based fast broadband. The opportunities need to be coordinated and then delivered by people who have a genuine interest in the outcomes. Volunteers will play a big role, not only in providing 'the feet on the ground' that give social contact to people, but also to help them understand and appreciate the potential of technology. Many will be digital champions, whilst others may be happy to be the foot soldiers who drive the vehicles that deliver services to a wider community.

The main conclusions for this chapter, and for the other chapters in this book, are that social care must change radically if it can deliver to vulnerable people the services that they need to give recipients worthwhile Quality of Life benefits. Indeed, the recipients must no longer feel that services are being delivered to them – they must believe that they are partners in a range of choices that are being offered to them. Everyone needs to understand that a consequence of extended life is both an increase in the number of people who need support, and a decrease in the number of younger people who are available to provide it. In such a world, care services are a valuable, and expensive, resource that may need to be rationed on the basis of need and want. Cost will inevitable be an additional factor when wealth is so poorly distributed, but that should become an incentive to create more innovative solutions that offer greater efficiencies. Prevention will also be a big driver as it offers a way of reducing demand. Ultimately, society should be aiming to give more life to years rather than extending the years of life.

Applications for technology already abound, and many have been used with success by service providers both in the UK and elsewhere. But there has been a reluctance to transform services by an increased use of technology, though in Scotland, the technology option has been placed first for a decade or more. Digital technologies can quickly make older approaches redundant, but the process of change can be slow and difficult. The switching-off of analogue telephone communications over the next 5 to 10 years will force more change. This will need to be planned and managed so that a' like for like' exchange will not be the preferred method. A digital roadmap approach is advocated, and this needs to be more detailed than the telecommunication technology version shown in Figure

48. Many of the new technologies are already available and described in the 7 waves model shown in Figure 51.

The technology cannot, alone, result in a digital transformation. That requires a more comprehensive change that involves commissioners, service providers and the people who will ultimately use these services. Commissioners will expect to see evidence of success before they commit to spending valuable resources on equipment that may work for only some of the people intended. Service providers don't want to be basing their propositions on the use of equipment that has a very limited lifetime, and which will expose them to commercial pressures if their competitors adopt a newer device or system that is superior in every way. End users need to learn that the technology can be their aid, in the same way as a vacuum cleaner, a washing machine or any other household appliance can reduce the physical strain on them. Our grandparents would be amazed (and might have trouble accepting) microwave ovens, colour television, mobile phones and handheld computers that we take for granted. We must prepare the narrative so that people who are not used to technology, or are scared of using it, learn to accept it and to embrace it.

Computers have transformed the office and the homes of people whose working lives covered some or all the 40 years from 1980 to the present day. Those who are older, or who were not in the workplaces that used computers, may be more reluctant to accept digital technology in the home, and definitely not if they replace devices or people that they are used to. Improving the acceptance of the technology, and its benefits, to the end users remains the greatest challenge to implementing changes in social care.

It will be the focus of Part 2 of this book which consider all aspects of transformation from the need to give people access to both the Internet through to innovative applications that can yield meaningful benefits. The process of transformation has already begun; there are examples of the technologies available already. But they have yet to be gathered into full service propositions that will pass the test of time due to their popularity with the majority of service users.

Work on Part 2 of this work has already begun.

There follow four appendices:

A. A series of vignettes – these are 8 enhanced case studies that can be used to test the reader's assessment skills, in understanding the needs and wants of the individual, and in offering them technology-based support that both mitigate risk and give them the potential for improving of maintaining their quality of life. Solutions are not provider – there may be many different ones – but some options will be discussed in Part 2 of the book, together with an analysis of the impact of digital transformation. Some summary points will be provided on-line.

B. List of suppliers – these organisations are the leading providers of equipment, devices and systems as described in Part A of this book

C. Useful reading – this list of books, academic papers, and online resources may be used to supplement the material found in this book.

D. Glossary of terms – a brief description of terms, expressions and acronyms used within the book.

Appendix A

Vignettes as Case Studies

1. Victor is 95 years old and lives with his 92year old wife, Rose, in a 3 bed semi-detached house on a former council housing estate in Hampshire. He has had Type 2 diabetes for the past 25 years but it is well managed with medication and diet. His wife is quite well but has high blood pressure and complains of pain due to osteoarthritis in her shoulder and her hip, which limits her mobility. She struggles to go upstairs, to the bedrooms and only bathroom. Victor walks with the aid of a cane, though he doesn't use it appropriately. He is able to go out on the bus most days, visiting the betting shop to meet with friends and to put a small bet on the horses. Rose goes out only twice a week, on Wednesday afternoons, to have her hair done, and to do the shopping with one of her friends, and on a Sunday morning when a neighbour takes her to church.

   Victor and Rose go out together only rarely, claiming that they don't like to leave the house empty because of potential intruders. The property has no security provision other than double-glazed windows and doors. They provide each other with companionship and ensure that they each take their medication on time, and attend appointments when necessary. All 3 children of their children have retired. Their only daughter, Belinda, lives locally and visits them twice a week, taking their laundry home to wash and iron. She also does some housework for them. One of their sons lives half an hour away from them but calls every week to keep the garden tidy, and to put out their recycling bins. He is not very good with technology, and has an old basic mobile phone. It may not be surprising that his parents have no digital technologies, though they have a large-screen television, and enjoy listening to the radio. They have also resisted all attempts to have central hearing installed, claiming that they cannot cope with the mess that it would involve.

   Their other son lives North of London with his family and his poorly mother-in-law. He rarely visits but telephones twice a week and ensures that his children also say a few words to their grandparents. He is frustrated that Victor and Mary refuse to have broadband installed, as he would like to monitor his parents remotely, and use video calls to communicate with them.

   Last month, Rose suffered a fall on the floor of the bathroom, and was unable to get up on her own. Victor heard her cries and went to her rescue. In trying to pull her up from the floor, he ended up falling also. He was able to crawl to the phone to call his son, who was able to rescue them before too long. He insisted that they needed to make changes to manage the risks of future accident and informed their GP who advised that the local authority should visit and make appropriate recommendations. They were assessed by a social worker and by an Occupational Therapist who they told that their priority was to stay together in their existing property, but to feel safer. They lead a frugal existence, even though their income per month, after taking away basic expenditure exceeds £1000 per month.

   What equipment and services could be introduced that Victor and Rose would accept?

2. Since the death of her husband 5 years ago Edna, aged 75, and has lived alone in a specious, centrally-heated and double-glazed bungalow close to local shops and a Post Office in a sea-side resort in North Wales. She has never been able to drive and now spends much of her time indoors, where she is able to smoke as many as 20 cigarettes a day while watching one of her many large screen TVs. Her diet is lacking in fruit and fresh vegetable though she otherwise eats and drinks well, even though she manages to drink at least a bottle of sherry and half a bottle of gin every week.

She has put on considerable weight in recent years and is short of breath when she walks even a short distance, which means that she will happily order a taxi to take her out to collect her pension, or to attend appointments at the local medical centre. She takes medication, twice a day for high blood pressure but considers herself otherwise to be fit and well but not as happy as she should be. She has a son who lives in Australia who she intends to visit but probably never will, and a daughter, Janet, who lives and works a few miles away with her husband and 3 teenage children. Janet visits her once a week to take her shopping to the local supermarket. She used to bring the children to see her every week, but they are now too busy with their after-school sports, their studying, and their busy social lives to visit their grandmother's home. She sees them twice a month when she visits their home for Sunday lunch.

Edna has her own occupational pension as well as money that she and her husband saved during their working lives, which she now uses to employ a cleaner 3 mornings a week, a a gardener who tidies up the lawn and performs odd jobs for her once a month. Money isn't an issue, and she has told her children that she would be happy to spend it all before she dies because they are both financially secure and comfortable. She has no problem with spending money on the latest gadgets or electronic equipment, though they are often left unused because she has nobody to show her how to use them.

Edna contracted influenza recently. She was unwell and, after having a home visit from her GP, was admitted to hospital with severe breathing problems. She needed oxygen and her condition was considered to be critical for a couple of days before she recovered enough to be moved to a local cottage hospital where she could be assessed prior to discharge to her own home. She was prescribed further medication and was offered a reablement package for 6 weeks, which included a basic pendant alarm. At the end of the reablement period, Edna felt much better and was able to perform most activities of daily living without help. However, her confidence was dented, and she realised for the first time in her life that she was vulnerable and would need to make some lifestyle adjustments if she was to survive another winter. She was visited by an independent living advisor from a local charity, who quickly understood that Edna wasn't lonely, but that she missed attending the social events that she had enjoyed going to with her husband when he was alive. She had loved her husband dearly but had no friends of her own. She was also visited by an assessor from a large national telecare service provider, who was able to show her a range of technology support applications that could help to reassure her and her family that she was safe and well. She saw how they could help her, and how they might enable her to avoid an early admission to a care home.

What might Edna's private care plan look like?

3. John is 42 and lives with his elderly mother, Mary, aged 80, in a small village in the Lake District. They have no other living relatives. He works part-time with a group of other people with learning disabilities for a social enterprise which runs an Internet café where people can be helped to use computers, and smartphone and with using online activities. He is an accomplished photographer and stores all his images efficiently on his iPad and home computer, and then on his own cloud computing account. He also likes music, both classical and pop, and always wears his iPod and headphones on his way to work. His personal communication skills have developed to match his digital skills, to the point where many customers ask specifically for his help when they come into the café.

   Mary has tinnitus and has to turn the television volume up to hear it. Consequently, both Mary and John sometimes miss the telephone and the front door bell when they are ringing. Mary's mobility has deteriorated slowly over the years also. She now used a cane for support, though she often forgets to take it with her if she suddenly remembers that she has to do task in the kitchen, such as switching off the cooker.

They were the victims of burglary last year. The details of the incident are quite vague, but it is possible the perpetrator(s) entered through the back door from the back garden either during the day when John was at work, or during the evening when he and his mother were watching TV. They tended to leave the back door unlocked until they go to bed at night at about 10pm. Few items were taken but Mary certainly lost some small items of jewellery that had sentimental value, while John lost his iPhone and charger.

John is now nervous about going out and leaving his mother alone, and also appears to be about going out alone on the bus, especially when it is dark, claiming that there is a lot of hate crime around. He had previously made use of a number of apps on his home that helped him to use public transport safely as well as to give him reminders about daily events, and sequences of activities that he could perform to help his mother in the house with domestic tasks. He has also abandoned some of his personal hygiene tasks that he had mastered in recent years, and now is at risk of losing his job in the internet café for poor attendance and time-keeping. Some of his colleagues have complained about his body odour and reluctance to shave daily.

Both his mother and his social worker are concerned about the changes in John and are prepared to try any new approaches, including an extended use of technology applications, to restore his confidence and to manage the issues that have been identified, including those relating to security. John has a personal support budget that could be used to help improve his well-being and to ensure that he feels safe. But there is a consensus that he needs to try innovative solutions alongside more conventional technology approaches. Meanwhile, John's main concern continues to be the safety of his mother, who he believes is at risk of accident, illness and injury because of the decline in her general health. He is happy to accept systems that support his mother but doesn't want to accept that she is worried about his welfare.

What combination of technology applications might play a role, and be acceptable to both John and Mary?

4. Pauline is a 90 year old widow who moved into a private retirement complex in Torquay when her husband of 60 years died 5 years ago. Her apartment is on the third floor, giving her beautiful views over the English Riviera. She has no children and little money once her service charges and food have been paid for. Fortunately, she has few needs but is sprightly and walks for miles every day along the promenade before returning home to watch quizzes and soaps on TV. Although Edna considers herself to be quite sociable, she is unlikely to do more than say good morning to the dozens of people she passes every day on her walks, though she does talk for a couple of minutes every day to the newsagent. When she and her husband owned a dog, they found that strangers used to stop them to stroke the dog and to talk to them. This is something that she misses.

She also likes to read and to listen to music either on the radio or on her CD player. Her singing voice is good, and she enjoys humming along to pop songs, though her favourites remain those from the 1950s and 1960s. When it rains, she spends time working through the daily cryptic crossword in her newspaper. She was given a tablet computing device by a nephew last Christmas so that she could read her newspapers online and read ebooks, but she struggled to master the device, and was disappointed to find that any savings she made on subscriptions were more than wiped out by the cost of having broadband to her home.

She has few vices; she would rather than spend any spare money on charities than buying fresh meat, fruit and vegetables for herself, though she will occasionally treat herself to a bar of Cadbury's Dairy Milk, which would last her 2 days. Before her husband died, she attended church regularly, but now attends only a few times a year, though she does support most of their fundraising events.

She prides herself on being no trouble to anyone, and only visits her GP for her annual flu jab and to renew her repeat prescription for antidepressants.

Then Pauline had an accident one Sunday morning. She saw a dog, off its lead, running loose along the promenade. She tried to grab its collar but was pulled off her feet, falling to the floor and banging her head against a lamp-post. She was knocked unconscious, waking up on a trolley at the Accident and Emergency Unit at the local hospital. The nursing staff were able to identify her by a letter in a handbag, but could find no details of next-of-kin or friends to contact. She had broken her hip, and stayed in hospital for a week, before being admitted to a care home for convalescence. She hated it there and was scared that she would be made to stay there, knowing that could not afford any additional fees, so was relieved when arrangements were made for her to return home with a package of reablement care and walking aids. However, the assessors felt that she could be discharged safely only if other provision was made to ensure her ongoing safety and to preserve her quality of life and well-being.

What combination of equipment, social prescriptions and advanced telecare services and applications might be proposed to help Pauline to overcome her immediate problems and then to thrive in her own home?

5. Carol is a 77 year old widow who lost her husband nearly 20 years ago following a long battle with cancer. She lives in a neat bungalow in Hereford, some 5 miles from the City Centre but conveniently close to a bus route. She uses the bus almost daily to visit the shops, the hairdresser, and the local library where she could have a cup or tea and meet up with some friends. She has two sons, One of her them lives with his family 200 miles away in Kent; he and his wife work in London in stressful jobs and have two daughters, one in a sixth form colleague, and another in York at university. They visit Carol about 4 times a year, and she goes to stay with them at Christmas. Her other son lives and works in Canada.

   Last year, she had a serious kidney infection and was hospitalised for 2 weeks. She was discharged home but required some help around the house and to keep the garden tidy. However, as her physical health improved, she had dismissed her home help and her gardener, telling them that she could look after herself and that home on her own. When her son arrived one weekend recently to see her, he was alarmed to find that her cat had died and had been left in the garden that had overgrown, and that his mother was not taking any of the four medications that had been prescribed; they were hoarded in a cupboard in the kitchen.

   He also found that the curtains were closed when they arrived at lunchtime. Neighbours told him that this was normal and that they were often closed all day and for many days at a time. One suggested that his mother regularly went out late at night, sometimes in the rain and without wearing appropriate clothing. There was evidence to suggest that his mother wasn't eating properly; there had also been a flood of water in the bathroom that had soaked carpets throughout the hallway that was causing an offensive smell throughout the property. He was unable to find a cause, while his mother claimed not to remember such an incident.

He contacted the local care agency only to be told that Carol had months ago physically assaulted her home help and had pushed her out, accusing her of coming on the wrong day and of not doing as she was told. They were not prepared to resume working for her without a formal care plan and measures to ensure that their staff were not intimidated. It also became clear to him that his mother's short-term memory had declined and that her awareness of day and time had also deteriorated; she had asked him to take her out to the shops as he was preparing for bed late on the Saturday night.

Before returning to Kent on the Monday morning, he contacted social services for an assessment of his mother's needs in the hope that a care plan could be put in place that would enable her to continue to live safely and securely at home, and without compromising the willingness of carers to visit her. He agreed to be present at the time of the assessment, but also to arrange for his mother to be assessed at a memory clinic. In the meantime, he took his mother back to Kent.

As he had expected, Carol was confirmed to have cognitive and memory problems that would compromise her ability to live safely at home, but also that she lacked activities to keep her occupied.

What might be included in her care plan to offer safety and improved quality of life?

6. Doris is a spinster and retired librarian aged 80 who has lived in the same terraced home for the past 75 years, a property that she inherited from her parents when they died in a car accident 30 years ago. She has no siblings, but has a pet golden retriever called Rover aged 10 and is both her guard dog and her constant companion by day and by night.

   She reads avidly, especially historic novels and non-fictional books about wars and voyages of discovery. Radio 4 is normally on all day and through the evening in Doris's home, but she gave up her television when she was told that she needed a new aerial for digital channels. Over the past 20 years, she has written a number of novels, the most recent of which she completed 2 years ago using her old manual typewriter; it is soon to be published, so she has invested her advance royalties on a laptop computer and a printer, to enable her to write more books, more quickly!

   Her health has always been good, something she considers to be the result of being a vegetarian, not mixing with sick people in GP surgeries and hospitals, and not wasting her money on heating, smoking, and other vices. She goes to a bridge club every Thursday evening in the home of one of her friends who has poor mobility. She accepts a small glass of sherry at these events, but only to be sociable. She also eats a small cake, but only because she believes it to have been baked by one of the other guests, though the cakes are always from a local supermarket!.

   It was on the way home from one of these evenings that she suffered an accident that may define the rest of her life. The cause is not known, but, according to the driver of the car that struck her, she and her dog were walking on the side of the road that didn't have a pavement. He claimed that the street lights had been turned off by the local authority to save money, and that she was in the middle of the road when he came round a bend at 28 miles per hour. He also claimed that she had followed her dog towards the middle of the road as the car approached, and he was unable to avoid dealing her a glancing blow.

   In hospital, she was found to be concussed, badly bruised, but otherwise very lucky in that she had no fractures. She was kept in hospital for 2 nights under observation but prior to discharge was found to have poor distance eyesight. It transpired that her distance vision had not been checked for many years, and that she also had cataracts. An assessment of her home was performed by an occupational therapist who expressed major concerns about clutter, the steep stairs, the state of the electrics, the age and functionality of her gas cooker and gas fire, and by her lack of social support.

Medically, she was considered to be fit enough for discharge but staff were worried that she might be at risk of further accidents if she return home. Doris's only concern was getting Rover back from the local dog charity that was looking after him. She certainly wouldn't entertain any suggestion that she should consider moving into sheltered accommodation or a care home.

What are the opportunities for introducing assistive technologies and a social alarm system to help Doris to continue to live independently at home? How long do you think she would be able to carry on living in this property?

7. Mary is 72 and has lived alone in her adapted bungalow since her husband died 2 years ago. She has Parkinson's disease and he was her main carer providing her with 24/7 support. He took her out in their car to go shopping and sight-seeing, as well as to visit friends and relatives using a portable wheelchair when appropriate; he also managed her medication, did the cooking and helped her with personal care tasks when she felt too weak to perform them herself.

Since his death, she has been suffering from depression and has struggled to adjust to a life on her own. Her only daughter, Jane, lives in Scotland with her family and has suggested to her mother that she either moves closer to them, or lives with them. But Mary worries about the cold weather up North, and doesn't want to impose on her daughter and her family. Instead, she has recruited Dorothy as a live-in carer who can provide her with the close personal support that she needs both during the day and during the night if necessary. Dorothy is a 64 year old spinster who recently returned to the UK after spending more than 25 years in Africa working as a nurse in a leprocy mission. She had few possession other than 2 suitcases of clothing, some jewellery, a small (old) car, and a radio which she listens to in the evening. Mary would prefer for Dorothy to come to her lounge to watch her 50 inch flat-screen TV with her, but Dorothy insists that she needs time on her own to pray and to meditate.

Dorothy is slightly built and eats far less than Mary, but Mary now weighs no more than 7 stone (40 Kg) and doesn't give her carer much trouble in helping her to mobilise. Nevertheless, on Dorothy's advice, Mary now sleeps in a single bed which has a controller for electrically raising either the bed head or the foot of the bed. It helps her to sit up, and also to get up or go to bed on her own when she is feeling well. Her bed is in the corner of the bedroom, leaving plenty of room for a commode next to the bed that she sometimes uses instead of going to her en-suite bathroom in the middle of the night.

When Mary's daughter visits, she is concerned about the slow detioration in her mother's condition, and her decreased mobility and likelihood of having a serious fall. Mary insists that she is coping well, and that the modifications that her husband had made to the bungalow would ensure that her quality of life would be maintained. Jane likes Dorothy and has been impressed by her devotion to Mary, and by her willingness to be at her beck and call on a 24/7 basis.

However, Mary suffered a fall while coming home from church with Dorothy one Sunday. An ambulance was called and she was taken to hospital with significant bruising but no broken bones. After a couple of days, the nursing staff wanted to discharge her to a nursing home for a couple of weeks of convalescence, but Mary wouldn't here of it. Instead, with the support of Dorothy she went home. But her daughter and her husband were furious and insisted on Mary being admitted to a care home. Mary resisted and told them that it was her life and her decision, and that Dorothy had agreed to look after her in her own home until the end of her life, but was willing to accept equipment and technology that would keep her safe.

What technology might help?

If Jane suspected that Dorothy was financially abusing Mary, could technology play a role in helping her monitor the situation?

8. Simon is 45 years of age and has been a wheel-chair user for the past 10 years following a traumatic road vehicle accident in which he received head and back injuries. Following a period of 18 months in hospital and in a specialist rehabilitation facility, he was able to return home supported by a large support package which has slowly been reduced as he has found new coping strategies. However, his marriage did not survive the transition; his ex-wife claimed that she was unable to deal with his panic attacks, general anxiety, and need for constant prompting to perform even the simplest of tasks. He now lives in a supported living environment where he has independence but also has some technology to enable him to operate appliances remotely as well as a telecare link to a 24 hour monitoring centre. He received 4 visits a day by a carer to supervise him getting up, preparing meals and taking his medication.

His care package has now been reduced to 3 visits a day with more flexibility around timings. The result is that he has more opportunity to get out and about but has become anxious at times when he believes that the carer is late or has forgotten about him altogether. He sometimes forgets that the carer has already been and will press his emergency call button several times seeking reassurance from the alarm centre. He also complains to social services that he is not receiving the care that he feels that he needs, and that he is insecure in his own home.

Simon manages to use a smart phone to link with his children and with other relatives, and has the capacity to make greater use of technology if it could be chosen appropriately to address his personal goals. These include being able to get out more and to meet with people who would empathise with him and help him overcome his intense feelings of loneliness. He is financially comfortable and receives income from a trust set up for him with money paid by the insurance company following his accident.

Nevertheless, Simon is depressed and feels very lonely every evening when he returns to his flat. He has spoken to friends and support workers that he is thinking of ending his life by overdosing on his medication. They are concerned that he may be serious and want to manage his medication so that he doesn't have access to more than a day's medication at a time, and can monitor both his activities and his medication adherence.

They inform his ex-wife who probably knows him better than anyone else. She believes that he is capable of suicide, and will find a way of doing it unless he is offered more to fill his life, including the friendship of women.

What can be done to improve Simon's mental health, whilst ensuring that he receives the support that he needs?

Appendix B

List of Suppliers of AT and Telecare Equipment

Below are listed a non-exhaustive list of organisations that are involved primarily in the manufacture and supply of Assistive Technologies in the most general sense. The focus is on products that are currently available and approved for use in the UK and Europe. The list will inevitably be out-of-date as soon as it is written. This is because of the dynamics of the market-place and a recognition that companies will cease trading either because their products have been overtaken by technical developments, or that they are so innovative that they are simply too early in what remains a conservative market-place. It is likely that there will also be considerable, further consolidation in the industry, both to produce economies of scale for marketing and sales purposes, and as manufacturers appreciate that there is a major shift from the supply of product to that of service. Such moves may be consistent with the digital change agenda where maintenance and repair are becoming less important in a world of subscriptions, upgrades and greater reliability. The need for installations by technicians may be replaced by volunteer services (perhaps led by 3rd sector organisations). This may be part of a broader transformation agenda which will be covered in Part 2 of this book.

3 rings - activity monitoring system for Internet of Things www.3rings.co.uk

Abilia Toby Churchill - manufacturers of stairlifts www.toby-churchill.com

Acorn - manufacturers of stairlifts http://www.acornstairlifts.co.uk

Acticheck – manufacturer of wrist-worn alerting device www.acticheck.com

Aico - intelligent environmental sensors solutions - https://www.aico.co.uk/

Alert it – manufacturers of plesiocare equipment www.alert-it.co.uk

Amazon - suppliers of Echo digital assistants https://www.amazon.co.uk/

Apple – manufacturers of iPhone, iPad and watches https://www.apple.com/uk/

Brain in Hand - apps for autistic spectrum disorders www.braininhand.co.uk

Britton Price Hoists – hoists for homes http://www.brittonprice.co.uk

Burton safes - suppliers of key-safes https://www.burtonsafes.co.uk

Buddi - GPS and mobile falls detection https://www.buddi.co.uk/

cair - supplier of telecoms systems linked sensors https://we-cair.com/

Canary care - activity monitoring systems https://www.canarycare.co.uk/

Chiltern Invadex – Assistive Technology providers - www.chilterninvadex.co.uk

Chubb Community Care – telecare systems www.chubbcommunitycare.co.uk

Closomat – toilets for disabled people http://www.clos-o-mat.com

Cm2000 – care management technology https://www.cm2000.co.uk/

CSL dualcom - secure connectivity specialists www.csldual.com

Docobo – integrated digital health solutions https://www.docobo.co.uk/

Doro Care - smart support products for older people www.doro.com/care

Essence – suppliers of smart AT https://www.essence-grp.com/smart-care

Everon - mobile care solutions https://everon.net/

Fitbit – activity and lifestyle tracking https://www.fitbit.com/uk/home

Google - home digital assistants https://www.google.co.uk/about/products/

Grandcare - activity monitoring and support systems http://www.grandcare.co.uk

Green Access - connected assisted living technology www.greenaccess.plc.uk

HTC – suppliers of smartphones and headsets https://www.htc.com/uk/

Hobbs Rehabilitation – neurotechnology https://www.hobbsrehabilitation.co.uk

Invacare - Independent Living Equipment www.invacare.co.uk

Just checking – systems for activity monitoring/audit www.justchecking.co.uk

Kemuri - wellbeing monitoring devices www.kemurisense.com

Keynetics – providers of Sentrikey lock boxes https://www.keyneticsltd.co.uk/

Legrand – manufacturer of Tynetec and Jontek equipment www.tynetec.co.uk

LG – smart appliances for the home https://www.lg.com/uk/appliances

Loc8or – wireless locator tags and finder devices https://loc8tor.co.uk

Medpage Ltd - manufacturer of electronic AT www.easylinkuk.co.uk

MindMe - suppliers of GPS location alarms www.mindme.care

MOMO – digital comms tools for young people https://mindofmyown.org.uk

Motech - digital platform service www.motechdevices.com

My Home Helper – tablet device for dementia https://www.myhomehelper.co.uk

Nottingham Rehab - supply of rehab equipment/services www.nrshealthcare.co.uk

Oculus – suppliers of virtual reality headsets https://www.oculus.com/

Orcha - App design and quality assurance www.orcha.co.uk

Ovenguard - fire safety devices for independent living http://www.ovenguard.co.uk

Ownfone - simple personalised mobile technology www.ownfone.com

Oysta - mobile technology services https://oystatechnology.co.uk/

Pivotell - medication dispensers and reminders http://www.pivotell.co.uk

Possum – providers of AT and telecare www.possum.co.uk

Protomed - suppliers of the Biodose Connect System www.biodose.co.uk

Remind Me Care – technology for dementia https://www.remindmecare.com/

Rewalk – supplier of exoskeleton https://rewalk.com/rewalk-in-uk/rewalk-in-uk/

Rise & Recline – manufacturers of reclining chairs/beds www.riseandrecline.co.uk

Samsung – designers of smart equipment/phones https://www.samsung.com/uk/

Sarabec - suppliers AT for hearing improvement www.accesstosound.com

SIMAvita - incontinence alerts systems http://www.simavita.com

Smartbox - communication aids, environmental control www.thinksmartbox.com

Solon Security – crime prevention specialists www.solonsecurity.co.uk

Speakset – TV linked remote consultation https://www.speakset.com/

Stannah - suppliers of lift equipment www.stannahstairlifts.co.uk

Stiltz - home lift equipment http://www.stiltz.co.uk

Telealarm Europe - suppliers of Bosch telecare equipment www.telealarm.com

Telecoms GRP - suppliers of easy to use phone http://www.easiphones.com

TextCare – suppliers of activity monitoring systems https://www.textcare.co.uk/

The Key Safe Company – suppliers of security products www.keysafe.co.uk

TelMeNow – retailers of AT and telecare http://telmenow.com

Touch Bionics – smart prosthetics and iLimb https://www.touchbionics.com

Tunstall - manufacturers of telecare equipment www.tunstall.co.uk

Tyto - devices for home medical consultation https://www.tytocare.com/

Unforgettable – products for people with dementia https://www.unforgettable.org/

Verklizan – software for monitoring centres www.verklizan.com

WHZAN – digital health and care https://www.whzan.com/public/Home.aspx

Appendix C

Further reading

"Three Generations of Telecare of the Elderly"; K. Doughty, K. Cameron & P. Garner; J. Telemedicine & Telecare, Vol. 2, No. 2, pp. 71-80 (1996)

"Continuous automated telecare assessment of the elderly"; K. Doughty & J.H. Costa; J Telemedicine & Telecare; vol. 3 no. 1, suppl 1 pp. 23-25 (1997)

"Reducing fall incidence in community elders by telecare using predictive systems"; K. Cameron, K. Hughes & K. Doughty; Proceedings of 19th Annual International Conference of the IEEE; Vol. 3 ; pp. 1036 – 1039 vol.3 (1997)

"A remote electronic monitoring system for the prevention of pressure sores"; G. Williams, L. J. Spicer , & K. Doughty; Proceedings of 19th Annual International Conference of the IEEE; Vol. 3 ; pp. 1076 – 1079 vol.3 (1997)

"Good Practice in Risk Assessment and Risk Management 2"; Hazel Kemshall and Jacki Pritchard; Jessica Kingsley. London (1997)

"A systems approach to achieving CarerNet-an integrated and intelligent telecare system"; G. Williams, K. Doughty & D. A. Bradley ; IEEE Transactions on Information Technology in Biomedicine ; Vol. 2 , No. 1, pp. 1-9 (1998)

"Continuous assessment of the risk of falling using telecare"; Kevin Doughty & Keith Cameron; J Telemed and Telecare; vol. 4 no. suppl 1 88-90 (1998)

"DIANA – a telecare system for supporting dementia sufferers in the community"; K. Doughty, G. Williams ; P. J. King ; R. Woods; Engineering in Medicine and Biology Society, 1998. Proceedings of the 20th Annual International Conference of the IEEE ; Vol. 4, pp. 1980 – 1983 (1998)

"TEMPEST – an intelligent pill reminder and delivery system for telecare applications"; K. Doughty, G. Williams & L. Holloway; Engineering in Medicine and Biology Society, 1998. Proceedings of the 20th Annual International Conference of the IEEE ; Vol. 3, pp.1206 – 1209 (1998)

"Distributed intelligent nodes as information filters in advanced telecare systems", G. Williams, K. Doughty & D.A. Bradley; 21st Annual Conference and the 1999 Annual Fall Meeting of the Biomedical Engineering Society] BMES/EMBS Conference, 1999. Proceedings; Vol. 2, (1999)

"MIDAS – Modular Intelligent Domiciliary Alarm System-a practical application of telecare"; K. Doughty, R. Isak, P. J. King, & P. Smith; 21st Annual Conference and the 1999 Annual Fall Meeting of the Biomedical Engineering Society] BMES/EMBS Conference, 1999. Proceedings; Vol. 2 (1999)

"Primary and secondary sensing techniques for fall detection in the home"; K Doughty, K Cameron – Proceedings of Hospitals Without Walls Conference, London: City University (1999)

"Telecare – New ideas for care and support @ home"; Puay Tang, David Gann and Richard Curry; Policy Press, Bristol (2000)

"The design of a practical and reliable fall detector for community and institutional telecare"; K Doughty, R Lewis & A McIntosh; J Telemed Telecare; vol. 6 no.2, suppl 1 150-154 (2000)

"FRED, EMMA and Ingrid – Watching over you"; Kevin Doughty; Old Age Psychiatrist, Millennium Issue No. 17, pp. 7-9 (2000)

"Safety and risk issues in using telecare"; G Williams, K Doughty & D A Bradley; J Telemed and Telecare; vol. 6, no. 5, pp. 249-262 (2000)

"Practical solutions for the integration of community alarms, assistive technologies and telecare", Quality in Ageing and Older Adults, Kevin Doughty, & Gareth Williams, Vol. 2, No. 1, pp.31 – 47 (2001)

"Extending the scope of community alarm services", Kevin Doughty, & Malcolm Fisk; Housing, Care and Support; Vol. 4, No. 2, pp.24 – 27 (2001)

"Social Alarms to Telecare"; Malcolm J. Fisk; Policy Press, Bristol (2003)

"Supporting independence: The emerging role of technology", K. Doughty; Housing, Care and Support, Vol. 7, No. 1, pp.11 – 17 (2004)

"Standards and the Dependability of Electronic Assistive Technology"; G. D. Baxter, A. F. Monk, K. Doughty, M. Blythe & G. Dewsbury; Chapter in Designing a More Inclusive World; Springer; pp 247-256 (2004)

"Alleviating the COPD Burden with ICT"; Tony Rice & Kevin Doughty; British Journal of Healthcare Computing and Information Management; Vol. 21, No. 1 pp. 27-28 (2004)

"Socially dependable design: The challenge of ageing populations for HCI"; Mark Blythe , Andrew Monk, & Kevin Doughty; Interacting with Computers; Vol. 17, No. 6, pp. 672–689 (2005)

"Assistive technology and telecare"; Simon Brownsell and David Bradley; Ed. Jeremy Porteus; Policy Press, Bristol (2005)

"Telecare – Practice and Potential"; Kevin Doughty; BT Community and Home Care (2007)

"The Digital Seventh Age"; Kevin Doughty; IET Engineering Management, Control & Automation, June/July (2007)

"Towards an evaluation framework for telecare services"; Gareth Williams & Kevin Doughty; Journal of Assistive Technologies, Volume: 1 Issue: 1, pp.42 – 47 (2007)

"Review of strategies for charging service users for telecare"; Kevin Doughty; Journal of Assistive Technologies, Vol. 1 Issue: 2, pp.22 – 25 (2007)

"Telecare, telehealth and assistive technologies — do we know what we're talking about?"; K. Doughty et al.; Journal of Assistive Technologies, Vol. 1 Issue: 2, pp.6 – 10 (2007)

"A systematic review of the benefits of home telecare for frail elderly people and those with long-term conditions
James Barlow , Debbie Singh , Steffen Bayer , Richard Curry; *Journal of Telemedicine and Telecare, 13*(4), 172–179 (2007)

"Lifestyle monitoring - extending telecare services into prediction and prevention"; Kevin Doughty; Journal of Assistive Technologies, Vol. 2 Issue: 1, pp.35 – 41 (2008)

"Accessing properties for the delivery of telecare services in an emergency"; Kevin Doughty & Keith Cameron; Journal of Assistive Technologies, Volume: 2 Issue: 2, pp.43 – 47 (2008)

"Home telehospice: new tools for end-of-life care services"; Audrey Kinsella & Kevin Doughty; Journal of Assistive Technologies, Volume: 2 Issue: 4, pp.47 – 50 (2008)

"Improving the quality of telecare services - the role of audit and training"; Kevin Doughty; Journal of Assistive Technologies, Volume: 3 Issue: 1, pp.24 – 28 (2009)

"The role of telecare in the protection of individuals from domestic violence"; Kevin Doughty & Joanna Clark; Journal of Assistive Technologies, Volume: 3 Issue: 4, pp.44 – 48 (2009)

"The use of telecare and community equipment to improve joint working across health and social care"; Kevin Doughty & Chris Steele; Journal of Assistive Technologies, Volume: 3 Issue: 4, pp.36 – 43 (2009)

"Safe walking technologies for people with mild to moderate cognitive impairments"; Kevin Doughty & Barbara Dunk; Journal of Assistive Technologies, Volume: 3 Issue: 2, pp.54 – 59 (2009)

"Preliminary results of a medical telecare pilot in Wrexham"; Kath Cooper & Kevin Doughty; Journal of Assistive Technologies, Volume: 3 Issue: 3, pp.36 – 42 (2009)

"The role of reminder aids and systems to support independence in people with memory problems"; Jan Costa & Kevin Doughty; Journal of Assistive Technologies, Volume: 3 Issue: 2, pp.60 – 64 (2009)

"Evaluation and value for money analysis of a new telecare service from a user perspective"; Ellen ap Dafydd, Janet Roberts & Kevin Doughty; Journal of Assistive Technologies, Volume: 3 Issue: 3, pp.43 – 49 (2009)

"Practical issues in the selection and use of GPS tracking and location technologies to support vulnerable people at risk of becoming lost or threatened"; Gareth Williams, Peter King, & Kevin Doughty; Journal of Assistive Technologies, Volume: 5 Issue: 3, pp.146 – 151 (2011)

"A systematic review of lifestyle monitoring technologies"; S. Brownsell, D. Bradley, S. Blackburn, F. Cardinaux, & M.S. Hawley *Journal of Telemedicine and Telecare*, *17*(4), 185–189 (2011)

"SPAs (smart phone applications) – a new form of assistive technology"; K. Doughty; Journal of Assistive Technologies; Vol. 5, No. 2, pp.88 – 94 (2011)

"The evolution of assisted living provider services (ALPS) to support twenty-first century health, social care and housing needs"; Dave Miles, & Kevin Doughty; Journal of Assistive Technologies, Volume: 5 Issue: 3, pp.140 – 145 (2011)

"Low cost connected healthcare systems in developing countries"; Kevin Doughty; Journal of Assistive Technologies, Volume: 6 Issue: 3, pp.216 – 220 (2012)

"Self-care, plesio-care, telecare and m-care: a new assisted living model"; Kevin Doughty, David Godfrey, & Billy Mulvihill; Journal of Assistive Technologies, Volume: 6 Issue: 4, pp.292 – 301 (2012)

"Systematic review of studies of the cost-effectiveness of telemedicine and telecare. Changes in the economic evidence over twenty years." Mistry, H*; Journal of Telemedicine and Telecare*, Volume *18* No.1, pp. 1–6 (2012)

"Digital reablement – a personalised service to reduce admissions and readmissions to hospitals and nursing homes"; Kevin Doughty, & Patrick Mulvihill; Journal of Assistive Technologies, Volume: 7 Issue: 4, pp.228 – 234 (2013)

"Imagining the Future: Enabling Technology"; Kevin Doughty; Report commissioned by IRISS (Institute for Research and Innovation in Social Service), Glasgow (2013)

"AT and telecare to manage fire risks in the homes of older and vulnerable people"; Kevin Doughty, & Mike Orton; Journal of Assistive Technologies, Volume: 8 Issue: 1, pp.35 – 43 (2014)

"The Patient Will See You Now"; Eric Topol; Basic Books, New York (2015)

"Wearable devices to support rehabilitation and social care"; Kevin Doughty, & Alistair Appleby; Journal of Assistive Technologies, Volume: 10 Issue: 1, pp.51 – 63 (2016)

"New models of assessment and prescription of smart assisted living technologies for personalised support of older and disabled people"; Kevin Doughty, & Gareth Williams; Journal of Assistive Technologies, Volume: 10 Issue: 1, pp.39 – 50 (2016)

"The New Mobile Age"; Joseph C. Kvedar; Partners Connected Health, Boston (2017)

"The Role of Technology in Addressing Health and Social Care Needs and Opportunities in a Rural Environment"; Kevin Doughty & Anne Livingstone; Journal of Corporate Citizenship; Issue 68, pp. 67-81, (2017)

"The Fourth Wave: Digital Health"; Paul Sonnier, Lexington (2018)

"Essentials of Dementia"; Shibley Rahman and Rob Howard; Jessica Kingsley, London (2018)

Appendix D

Glossary of terms and phrases

i. Acronyms

AAC	Alternative Augmentative Communication
AAL	Ambient Assisted Living
AD	Alzheimer's disease
ADL	Activities of Daily Living
ADSL	Asymmetric Digital Subscriber Line
AI	Artificial Intelligence
API	Application Programming Interface
AR	Augmented Reality
ASD	Autistic Spectrum Disorders
CCG	Clinical Commissioning Group
CHD	coronary heart disease
COPD	Chronic Obstructive Pulmonary Disease
CUHTec	Centre for Usable Home Technologies
ECR	environmental control systems
EHR	electronic health record
GDPR	General Data Protection Regulation
GUI	General User Interface
IADL	instrumental activity of daily living
ICT	Information and Communication Technology
iCUHTec	International Centre for Usable Home Technologies
IDD	intellectual and developmental disabilities
IoT	Internet of Things
LA	Local Authority
LTC	Long Term Condition
NHS	National Health Service
NCD	Non-communicable Disease
PACS	Picture Archiving and Communications Systems

PBX	Private Branch Exchange
PERS	Personal Emergency Response System
PHR	personal health record
POTS	Plain old Telephone System
PSTN	Public Switched Telephone Network
PTSD	post-traumatic stress disorder
RCT	randomized controlled trial
RF	Radio frequency
RPM	Remote Patient Monitoring
TEC	Technology Enabled Care
TENS	Transcutaneous Electrical Nerve Stimulation
TS	Technology Support
TSA	Telecare Services Association
UKTH	UK Telehealthcare
VR	Virtual Reality
WISER	Watchful Intelligent Safe Empowering and Responsive (Home)

ii. Terms and Phrases

	Terms	Meaning
A	**Accessibility**	Physical and sensory access to buildings, vehicles, services, information and web-sites, and relating to matters such as design, appearance, lighting, colour, texture, and visual or audible cues
	Accreditation	Process for acknowledging that a service is being delivered to agreed standards of quality
	Activities of Daily Living	key tasks such as walking, eating and toileting
	Activity Monitoring	collection of data on the performance of tasks using technology
	Asymmetric digital subscriber line (ADSL)	a type of "always-on" telephone service that enables faster data transmission over standard copper telephone lines than a conventional dial-up modem.
	Ambient Assisted Living	background support for independent living using technology
	Android	a mobile operating system developed by Google, and designed primarily for touchscreen mobile devices such as smartphones and tablets.
	API	a set of routines, protocols, and tools for building software and applications.
	Apps	software applications that can be run on smartphones and tablet devices
	Assessment	the processing of understanding wants, needs and risks, and of identifying potential solutions
	Assistive Technology -	Devices or Systems that can help people or carers to support themselves or others to perform domestic or personal tasks more easily or safely or to do things that they couldn't otherwise manage
	Augmentative and Alternative Communication	Means of communication that supplements conventional text or speech and used when one or the other is impaired due to disability or illness
	Augmented Reality	Computer system for adding elements to an actual image in order to overlay objects or ideas
	Autistic Spectrum Disorder -	describes a range of conditions classified as neurodevelopmental disorders and including features such as social deficits and communication difficulties, stereotyped or repetitive behaviours and interests, sensory issues, and in some cases, cognitive delays.
	Awareness	understanding the potential to do things differently
	Asset-based	Drawing on skills and abilities of individual

B	**Barrier free design**	Access to environments, products or services for the widest possible group of people
	Big data	a term for data sets that are so large or complex that traditional processing applications are inadequate, but which have major uses in predictive analytics.
	Bluetooth	a wireless technology standard for exchanging data over short distances using short-wavelength radio waves from fixed or mobile devices
	Bogus caller	a caller at a property who attempts to gain admission or to sell products under false pretences, often claiming to represent a utility or repairs organisation
	Bot	A computer system that has been compromised by malware
	Broadband -	telecommunications bandwidth capable of supporting video interactions
	Bucket list	personal aims and ambitions to be achieved one day
C	**Carephone**	A dispersed alarm unit that connects to the telephone network
	Carer	Professional staff or family members who provide care
	Chronic disease	health conditions that need to be managed for ever
	Cognitive deficits	the result of brain conditions, such as dementia, which lead to short term emory loss and difficulties dealing with time and space
	Co-design	Planning and working in partnership with stakeholders to ensure that all pinions and expertise are considered
	Commissioning	Giving instruction to an organisation to undertake specific tasks or sponsibilities based on a specification
	Community alarms	basic communication system consisting of pull cords, alarm buttons and tercom systems linked to a telephone monitoring centre or housing warden
	Connectivity	the means by which individuals, their computers, mobile devices, and local ea networks connect to the global Internet or to specific monitoring centres
	Consumer	Individuals buying or using products, systems or services
	Culture	Behaviours and beliefs that derive from societal norms
	Cyber security	Technology processes and systems that protect from unsolicited electronic tacks or access
D	**Delayed Transfer of Care**	a patient who is medically stable who must remain in hospital because there no safe discharge option available
	Digital care	support processes, systems and devices that employ digital, communication electronic technologies to help provide care or support
	Disability	The result of long-term physical, mental, intellectual or sensory impairments hich may hinder people in participating in society on an equal basis

Disease management The continuous coordinated health care process to manage and improve ealth

Duty of care a responsibility of an organisation to recognise and help manage risk

E **e-healthcare** the use and sharing of electronic records and communications to help ovide people with more coordinated and responsive healthcare services

Electronic Assistive echnology Support systems that allow electronic communication between the user and sistive devices

Electronic Health Record A record that can be maintained directly by a patient to securely hold formation relating to health history, treatment and medication

Electronic Patient Record A record with information and data relating to the user as a basis for eatment options

Emergency response protocols and practical support to help someone in their home in the event them suffering a health or social care crisis

Empowerment A process that offers greater control over decisions that affect the way that they live

Enablement – help to restore an individual to achieve independence and confidence in performing practical everyday tasks such as shopping and managing their medication

Environmental Control Systems electronic systems that enable people with profound physical disabilities to control and access everyday electronic devices, doors and windows through switches that are activated by touching an icon or using eye gaze or blowing down a tube for example.

Environmental monitoring and alarms the sensing of the home environment for dangerous levels of smoke, water, temperature (high and low) and air quality including gases (such as carbon monoxide and methane), and the raising of an alarm or an alert if they exceed recognised threshold values

Epilepsy a group of neurological diseases characterized by episodic seizures that can vary from brief and nearly undetectable to long periods of vigorous shaking

Ethics the branch of knowledge that deals with moral principles regarding actions

Exoskeleton a wearable powered suit with a system of motors, pneumatics, levers or hydraulics that allow for limb movement, increased strength and endurance

F **Family carers** close relatives who provide informal care and support

Fibre to the cabinet - a telecommunications system based on fibre-optic cables run to a platform on the street that can serve several households and give them all high-speed access to the Internet

Frail A clinical syndrome characterised by several characteristics such as recent weight loss, exhaustion, weak hand grip, slow walking speed and low physical activity

	Term	Definition
G	**2G**	short form for second-generation wireless telephone technology launched on the GSM digital standard in Finland 1991 and the basic form of mobile telephony in much of the UK before 2012
	3G	short form for third generation of mobile telecommunications technology supporting network services that provide a higher information transfer rate enabling the viewing of internet pages
	4G	short for fourth generation of mobile telecommunications technology and capable of supporting amended mobile web access, IP telephony, gaming services, high-definition mobile TV, video conferencing, 3D television, and cloud computing
	5G	short for the next (fifth) generation of mobile telecommunication technology and supporting data rates nearly equivalent to broadband for tens of thousands of users within a small area
	Geofence	A virtual boundary that defines a safe zone
	Governance	The function determining an organisation's direction, objectives, policy and practice frameworks to effect efficient service delivery
	Global Positioning System	A system of satellites that can provide location information within a few metres
H	**Hazards**	Anything or factor in the home (or any other environment) that has the potential to harm people
	Health coaching	Techniques to help people change their behaviours that can lead to ill-health
	Home automation	System to provide user with more control over their home environment enabling improvements in usability, security and comfort
I	**International Centre for Usable Home Technology**	a virtual organisation supporting the training and support needs of organisations with telecare interests
	Impairment	Reduced capacity due to congenital or acquired physical or sensory impairment, ill-health, age and frailty, giving rise to disability
	Implants -	a medical device manufactured to replace or enhance a missing or damaged biological structure and, increasingly, capable of communicating sensed and performance information.
	Inclusion	Harnessing of views and opinions of a range of individuals in relation to accessibility of designs, building and products
	Independence	The circumstances by which individuals remain, with or without care and support, in control of important aspects of their lives
	Information and Communication Technologies	Umbrella term for communication devices and systems, and the service and applications that are associated with them e.g. computer networks, video conferencing, remote consultations

	Term	Definition
	Ingestible	a biomedical device that can be swallowed so that it can become active as a sensor or actuator within the body
	Innovation	New processes, devices or materials that result in improved outcomes
	Instrumental Activities of Daily Living	tasks, including shopping, managing finances and grooming, that need to be performed to live independently
	Integrated Care	the principle for care delivery to improve a person's experience and achieve greater efficiency and value from health and social care delivery
	Internet	this consists of millions of linked computers all around the world enabling the World Wide Web to provide the information and services that we can use
	Internet of Things	a creation in which everyday objects, including appliances in the home are connected to a network that allows them to send and receive data
	Internet Protocol	the method and addresses by which data is sent from one computer to another device on the *Internet*
	Interoperability	the ability of different information technology systems, software applications and digital devices to communicate with each other, exchange data, and use information
K	**Kinect**	a line of motion sensing input devices by Microsoft for use with their video game consoles, and using a webcam-style add-on peripheral to enable users to control and interact with console games without the need for a controller using gestures and spoken commands
L	**Lifestyle Monitoring**	Monitoring of environmental conditions, activities and physiology to identify circumstances that reflect adverse changes to well-being
	Local authority	Councils that may have social care responsibilities
	Long term conditions	illnesses and health conditions that may not in themselves be acute, but which can impact on life for several years or until death
	Loneliness	a feeling of being cut off and/or alienated from other people, so that it feels difficult to have any form of meaningful human contact.
M	**mCare -**	care and support functions that can be managed at a distance using mobile communication technologies and devices
	Medical alarms	an *alarm* system designed to signal the presence of a potential emergency of a medical nature and which requires the urgent attention of *medically trained and qualified* personnel
	Medication dispensers	Devices that can be filled with prescribed pills by a suitably qualified or authorised person, and then programmed to release appropriate dose (or make the pills available) at set times to help with medication adherence.

	mHealth	an abbreviation for **mobile health**, a term used for the practice of medicine and public health when supported by mobile devices including smart phones and tablet devices
	Monitoring centre	A telephone call and coordination centre, usually operating 24/7, that is available to connect with people who have health and support needs
	Motivational interviewing	Encouragement of behaviour change through individual support
N	**Non-communicable disease**	Illness not caused by an infection, and therefore cannot be transmitted to others
	Nursing home	Residential care facility with 24-hour nursing staff
O	**Outcomes**	the impacts or end results of personalised services on a person's life.
	Outcomes-focused	an intention to aim support at delivering the necessary services in ways that enable everyone to achieve their agreed and preferred outcomes
	Outcome based commissioning	Method of recruiting and paying for support services based on achieving desired results such as avoidance of hospitalisation or admission to residential care
	Out-of-hours	times and days that are outside the normal working week
P	**Peripheral devices**	Sensors, actuators and elements of a computerised system that increase its capability and/or usability
	Person Centred Care	a way of doing things that sees people using care services as equal partners in planning and developing *care* to make sure it meets their needs.
	Personal assistant	Someone employed by someone with a disability to help them to live a fuller and more independent life
	Personal care	Support such as feeding, dressing and toileting of people who are unable to perform Activities of Daily Living on their own
	Personalisation	
	PERS	North American term for social alarms and community alarm systems.
	Personal technologies	Consumer rather than enterprise devices and system, and would include electronic gadgets, smart phones, tablets, apps, home computers, consoles, video games, GPS devices and, increasingly, wearable fitness or well-being devices including smart watches.
	Physiological monitoring systems	Monitor parameters such as pulse rate, respiration rate, blood pressure, level of oxygen absorbed in the blood and basal temperature, often through peripheral sensing devices, so that clinicians can be informed of changes in a patient's condition.
	Plesiocare	support provided by people who are located close (usually in the same room or dwelling) to the person needing it

	Prevention	Strategies and approaches aimed at maintaining good health by avoiding accidents and risky behaviour and the contracting of illness and disease
	Procurement	Identifying and purchasing products or services based on a specification
	Profiling -	Collection of information about a person to ensure that support choices are appropriate to them, their circumstances and what's matters to them
	Protocol	A set of rules that determine a course of action on receipt of information
	Public health	The health and well-being of the community as a whole
	Pulse oximeter	a medical device that indirectly monitors the oxygen saturation of a patient's blood and changes in blood volume in the skin
Q	**Quadstep**	stair-climbing or descending aid which consists of a walking stick combined with a low step which makes it easier and less painful for people who have problems bending their knees to use the stairs.
	Quality assurance	Process by which services are deemed to meet appropriate standards
	Quality of Life	A highly subjective measure of happiness that includes several components including security, job satisfaction, family life, health and safety
R	**Reablement**	Approach to supporting people to relearn their independence skills following hospitalisation or change in circumstances
	Rehabilitation	Restoration of skills for a person following accident or illness aimed at regaining as much function as possible
	Retirement village	US and Australian model of supported living which can offer care and leisure facilities
	Risk	Probability or threat of damage, injury, or other negative occurrence caused by external or internal factors, which may be avoided through pre-emptive action
	Risk assessment	Process of evaluating hazards and likelihood of harm due to exposure
	Risk management	Mitigation of risks in order to reduce the chances of danger or injury
	Robots	electro-mechanical device that is guided by a computer program or circuit to perform complex tasks or to help a person to perform them
S	**Safe walking**	Supporting people to go out with support from GPS location devices
	Safeguarding	Protection of vulnerable adults or children from forms of abuse
	Sensors	Device that responds to or measures a stimulus or parameter
	Sheltered housing	Accommodation provided by councils or housing associations mainly for older people who need ready access to support
	Skype	a popular example of an application that provides video chat and voice call services on computers, tablets and smartphones

	Smart homes	a home equipped with lighting, heating, and electronic devices that can be controlled remotely by smartphone or computer.
	Smart phone	mobile phone with many functions of a computer, with a touchscreen, Internet access, and operating system that runs downloaded apps
	Smart watch	Mobile device with a touchscreen display, designed to be worn on the wrist
	Soft robotics	Design of intelligent non-industrial devices that can perform domestic tasks or provide support for humans in the home
	Social alarms	Also known as personal alarms or community alarms, they are the most basic form of telecare and usually consist of a pendant unit with an alarm button for use in an emergency, and a base unit that works with the telephone system, to provide a connection to a 24-hour monitoring centre.
	Social Capital	the network of social connections between people, and their shared values and behaviour, that encourage mutually advantageous cooperation
	Social exclusion	The state of being disadvantaged by lack of access to social, political and/or economic activities required for normal living in the community
	Social Media	websites and online means of communication used by large groups of people to share information and to develop social and professional contacts
	Social Isolation	state in which persons, groups, have little contact, communication or cooperation with one another, and which can lead to loneliness
	Sphygmo-manometer	a device used to measure blood pressure, composed of an inflatable cuff to collapse and then release the artery under the cuff in a controlled manner, and used in conjunction with a means to determine the pressure blood flow starts, and at what pressure it is unimpeded
	Spirometer	a device for measuring the volume of air inspired and expired by the lungs
	Stairsteady	a special handrail with a supporting rail that slides when pushed but locks when weight is applied, and which acts as a support aid when climbing and descending the stairs
	Sundowning	Tendency of people with dementia to become agitated and leave their home when it becomes dark
	Supported housing	Accommodation with various forms of support including carers and technology provided for people with disabilities
T	**T-Cubed**	a specialist telecare and smart assistive technology company providing evaluation, training and support services to public, private and third sector organisations
	Telecare	The means by which technology-based applications and services at a distance are provided for vulnerable people needing support
	Technology Enabled Care,	Recently created term (by Dept of Health in England) to cover e-healthcare, telehealth, telecare and telemedicine i.e. anything associated with healthcare which requires digital technology

	Term	Definition
	Technology Support	Term describing all applications of technology to directly help people to live successfully in their own homes
	Telecare Services Association	A trade body representing telecare equipment providers, service providers, commissioners and other related organisations
	Telecoaching -	A range of support methods delivered online and/or by telephone or video consultation providing clinical or self-management support to people, empowering them to improve their own care and well-being
	Teleconsultation	Video-based meeting between patients, carers and professional staff
	Telehealth	The means by which technologies and related services concerned with health and well-being are accessed by people or provided for them irrespective of their location
	Telemedicine	Remote diagnosis and treatment of patients using medical information, including scans or video, transmitted over potentially long distances
	Telemonitoring	Use of telecommunication technologies to remotely collect data relevant to health and well-being of an individual
	Telepresence	Robotic and instrumentation that enable physicians to perform procedures or interactions from remote locations
	Transcutaneous electrical nerve stimulation	use of electric current produced by a device to stimulate the nerves for therapeutic purposes
	Tricorder	A medical device used by doctors to help diagnose diseases and collect bodily **information** about a patient
U	**Ubicare**	Providing care and support in any and all locations
	UK Telehealthcare	Membership organisation arranging Market Place events
	Usability	The extent to which a product, system or service can be used to achieve the intended purpose
V	**Videoconferencing**	Real-time two-way transmission of audio and video images between two or more distant locations
	Vignette	a short description of an event, behaviour or a person used to provide information for participants in a course of exercise to consider
	Virtual reality	An artificial environment that is created with software and presented to the user through a headset so that the user accepts it as a real environment
	Virtual visits	A means of presenting remote care staff or family with the information that they might have collected during a physical visit to the home of an individual
	Vital signs	Clinical measurements, specifically pulse rate, temperature, respiration rate, and blood pressure, that indicate a patient's essential body functional state

	Term	Definition
	Vivo Guide	An on-line portal giving the public and professionals decision support tools and access to whole of market assisted living technology products
	Voids	Rental properties that the landlord finds difficult to let, usually due to a poor location, a lack of good amenities or disruptive neighbours
	Volunteers	People who support vulnerable people at their own expense usually acting through a third sector organisation
W	**Wandering**	Non politically-correct expression for dementia sufferers who go out and forget where they are going
	Wearables	– sensors or actuators that support more independent living which can be worn or attached to the body
	Website	- a virtual location on the World Wide Web containing several subject or company related webpages and data files accessible through a browser
	Well-being	- a complex combination of a person's physical, mental, emotional and social health factors which are strongly linked to happiness and life satisfaction.
	Wi-Fi	A local area wireless computer networking technology that allows electronic devices to connect to the network
	WISER Homes	Properties with linked sensors, actuators and interfaces.
Z	**Zimmer Frame**	A walking frame for disabled or elderly people who need additional support to maintain balance or stability while walking

About the Author

Kevin Doughty was born in Caernarfon, in a largely Welsh-speaking area of North Wales. He became the Head Prefect of the local grammar school and captain of the rugby team, but he suffered a complex double fracture of his leg playing football as a 17-year-old. This frustrated his ambition to become a medical doctor, so he chose instead to study physics at university, which subsequently revived in him his interest in health and medical matters. He went on to study Electronic Engineering for his Ph.D. which focused on applications for thin film electro-active sensors that included high frequency ultrasonic imaging, and body-worn devices for acoustic and vibration detection and analysis. Working with the consultant who had treated his own leg fracture, he invented a system for non-invasively monitoring the healing of broken long bones (such as the tibia) within a plaster cast, and a novel spectral analysis system for monitoring the joints of people who had suffered sports injuries.

After a further 6 years as a post-doctoral researcher, he became a lecturer in wireless Telecommunications Systems at the University of Wales, in Bangor where he combined his expertise in digital radio with an interest in the newly emerging field of optical and fibre communications. This led to a decade-long collaboration with BT's Research Centre at Martlesham, where he was persuaded to use his experience in the long-term care of older people to help them develop projects in the new field of telecare, in order to exploit the company's existing links into people's homes. He established the UK's first academic Telemedicine and Telecare Research Group in 1993. In 1996 he published, jointly with BT collaborators, the paper "Three Generations of Telecare of the Elderly" which proposed, for the first time, the use of social alarm systems as a platform for developing telecare services. It also introduced the concepts of lifestyle, activity and physiological monitoring using new sensors and dashboard arrangements to help manage chronic disease and disability, together with video telephony to reduce demands on travel, and for enabling virtual visits.

In 1998, he spun out 'Technology in Healthcare' as Europe's first specialist telecare company. He recruited several his former star students to develop and fabricate a range of smart sensors that could be used within existing social alarm systems. These included environmental sensors, included novel devices that looked at temperature extremes, detect falls, inappropriate exiting of homes or beds, and hybrid devices with integrated programmable interfaces. The sensors were installed in one of the country's first demonstrator smart homes. He then worked with Tunstall Telecom, the market leader, to pilot devices with forward looking housing associations and local authorities, including West Lothian who were transforming their provision for older people using technology. The sensors gained Millennium Product status as well as winning international innovation awards.

Subsequently, Tunstall took over his company and appointed Kevin their Director of Technology in Healthcare, a position that he held until 2009. During this period, he was diagnosed and treated for colorectal cancer, after which he was seconded to the University of York's Centre for Usable Home Technology (CUHTec) that was set up with funding from the Joseph Rowntree Foundation. CUHTec attracted several hundred organisations as members and offered focused training over several years. An advanced user group was also established to promote best practice, and to inspire further research into future needs and developments.

In 2006, Kevin co-established T-Cubed, the Telecare Think Tank, which he now chairs, and which provides a vehicle to support stakeholders across the UK and Ireland to improve service specifications and quality, and to offer new developers a platform for demonstrating their innovations. He has worked as a specialist for the Telecare Services Association and for UK Trade and Investment as they aimed to exploit the UK's technology excellence during the London Olympics. This led him to undertake visits to the Far East and to Australia, where he has returned several times to lecture and offer workshops, as well as to be part of a roadshow organised by Community Resourcing on behalf of the Queensland Government. He continues to publish widely and is a frequent invited speaker at both local and national seminars and conferences. He is a Visiting Professor in the Digital Transformation of Care Services at the University of Cumbria and performs occasional lecturing to occupational therapy and physiotherapy classes.

Kevin continues to work as the executive chairman of T-Cubed Ltd., a leading independent consultancy specialising in digital care, assisted living technologies and technology enabled care services. It works with service providers, monitoring centres, service commissioners and product developers to help them deliver improved products, services and outcomes for their customers. Its customers include national government agencies, local authorities, NHS organisations and housing associations to develop strategies, road-maps and hard-hitting reports on the future direction of care services. It also develops and delivers bespoke training packages to support assessors and front-line staff is extending the role of technology to support people with care needs and ambitions to improve their lives.

Kevin can be contacted at: kevin@t-cubed.co.uk

www.t-cubed.co.uk

About Part 2

The first part of this booked has described a journey of over 50 years from the establishment of sheltered housing with wired access to warden services, through the development of monitoring centres and carephone systems, to a situation where telecommunications have revolutionised the way that we connect to each other socially and in an emergency. The journey is far from over, and the destination remains unclear, perhaps limited only by our imaginations, what we really want from health and support services, and by the inevitable cost of provision. The austerity years, 2010 to 2018, have done little to answer the question of who pays. Perhaps this isn't surprising in a world of such great equalities where the poorest poor have no money other than the state pension and benefits, while more older people than ever have both assets, in the form of expensive homes, and occupational pensions that may be considerably higher than the working wage.

The context for the second part of the book is one of presenting inevitable opportunities for changing the landscape of support, against a backdrop of cultural acceptance for doing things differently, and for many individuals choosing to pay for them too rather than expecting the state to provide for them despite their relative wealth. It has perhaps taken the best part of 30 years for the public to realise that social care is not free at the point of access, and that it is both expensive and increasingly necessary. Without good and accessible support in the community, the NHS will not survive because the system will grind to a halt if patients cannot be safely discharged to their own homes after treatment. But solving the discharge problem on its own won't be enough, we need also to prevent many emergency admissions taking place in the first place. That is perhaps the biggest long-term problem, and one that needs to be tackled in lots of different ways, including enabling and empowering people to become co-carers in their own support plans. That means us all having to take more responsibility for our lifestyles and the risky behaviours that we choose, and will may lead us to be a lasting drain on the health service. Of course, people who have health needs also tend to have more disabilities, and then more social care needs. The message must be for us all to be more considered with a public health agenda that will involve us taking more exercise, eating less, having better diets, and supporting each other to maintain our health and well-being.

Technology must play a big role in this enormous agenda. It provides a vehicle for support but individually we must be the drivers. Society can provide the tools and the infrastructure, and we, as individuals, need to use them wisely and continuously. Part 2 of the book begins with the practicalities of communication, and the role of the Internet to provide information and access to resources irrespective of location, but in the context of keeping safe on-line. People must feel safe and must not worry that their data will be harvested for harm. The book will describe the digital transformation journey in terms of applications for the Internet of Things, and a world where every electronic device from a fridge to a television, connects to each other and to the user. Health and social care will be beneficiaries of these changes, most of which have not yet been developed as applications that will benefit older and more vulnerable people.

We must be prepared to maximise opportunities for people to use their own devices, because they are more likely to use and continue to use technology that they are familiar will. Only then can we expect people to embrace apps and peripherals that can, in their ow hands, become powerful machines that can be personalised to their own needs. Apps are emerging in every field and will play an enormous role in health and social care if everyone can embrace them in the same way as they are prepared to accept medication, surgery and other forms of treatment. Opportunities will be present throughout Part 2.

It follows that Part 2 will need to consider how technology may be used as a form of glue that links families, formal support agencies, the NHS and vulnerable individuals, wrapping those who are in greatest need in a safety blanket of support. The shape and size of these blankets need to be worked out for each population group and then offered in an appropriate manner so that people can maintain their independence rather than be nannied by the state. There will, of course, be a big role for the state, especially in enabling town, cities and communities to be friendlier towards those with disabilities and long-term conditions, and in ensuring that care homes and nursing homes offer a quality environment that doesn't rely on an antiquated regime of inspection. Digital devices and services allow different ways of measuring outcomes and the factors that matter to staff and to residents of these establishments. They allow comparisons to be made in real-time, and interventions required immediately.

Finally, as the western world comes to terms with a birth rate that is insufficient to allow the population to grow, the inevitable increases in the number of older and very old people will result in there being insufficient people to provide the required level of care, and at a price that individuals and the country can afford. It means relying on greater efficiencies that technology can support or finding radical alternatives. Either way, volunteers must play a much bigger role either in providing hands-on care or, more likely, showing people how they can perform more self-care with the aid of digital technology. Volunteers may continue their role as digital champions, showing others how to go safely online, to make use of the Internet and to maximise the potential of emerging Internet of Things technologies. They might also become the first line of response when something goes wrong in the digital world. Maintenance and repair may be left to the (expensive) experts, but resetting of passwords, and solving minor connectivity issues may be an area where volunteers can provide more responsive support.

Eventually, more direct support functions will be possible using soft or social robots. Technically, they can be manufactured to help with many personal care tasks and with many of the responsibilities currently enjoyed by domiciliary carers. However, their acceptability has yet to be proved, nor the form that will make them commercially viable. Some may be fixed machines, perhaps kept at the bedside or in the bathroom, while others will be able to move about the home, providing companionship and connectivity as well as functional support. Robots may be the final frontier of care, but a failure to consider their potential and cost may be an expensive failure.

The aim is to complete Part 2 during 2019.

15922282R00136

Printed in Great Britain
by Amazon